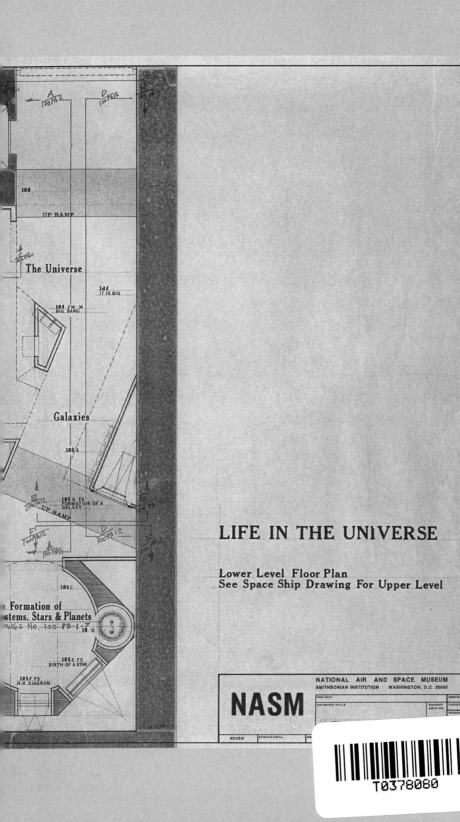

LIFE IN THE UNIVERSE

Lower Level Floor Plan
See Space Ship Drawing For Upper Level

INSPIRED
ENTERPRISE

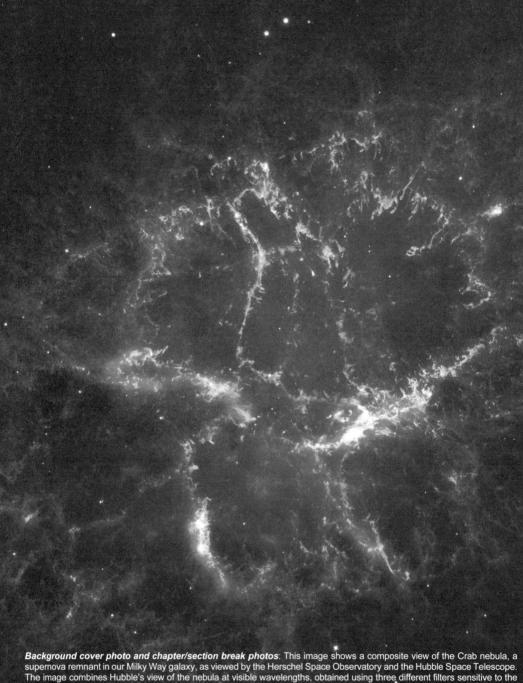

Background cover photo and chapter/section break photos: This image shows a composite view of the Crab nebula, a supernova remnant in our Milky Way galaxy, as viewed by the Herschel Space Observatory and the Hubble Space Telescope. The image combines Hubble's view of the nebula at visible wavelengths, obtained using three different filters sensitive to the emission from oxygen and sulfur ions and is shown here in blue. Herschel's far-infrared image reveals the emission from dust in the nebula and is shown here in red. The Herschel image is based on data taken with the Photoconductor Array Camera and Spectrometer (PACS) instrument at a wavelength of 70 microns; the Hubble image is based on archival data from the Wide Field and Planetary Camera 2 (WFPC2). *Photo courtesy ESA/Herschel/PACS/MESS Key Program Supernova Remnant Team; NASA, ESA and Allison Loll/Jeff Hester (Arizona State University)* ***Book Front and Back Cover Endpages***: Here is the architectural floor plan of the *Life in the Universe?* exhibit that opened in the National Air and Space Museum building in 1976. Note in the upper left corner of the drawing is the 11½-foot filming model *Enterprise* from *Star Trek* hanging above the exhibit exit. *Photo courtesy of the Smithsonian Institution Archives, image SIA2024-00619*

INSPIRED
ENTERPRISE

How **NASA**, the Smithsonian, and the Aerospace Community Helped Launch **STAR TREK**

 **GLEN E. SWANSON**

Foreword by Margaret A. Weitekamp, PhD
Curator and Chair, Department of Space History
National Air and Space Museum, Smithsonian Institution

4880 Lower Valley Road · Atglen, PA 19310

Inspired Enterprise: How NASA, the Smithsonian, and the Aerospace Community Helped Launch Star Trek and Schiffer Publishing, Ltd., are not endorsed by or associated with *Star Trek*, NBC, CBS, or Paramount and its affiliates. For the purposes of information, commentary, and history, this book may reference various *Star Trek*, NBC, CBS, and Paramount copyrighted material, trademarks, and registered marks owned by *Star Trek*, NBC, CBS, or Paramount and its affiliates. Copyrighted products, titles, and terms and other such names are used in this book solely for editorial, scholarship, and information purposes under the Fair Use Doctrine. Neither the author nor the publisher makes any commercial claim to their use, and neither is affiliated with *Star Trek*, NBC, CBS, or Paramount and its affiliates. All photographic images and pictorial matter shown in this book are individual works of art and are not sold by the author or publisher. Items reproduced herein is by arrangement with various public and private archives and institutions, as well as from the voluntary, non-compensated contributions from private collections of both the author and individual contributors. See the associated photo credits for all applicable attributions. All non–*Star Trek* trademarks referred to or depicted in this book and other names mentioned in the text are the property of their respective owners and are used or noted solely for editorial, scholarship, and information purposes.

Copyright © 2025 by Glen E. Swanson

Library of Congress Control Number: 2024941498

All rights reserved. No part of this work may be reproduced or used in any form or by any means—graphic, electronic, or mechanical, including photocopying or information storage and retrieval systems—without written permission from the publisher.

The scanning, uploading, and distribution of this book or any part thereof via the Internet or any other means without the permission of the publisher is illegal and punishable by law. Please purchase only authorized editions and do not participate in or encourage the electronic piracy of copyrighted materials.

"Schiffer," "Schiffer Publishing, Ltd.," and the pen and inkwell logo are registered trademarks of Schiffer Publishing, Ltd.

Designed by Jack Chappell
Type set in Inter/Avenir

ISBN: 978-0-7643-6936-0
ePub: 978-1-5073-0514-0

Printed in India
10 9 8 7 6 5 4 3 2 1

Published by Schiffer Publishing, Ltd.
4880 Lower Valley Road
Atglen, PA 19310
Phone: (610) 593-1777; Fax: (610) 593-2002
Email: Info@schifferbooks.com
Web: www.schifferbooks.com

For our complete selection of fine books on this and related subjects, please visit our website at www.schifferbooks.com. You may also write for a free catalog.
Schiffer Publishing's titles are available at special discounts for bulk purchases for sales promotions or premiums. Special editions, including personalized covers, corporate imprints, and excerpts, can be created in large quantities for special needs. For more information, contact the publisher.
We are always looking for people to write books on new and related subjects. If you have an idea for a book, please contact us at proposals@schifferbooks.com.

THE BOOK IS DEDICATED TO MY MOM WHO HAS BEEN THE GUIDING INSPIRATION IN MY TREK THROUGH LIFE EVER SINCE SHE FIRST INTRODUCED ME TO MR. SPOCK IN 1971

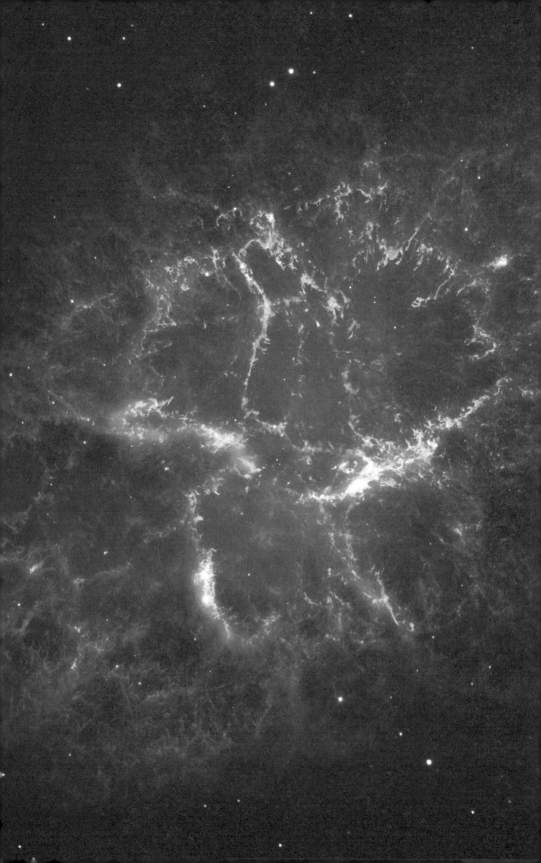

CONTENTS

Foreword 9
Introduction.. 13
Acknowledgments 17

CHAPTER 1   The RAND Corporation and Harvey P. Lynn Jr.....23
CHAPTER 2   Professional Nitpicker, Kellam de Forest 43
CHAPTER 3   Stephen Whitfield, AMT,
               and *The Making of Star Trek* 55
CHAPTER 4   The Military and Nationalism in *Star Trek*............79
CHAPTER 5   The Aerospace Industry and *Star Trek* 95
CHAPTER 6   NASA and *Star Trek* 113
CHAPTER 7   The Smithsonian and *Star Trek* 129
CHAPTER 8   Vindication through Syndication 157

Afterword 167
Endnotes 173
Bibliography 225
Index 233

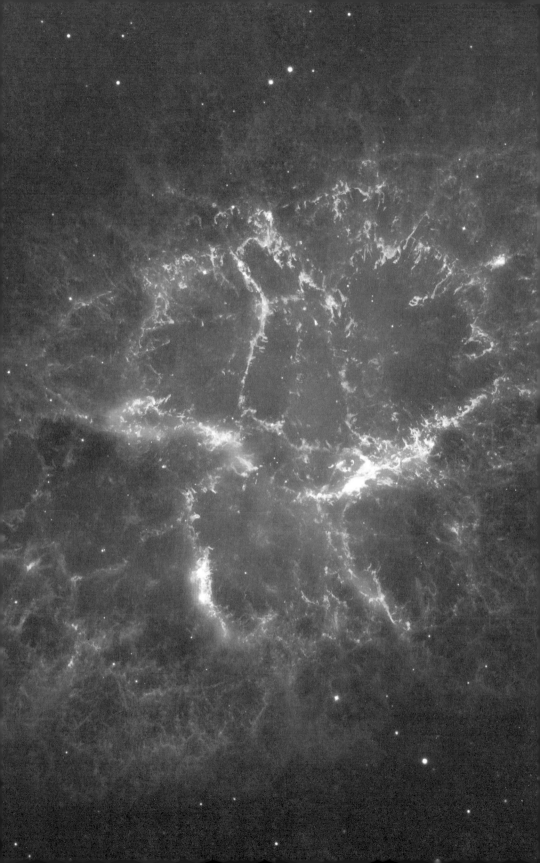

FOREWORD

A scholar's goals are often complex and sometimes just a bit contradictory. Just as news reporters try to avoid being a part of the news itself, curators and other museum professionals seek to be careful researchers, providing context and explanation, even as we also recognize our institutions' inextricable presence in the making and preservation of history. An organization as wide ranging as the Smithsonian Institution—the world's largest museum, education, and research complex—necessarily has an impact by its very presence and actions. In addition to being an artifact/specimen library for the nation and the world, the Smithsonian seeks to provide useful frameworks for understanding, adding depth to explanations, whether for scientific knowledge, cultural appreciation, or historical events. So, in my role as a curator at the Smithsonian Institution, let me try to provide a little context and some scholarly frameworks.

As Swanson explains in the following volume, in addition to the various scientific influences that *Star Trek* creator Gene Roddenberry sought for his television program, the Smithsonian's National Air and Space Museum played a direct role in sustaining the young television series by acquiring for the collection a copy of the show's pilot episode. Then, in 1974, the Smithsonian's National Air and Space Museum collected the 11-foot studio model of the *Star Trek* starship *Enterprise*. Preparing for a new building that would provide a showcase for the museum's vast holdings, the curators at the time sought to illustrate the imagination and inspiration that fueled flight, in the air and in space.

Stepping back and reflecting on the cultural moment of 1974, however, it's a little startling to realize what a different environment it was for the *Enterprise* model then. When Paramount shipped the 11-foot shooting model to the Smithsonian, it was a leftover prop from a canceled three-season television show. The program had become a syndication staple, appearing as reruns on screens across the country. Two seasons of *Star Trek: The Animated Series* (1973–75) had continued the franchise's life. And Roddenberry was writing and pitching other ideas. But *Star Trek*'s fan community was scattered and disjointed. Fans gathered occasionally at conventions or more frequently at fan clubs—but their cultural clout had not yet coalesced. Moreover, it would be several years before George Lucas's *Star Wars* (1977) revived the somewhat somnolent category of science fiction films, reinvigorating the genre. When the museum inquired about the *Star Trek* starship *Enterprise* studio model, executives at Paramount Pictures really did not see any reason to keep it. For the museum, it seemed like a great way to add a bit of Hollywood glitz to a new exhibit. And it fit with the mix of older and more-recent materials that were being assembled for the museum's artifact-rich exhibits.

As Swanson describes, the *Enterprise* studio model first went on exhibit in the *Life in the Universe?* exhibit mounted in the Smithsonian's Arts and Industries building, before that exhibit was disassembled and reconstituted (with some adjustments) in the museum's new building on the National Mall that opened on July 1, 1976. But the *Enterprise* model was not alone in illustrating popular cultural imaginings about spaceflight in the museum's new building. The inaugural *Rocketry and Spaceflight* exhibit, for instance, included the history not only of rockets going back to early Chinese fireworks but also commercially available toys from *Buck Rogers* and *Flash Gordon*, a mockup of the space projectile that Jules Verne described in his 1865 novel *La Voyage a la Lune* (*A Trip to the Moon*), and a model of a spacecraft imagined by Russian theorist and writer Konstantin Tsiolkovsky. Having the studio model of the *Star Trek* starship *Enterprise* right down the hall added an illustration of what it might look like to have an immense crewed starship flying to different solar systems.

No one associated with the studio model acquisition in the 1970s could have predicted that *Star Trek* would survive in various and evolving forms for decades to come, gaining strength and popularity well into the twenty-first century. *Star Trek* continues to grow, with new creative productions, seemingly endless merchandise, and a fandom that is legendary for its passion, attention to detail, and community building. No one could have seen that coming. Nor could those curators have anticipated how generations of scholars would analyze *Star Trek*.

When people visit the museum today, they often notice the popular-culture touches in the new renovated exhibits in the building on the National Mall and assume that these are fresh additions inspired by recent scholarly attention to the subject. As I have noted above, those kinds of exhibit inclusions are not new. But the academic study of popular culture has flourished in recent decades. On the basis of years of scholarly work dating back to the 1970s, researchers have considered television, comics, movies, and other forms of popular culture as significant cultural texts that are worthy of academic analysis and attention.

In particular, historians, media-studies scholars, and other academics have found *Star Trek* to be a rich source for careful analysis. Books and articles have dissected the original television program, its spin-off successors, and the motion pictures. Even the recent streaming series have collectively provided rich

FOREWORD

sources for academic critique. Indeed, given the efforts by *Star Trek* creators and fans to interweave all parts of *Star Trek* into one canon, media-studies scholar Daniel Bernardi has described the *Star Trek* universe as a "mega-text," a body of work so unified that it can legitimately be studied as one text.[1]

Any summary of the many books and scholarly articles can only touch on the variety that exists. There have been analyses of *Star Trek* and race, as well as gender, religion, and the use of real history.[2] *Star Trek*'s fandom has also merited critical attention in its own right.[3] Scholarly books like this one, about the production side of *Star Trek*, have been rarer than analyses of how *Star Trek* has been received or interpreted. Indeed, most books about *Star Trek*'s production have been insider's guides or reference books aimed at fans.[4] Production has only just begun to receive scholarly attention rooted in analyses of the greater historical context and grounded in deep archival research.[5]

Such analyses constitute a decades-long conversation carried out through books and articles, in which researchers answer previous scholarship with new evidence and new arguments. Likewise, museum work is also a kind of cathedral building that takes place across generations. Curators bring artifacts into museum collections in order to document historical (or cultural, technological, or scientific) achievements. But then the artifacts have a continued presence in the collection as interpretations, contexts, and understanding change over time. New curators find fresh ways to showcase the importance of those objects, telling stories that offer different perspectives on the history. And curators work with the rest of a museum's staff to maintain and preserve the collection so that it can be handed off to the next generation for *their* interpretations and care.

There are good reasons that historians revisit subjects. There are new materials to unearth, and new interpretations to be made with additional information. Much has been written about *Star Trek*, but there is still more to learn. Swanson's research has gathered together images, documents, letters, blueprints, photographs, and plans that shed new light on the history of *Star Trek*'s many early influences. Included in that story is the Smithsonian Institution's direct role. Along with gaining an appreciation for the many influences that shaped the beloved space science fiction franchise, I hope that readers will also get some new insight into the work that the Smithsonian does as caretakers of historical artifacts on behalf of the nation and the world.

Margaret A. Weitekamp, PhD
Curator and Chair, Department of Space History
National Air and Space Museum, Smithsonian Institution

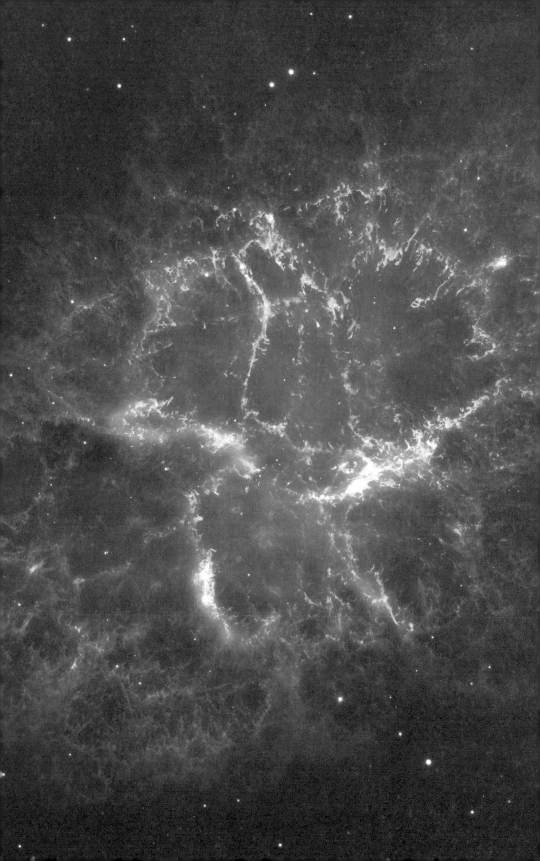

INTRODUCTION

The genesis of this book began in 1998, when I became the chief historian for NASA's Johnson Space Center in Houston. As a full-time civil servant, I had access to various archives and records that chronicled the history of our nation's space program. It was during this time that I began finding evidence not only of NASA's support, but that of others who served to inspire the creation of the original *Star Trek* television series. I found memos, letters, and other documents that collectively began forming a narrative showing how individuals within NASA and other government institutions, such as the military and the Smithsonian, influenced the development of Gene Roddenberry's creation. I found this, as Mr. Spock would say, "fascinating."

As a baby boomer, *Star Trek* was a big part of my childhood. I was six years old when the last episode aired on TV back in 1969 before it was cancelled by NBC. I can't even remember if my parents let me watch it back then because it aired around my bedtime. But things changed in the 1970s with the show's syndication. Suddenly I, along with millions of others, was able to watch *Star Trek* not just one night a week, but five nights a week, and sometimes even twice a day!

It's crazy to think that back then there were only a few books, toys, and licensed products for fans of *Star Trek*. And to be honest, the ones that did exist were low quality. But there were some gems, like the AMT *Enterprise* and Klingon model kits. Then there was Stephen Whitfield's book, *The Making of Star Trek*, which became our bible. My friends and I would spend hours poring over that book while talking about the show and playing with our model kits.

Even as I grew up, my interest in *Star Trek* never waned. In the early 1990s I founded the world's first peer-reviewed journal dedicated to the history of spaceflight. Called *Quest: The History of Spaceflight Quarterly*, it is still in print today. Around that same time, I also published another magazine called *Countdown* that focused on the space shuttle program. Both publications often contained articles about *Star Trek* because the show was so intimately connected to space.

Over the years, my *Star Trek* model kits disappeared but I kept my original copy of Whitfield's book. Yellowed and falling apart from age, I took the book with me when I moved to Houston to work for NASA. I remember seeing a photo in the book of Gene Roddenberry and the actors that played the show's characters Mr. Scott and Dr. McCoy. The picture showed them standing next to an experimental aircraft affixed with a large NASA logo. I always wondered what the story was behind that photo and when I worked for NASA, I finally found out.

After leaving NASA I continued researching the early influencers of *Star Trek*. Much has been written about how Roddenberry's show inspired others, but little has been documented about what inspired *Star Trek*. Who exactly were the individuals that helped make the show? People like Harvey Lynn, the RAND scientist who was hired as the first technical consultant, and Kellam de Forest, who was paid to do fact checking that ultimately gave *Star Trek* the Gorn, Stardates and Spican Flame Gems. And Stephen Whitfield, the man who helped give us the first AMT *Star Trek* model kits and who first told us about these early influencers through his book *The Making of Star Trek*.

When the COVID-19 pandemic hit, I finally had time to begin writing my book. I published several articles to help gauge reader response to the topic. The first was a piece that appeared in the January/February 2021 issue of *Michigan History* magazine, which focused on the history of the original AMT *Star Trek* plastic model kits. The second was a March 2021 article about how *Star Trek* helped NASA and how NASA helped *Star Trek*, which ran in the Smithsonian's *Air & Space* magazine. Both articles were well received and inspired me to write a third, more extensive piece for the online publication the *Space Review*. That article generated a lot of feedback which included suggestions to write this book

In October 2023, one of the most extensive collections of *Star Trek* props, filming miniatures, and costumes ever gathered in one location occurred in Dallas, Texas. The assembled items were part of Heritage Auction's Greg Jein Collection. Screen-used *Star Trek* memorabilia from the original television series, including the films and later television spin-offs, were all sold to the tune of a record-breaking $13.6 million.

Greg Jein was an Oscar- and Emmy-nominated model-maker and visual-effects artist who got his start working on the original *Star Trek* television series when he was hired back in 1968 to help with Roddenberry's Lincoln Enterprises merchandising company. Over the years, Jein rose to fame as he led the team that built the Mother Ship from *Close Encounters of the Third Kind*, which, like the original production model *Enterprise*, is now part of the permanent collections of the Smithsonian.

I was fortunate enough to be able to attend the Jein auction. Sitting among the on-site bidders (most participated remotely), I observed that most were middle-aged or older males like me who came to relive part of their childhood and, if they were lucky, walk away with a piece of it. Some were career model makers who came not only to bid on items, but also to show their respects for Jein. If one wanted to relive *Star Trek*, this auction was the event to attend. As pieces of *Star Trek* history sold, I was able to get a good look at many of the

INTRODUCTION

most-valuable items. These included screen-used props from the original series as well as filming miniatures and costumes. Highlights included the original Botany Bay model seen in the classic *Star Trek* episode "Space Seed" and the 22-inch filming-miniature shuttlecraft.

As a *Star Trek* fan, I have one foot firmly stuck in the past because I like nostalgia, while, at the same time, I have the other foot placed in the future because I like the positive message that the original series brought to its viewers. While watching others compete at the auction for "a piece of the action," to quote a line from a classic *Star Trek* episode, I thought about the show's influence. *Star Trek* gave viewers doses of serious science, politics, and commentary. But the show could also be just plain campy, which is what made it fun television to watch.

Talking with others at the auction, I learned that many were drawn to *Star Trek* for the same reasons as I was—we all liked the show. Good writing, great chemistry and, above all, a positive message made *Star Trek* stand out. Even if such a future never materialized exactly the way the show had predicted, we all agreed that the series was influential even to the point of determining some of our careers.

In writing this book, I learned a lot, especially about the show's past. The *Star Trek* franchise of today is very crowded, with countless publications and other products competing for attention. A lot has been written about how *Star Trek* has inspired others. But little has been published about how *Star Trek* itself was inspired. It is my sincere hope that this book will present new information to show that there is still much to learn about the inspired origins of one of the most beloved science fiction series of all time.

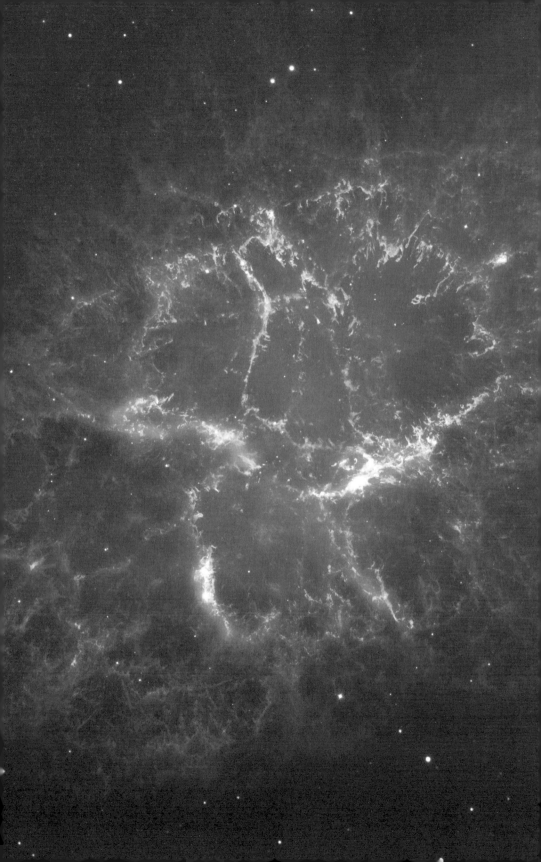

ACKNOWLEDGMENTS

Writing a book is a formidable task. I began working on *Inspired Enterprise* nearly 25 years ago and, during the journey, amassed a large number of individuals and institutions that proved invaluable in helping complete it. It is hard to keep track of everyone who contributed over the years, but I've attempted to do my best. If I've forgotten anyone, it was not intentional.

During my formulative years beginning in the mid 1970s, I was fortunate enough to grow up in a neighborhood that nurtured imagination and creativity. Godwin Heights in Wyoming, Michigan was where I went to school and, thanks to the generous tax support of the nearby General Motors plant, we had a new high school built that included a planetarium. It was under the dome of this amazing learning tool that I fell in love with all things relating to space.

During this time, I also became a member of our local *Star Trek* club. Called the "*Star Trek* Club of Grand Rapids," or simply "STCOGR," this is where I met others who were like me. I widened my circle of friends through this club and, to this day, still keep in touch with many of them. These include Eric Janulis, Ed Nelson and Russ Venlos. Sadly, some have passed away including Dave Hoffman who was the director of our planetarium.

Another friend since elementary school was Jerry Fellows, who passed away in 2011 at age 47. He loved *Star Trek* as much as I did and, during our afterschool walks home, we often talked about the show. I also can't forget the late Dave Marshall, another friend in school whose interest in science fiction and visual effects inspired me to a greater appreciation of the genre and movies in general.

In writing this book, one of my biggest supporters throughout this long and arduous process has been Karl Tate, whose encyclopedic memory of all matters pertaining to the *Star Trek* franchise proved invaluable as well as his talents as a graphic artist. I first met Karl at Wonderfest back in 2019 and since that time, we have embarked on various projects and adventures. Karl read through my drafts to make sure my *Trek* facts were correct. He also contributed many of the graphics as well as provided input on the book's cover and its overall design.

Jim Banke, a fellow *Star Trek* fan whom I've known since the 1990s, came along when I needed help the most. Jim covered the space beat for *Florida Today* and later Space.com. He read my manuscript and provided invaluable editorial advice. In addition, long-time friend and fellow spaceflight historian Dwayne Day was also very helpful in steering me on the right path.

Other contributors included David Arland whose knowledge of not only *Star Trek* but also vintage television and early broadcast history has proven helpful. Kipp Teague offered some great snapshots that helped show what it was like growing up with *Star Trek* in the 1960s. Christopher Beamish and his extensive *Star Trek* trims and original photos also proved to be a great asset. William Neff whose access to various newspaper databases at his employer, the *Washington Post*, helped me find articles that supported my narrative. Bill Kobylak provided me with some rare and unique items in his collection that included a bunch of original screen-used tribbles. Gary Kerr, also known as "Mr. Starship *Enterprise*," who worked on the 2016 Smithsonian restoration of the original 11-foot filming model, offered open access to his knowledge and personal image collection that helped me fill in details about the ever-changing history of our favorite starship. I also can't forget Matt Cushman and his brother Chris whose detailed cutaways of various "ships of the line" help make the *Star Trek* universe seem more real to us all. Others include Gerald Gurian, whose collection of original *Star Trek* trims helped reveal things that I never knew existed. Larry Nemecek also helped me during my book's Kickstarter by allowing me to be a guest on one of his podcasts. That appearance served to boost my fundraising campaign toward a successful completion.

Speaking of Kickstarter, a big thank you to all of my contributors. Special callouts to Rhet Topham and Javier Bonafont whose contributions not only allowed me to reach my funding goal but to exceed it. Thanks to their financial support, these two are the proud owners of original concept drawings of the early *Enterprise* and Klingon ships done by Matt Jefferies. I also need to thank those who contributed to my second fundraising effort. Through their support I was able to raise funds that helped me fund my Indiegogo campaign.

A book about *Star Trek* would not be complete without acknowledging the support of Doug Drexler. Doug remains an inspiration to all fans with his bountiful enthusiasm even through tough personal times that were especially challenging. Doug is the fan that all of us wish we could be as he is the rare individual who has taken his interest in *Star Trek* and made it into a successful career. Doug remains approachable to anyone, and is willing to help no matter what. His stories are amazing, and I hope he will preserve them all in his forthcoming autobiography that will surely inspire others to grab hold of life, pursue one's passion and never give up. There is also Rick Sternbach, another very approachable *Star Trek* celebrity from the era of *Star Trek: The Motion Picture* and subsequent television spinoffs. Rick never hesitated to respond to my questions no matter how trivial they may seem.

ACKNOWLEDGMENTS

Thanks to Mike Okuda, who paused from his busy schedule to help and, along the way, encouraged me to keep on going so that readers will learn something new about the making of the franchise.

Marc Cushman and the contributions made by his six-volume *Star Trek* history *These are the Voyages* cannot be overlooked. His work taught me the value of recognizing one's sources. The Gene Roddenberry Papers, housed in the Charles E. Young Research Library at the University of California in Los Angeles, provided the bulk of my primary source material. I am especially grateful to Simon Elliott and Peggy Alexander, members of their research library staff who accommodated my multiple visits and requests to use their collection. In addition, the University of California at Riverside, which houses the Jay Kay Klein Papers in their Special Collections and University Archives Eaton Collection of Science Fiction and Fantasy, is a place in which you can easily lose oneself while examining their more than 60,000 photographs chronicling the history of science fiction conventions. The keepers of the gate to this wonderful resource are Phoenix Alexander, Jessica Geiser, Andrew Lippert, and Melissa Conway, all of whom helped me in using their material for my book.

Robert Luke Kelly, Vicki Lynne Glantz and Nora M. Plant of the University of Wyoming's American Heritage Center helped me explore their *Star Trek* collections, which include the Gene L. Coon's papers along with those of Forrest J. Ackerman and Martin Caidin. Wayne State University's Walter P. Reuther Library and their Labor and Urban Affairs Archives also proved to be a good resource for material on AMT's employment history, which their staff members Gavin Strassel and Shae Rafferty helped me find. Mark Bowden of the Detroit Public Library Archives was also helpful as was Jeremy Dimick from the Detroit Historical Society. These folks and others helped me track down John Mueller, also known as Mr. AMT who is alive and well. John began work at AMT in the early 1960s and stayed employed with them in either a full or part-time capacity for over forty years. Now retired, I visited him and his wife in Iowa where he allowed me full access to his extensive files and photo collection which proved invaluable.

Edda Manriquez of the Academy of Motion Picture Arts and Sciences (AMPAS) Film Archive's Pickford Center for Motion Picture Study was incredibly patient while she tracked down several reels of obscure original special effects footage from *Star Trek*. In addition, the Academy of Motion Picture Arts and Sciences' Margaret Herrick Library holds a wonderful series of photos in their Linwood Dunn collection that features the 11-foot original filming model *Enterprise* and the 22-inch miniature shuttlecraft. Many thanks to Taylor Morales, Louise Hilton, and Faye Thompson for helping me get access to this material.

A special thank you to Cathy Le, the librarian at Golden West College in Huntington Beach, California, who helped uncover photos and documents pertaining to the 1972 and 1973 Space Expos held on campus. This information along with interviews with Ron Yungul who attended both of these events, helped confirm details on the first public display of the original 11-foot production model *Enterprise* from *Star Trek* before it was shown at the Smithsonian.

The folks at NASA also cannot be overlooked for their invaluable contributions. This includes Leslie Williams, the now retired news chief at NASA's Armstrong Flight Research Center, who helped locate a series of NASA photos that show the original production cast and crew of *Star Trek* visiting, what was then called, the Dryden Flight Research Center, on April 13, 1967. The folks at NASA/JPL, including James Fanson, Marc Rayman, Victoria Castañeda, Julie Cooper, and Erik Conway, helped me find records of *Star Trek* in their collections. Barbara Scott, Ken Carpenter, Holly McIntyre, and Christine Stevens, all from NASA's Goddard Spaceflight Center, were also instrumental in helping me connect with the late Alberta Moran and her family to find photos documenting Leonard Nimoy's 1967 visit in conjunction with the National Space Club's Goddard Memorial Dinner.

Other individuals that helped include Greg Weir and Karen Schnaubelt, both of whom provided photos and background material on Karen's father, Franz Joseph Schnaubelt, author of the original *Star Trek Technical Manual* and the *Star Trek Blueprints*. Jonathan Ward's extensive knowledge of astronomy helped me identify the many astronomical images that are seen throughout the original *Star Trek* episodes. I also cannot forget the help of John Fahey of the US Naval History and Heritage Command and USAF Academy command historian Brian Laslie, both of whom provided great material to help verify the military connections to *Star Trek*.

Jamie Hood of AMT/Round 2 models contributed his time and allowed access to early drawings of AMT's first *Star Trek* models; Michael Kmet and Maurice Molyneaux, the duo responsible for the *Star Trek* Fact Trek blog also were helpful as was Jeff Bond who steered me clear of faulty conclusions regarding the music of *Star Trek*. Others include Scott D. Swank and Patrick Cutter of the National Museum of Dentistry; *Space Review* editor Jeff Foust; Michigan History editor Emily Allison; Smithsonian *Air & Space* editor Christopher Klimek and his assistant Caroline Sheen; and Todd Crumley and Heather Sulier, along with the folks at the National Archives and Records Administration (NARA), who helped with my in-person visits.

Many of the stories shared in my book would not have been possible without the contributions of those interviewed. Sadly, some of these folks passed away before they could see the finished book. I want to thank those that generously gave of their time to talk to me. These include Peggy Alexander, Tom Crouch, Steve Durant, Ann de Forest, Jackie Wiley Hartley, Patricia Hartley, Greg Kennedy, Mike Mackowski, Alberta Moran (deceased), Pamela

ACKNOWLEDGMENTS

"Penny" Jeanne Moran, John Mueller, Paul Newitt, Bobbi Moore Poe, Meagan Prelinger, Bannon Preston, Larkin Preston, Richard K. Preston II, Herbert S. Schlosser (deceased), Karen Schnaubelt, Melvin Schuetz, Peter Sloman, Amy Stamm, Scott Steidinger, Richard Van Treuren, Gregory Weir, Susan Kathleen Whitfield and Ron Yungul. Of those interviewed, three opened their homes to allow me to meet with them in person to not only share their recollections but also their personal photos and documents. These were Bobbi Moore (Stephen Whitfield's sister) Kellam de Forest (deceased) and Dennis Lynn (Harvey Lynn's surviving son).

I also need to thank the wonderful folks at the Smithsonian. This includes Carolyn Russo, Marguerite Roby, Kate Igoe, Deborah Shapiro, Margaret Weitekamp, Erik Satrum, Heidi Stover, Brian Nicklas, Elizabeth Borja, and Amanda Buel. All these people were highly resourceful and extremely helpful, not only in granting me permissions to material used in this book but also in going out of their way to hunt down some of the more obscure *Star Trek* items in their holdings.

Thank you to Schiffer Publishing for taking on the task of getting this published.

Finally, I must thank those closest to me for all their support. This includes my amazing wife Deana who I was fortunate enough to meet after leaving NASA. She had just moved from Southern California to West Michigan to begin a teaching career as an assistant professor at Grand Valley State University. We began dating in 2003 while I was deciding if I should stay in Houston or move back to my hometown in Grand Rapids. Now, after nearly twenty years of marriage, I never regretted my choice. She has been the best soul mate I could hope for who loves so many of the same things that I do, especially space. Now a full professor, she has helped me intellectually by being an excellent editor, a creative promotor, and patient counselor. She is accomplished scholar in her own right, a published author and my goddamn partner who fully understands that being around someone who writes about the past all while trying to live in the present, isn't easy. I also want to thank our son Luke who refused to stop growing into the fine young man that he is during all the time I spent writing this book. And finally, I can't forget my mother, who is the adventurous soul who first introduced me to *Star Trek* over half a century ago.

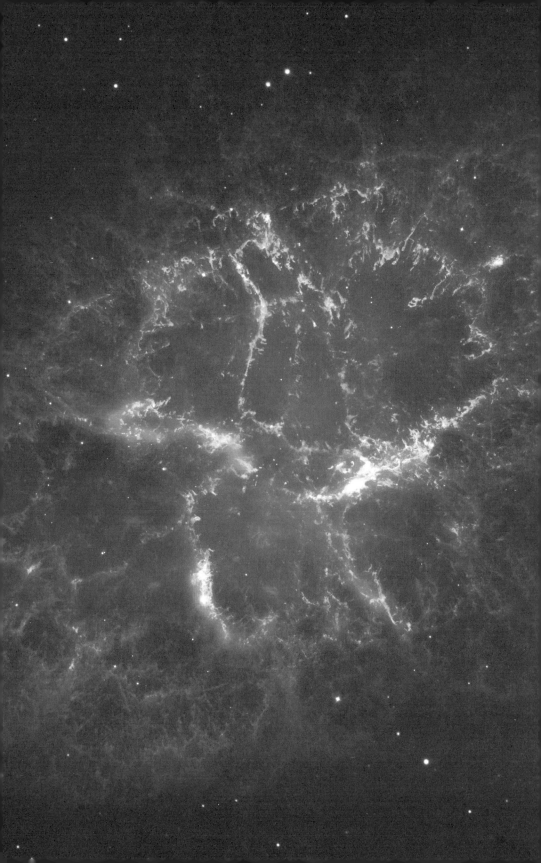

CHAPTER 1

The RAND Corporation and Harvey P. Lynn Jr.

To help sell his vision of *Star Trek* as a new television series, Gene Roddenberry put together a pitch designed to get the attention of network executives. In *The Making of Star Trek*, author Stephen Whitfield described how all writers for television assembled something called the "Series Format." He explained that this outline or framework served as a sales tool with which writers presented their ideas as "the skeleton on which a new concept in television science fiction was to be built."[1] Later this would evolve into the famous *Star Trek Writers/Directors Guide*, or simply the *Star Trek Guide*, which would serve as the bible or primer for those wishing to write for the show.

To help convince producers skeptical of anything having to do with science fiction, Roddenberry began his studio pitch in a very nontraditional way. On the second page of his pitch, he wrote the following:

> *Star Trek* offers an almost infinite number of exciting Science Fiction stories, thoroughly practical for television. How? Astronomers express it this way:
>
> $$Ff^2 (MgE)-C^1Ri^1 \times M = L/So$$
>
> Or, to put it in simpler terms; by multiplying the 400,000,000,000 galaxies [star clusters] in the heavens by an estimation of average stars per galaxy [7,700,000,000,000,000,000,000,000,000,000], we have the approximate number of stars in the universe, as we now understand it. And so . . .

... if only one in a billion of these stars is a "sun" with a planet ...

... and only one in a billion of these is of Earth size and composition ...

... there would still be something near 2,800,000,000,000,000,000,000,000,000 worlds with a potential of oxygen-carbon life ...

... or (by the most conservative estimates of chemical and organic probability), something like three million worlds with a chance of intelligent life and social evolution similar to our own.[2]

"In a language of television," Roddenberry explained that *Star Trek* was "built around characters that travel to worlds 'similar' to our own and meet the action-adventure-drama which becomes our stories." He was not clear on when exactly the show takes place, only that "the time is somewhere in the future," but he did tease that it is still "close enough to our own time for our continuing characters to be fully identifiable as people like us."[3]

Roddenberry knew that to try to sell a serious science fiction television series, he had to break the mold that resulted in mediocrity. To grab the studio's attention, he emphasized very early in his studio pitch the "science" part of "science fiction," taking a gamble that this approach might work. During one early interview with Roddenberry about *Star Trek*, the interviewer mentioned the words "science fiction." "Science fiction nothing," snapped Roddenberry in reply to the interviewer's question. "It's science fact. . . . It's as true as science . . . er . . . fiction can be. Everything we use is based on some scientific fact. Or some theoretical scientific fact."[4]

Roddenberry felt that there was no better way to impress upon a skeptical public the scientific nature of his new show than to use an equation, even if he had to make up one. Fans have noted that Roddenberry's equation is similar to another equation first developed by American astronomer and astrophysicist Frank Drake.

During the late 1950s and early 1960s, Drake performed the first modern SETI (Search for Extra Terrestrial Intelligence) research. During this same time, physicists Giuseppe Cocconi and Philip Morrison published a seminal paper titled "Searching for Interstellar Communications." In this paper, the authors described how communications over galactic distances were possible by using radio waves, and that existing equipment could be used to send messages to nearby stars as well as receive them. The authors noted, "Few will deny the profound importance, practical and philosophical, which the detection of interstellar communications would have. We therefore feel that a discriminating search for signals deserves a considerable effort. The probability of success is difficult to estimate; but if we never search the chance of success is zero."[5]

CHAPTER ONE

Shortly after this paper appeared, Drake was approached by the National Academy of Sciences to convene a small meeting to examine the questions raised by it and to propose where they should go next. "As I planned the meeting, I realized a few days ahead of time we needed an agenda," explained Drake during a NASA forum held in 2003 that explored the search for extraterrestrial intelligence, "and so I wrote down all the things you needed to know to predict how hard it's going to be to detect extraterrestrial life."[6] This then led to an equation used to estimate the number of active and communicative extraterrestrial civilizations in the galaxy.

$$N = R^* \times f_p \times n_e \times f_l \times f_i \times f_c \times L$$

Known as the "Drake equation," it is a probabilistic argument more than a precise number created to help stimulate discussions about the possibility of intelligent life in the universe.

Roddenberry was at least aware of something like the Drake equation when he began assembling his vision for *Star Trek*. Whitfield, who worked with Roddenberry during production of the series, wrote in the introductory pages of his book *The Making of Star Trek* that

> the technical achievement some eight years ago of the ability to detect reasonable manifestations of intelligent life over interstellar distance has stimulated far-reaching theoretical studies of these matters. Some minor attempts at the detection of possible radio signals have been carried out in the United States and the Soviet Union. Although the conclusions to be drawn from the theoretical studies are still controversial, the most widely accepted key points to emerge are as follows:
>
> (1) There has probably been and is at present in the universe an enormous number of life-bearing planets, most of which have evolved an intelligent, technical species . . .

At the end of the above copy, Whitfield included the following: "Excerpt from *Prospectus in the Search for Extraterrestrial Civilizations* by F. D. Drake, Center for Radiophysics and Space Research, Cornell University, Ithaca, New York."[7]

Here the evidence seems to favor that Roddenberry knew of Drake's work. But Drake wasn't alone in his formulations of equations that helped us consider the probability of intelligent life outside Earth.

◀ INSPIRED **ENTERPRISE** ▶

The RAND Corporation and *Habitable Planets for Man*

The Manhattan Project was a huge collaborative effort. During this time, many of the greatest minds worked alongside the military to unlock the secrets of the atom and develop the world's first atomic bomb, which helped end the war in the Pacific and bring a close to the Second World War. That effort brought attention to the importance of how science, technology, and research could work to help ensure success on the battlefield.

After World War II, the military took notice of this successful collaboration. Individuals within the War Department and the Office of Scientific Research and Development began to look at how they could form a private organization that would connect the military with research and development.

On October 1, 1945, the Douglas Aircraft Company established Project RAND (**R**esearch **AN**d **D**evelopment). Less than a year later, they issued their first report, *Preliminary Design of an Experimental World-Circling Spaceship*, which laid out a plan for the design, performance, and use of artificial satellites. This was thirteen years before the launch of Explorer I, America's first artificial satellite, and the creation by Congress of the National Aeronautics and Space Administration (NASA).

By 1948, Project RAND expanded to include researchers and expertise in a growing number of fields, including mathematics, engineering, aerodynamics, physics, chemistry, economics, and psychology. After the US Air Force encouraged Project RAND to become an independent nonprofit corporation, Douglas Aircraft did just that. On May 14 of that same year, RAND became a nonprofit corporation in California to "further and promote scientific, educational and charitable purposes, all for the public welfare and security of the United States of America."[8] Out of RAND emerged another scientist whose work was to be noticed and used by Roddenberry.

Around the same time as Drake's equation was developed, there appeared another study that proposed a slightly different formula. This was done by Stephen H. Dole, a scientist who worked for RAND. Like with Drake's work, Roddenberry used Dole's research to help connect his fictional world of *Star Trek* to the real world of science.

Dole was born on June 26, 1916, in West Orange, New Jersey, and received his bachelor's degree in chemistry cum laude from Lafayette College. He went on to study at the US Naval Academy Postgraduate School at Princeton University and then at UCLA. Dole served in the Pacific theater as a US Navy communications officer during World War II. After the war he joined the RAND Corporation, where he was a member of the senior staff's engineering science department. As head of RAND's Human Engineering Group from 1959 to 1968, Dole conducted studies on systems and assessments of the radiation environment in space. During

CHAPTER ONE

this time, he also directed a number of studies sponsored by NASA, including reports on contingency planning for spaceflight emergencies.

In 1962, Dole prepared a study for the US Air Force that estimated the probabilities of finding habitable planets in the universe. Recognizing that "the space age is still very much in its infancy,"[9] Dole focused his study not on the methods of propulsion or the technical problems on how man might travel in space but rather on "Where will man eventually want to go and what will he find when he gets there?"[10] This was the kind of work that appealed to Roddenberry as he sought to build his *Star Trek* universe on the most-current scientific research. When Dole's report made it out of the RAND think tank and into the popular press under the title *Habitable Planets for Man*, Roddenberry made sure to get a copy.

In *Habitable Planets for Man*, Dole revealed the nature of worlds that may support life in the universe, the probability of their existence, and ways of finding them, including assessments of nearby stars that have a relatively high probability of containing habitable planets. His work included equations, but unlike those of Drake's, which focused on intelligent life, Dole's were designed to help estimate the number of planets in our galaxy that may harbor life, whether it be intelligent or not.

In addition to giving Roddenberry evidence that life on other planets was highly probable, Dole's RAND report discussed the system used by astronomers to help classify stars. Most stars are classified by using a series of letters, O, B, A, F, G, K, and M, with the hottest stars being O type and the coolest M type. This system may have inspired Roddenberry with the idea to apply a similar classification system to planets. Even though our own sun is a class G star, Roddenberry chose to refer to those planets that are most like Earth as "Class M" planets and therefore among the ones most visited throughout the series. The third revision of the *Star Trek Writers Guide*, dated April 17, 1967, stated, "Where possible, you will confine your landings and contacts to Class 'M' planets approximating Earth-Mars conditions."[11] Perhaps Roddenberry chose "M" because it falls in the middle of the alphabet, reasoning that half the planets encountered might be like Earth while the other half were something else.

Roddenberry also realized that establishing the concept of "Class M" planets was a production compromise. As Whitfield noted in his book, "While billions and billions of planets scattered throughout the galaxy undoubtedly contain many weird life forms, the cost, in makeup and special costumes, makes their frequent appearance on a weekly television show prohibitive. Class M planets, on the other hand, will contain aliens, but generally of a humanoid type, bringing makeup and costuming costs within the realm of reason."[12] The idea of using earthlike planets was also built into the original series in the form of "parallel worlds." In the first draft of his studio pitch, Roddenberry wrote, "The 'Parallel Worlds' concept makes production practical by permitting action-adventure science fiction at a practical budget figure via the use of available 'Earth' casting, sets, locations, costuming, and so on."[13] Roddenberry knew that in order to help sell his show to studio executives, he had to keep costs down. Gallivanting through the Galaxy each week exploring strange new worlds would be a deal breaker if most of the production costs went into building them. The use of earth-like "Class M" planets was the solution. This allowed television viewers to still explore the final frontier without breaking the bank.

Roddenberry was a good visionary. He saw the potential of the future and what it might bring, but he was not a scientist or a mathematician. A close look at his equation at the beginning of this chapter quickly reveals that his version simply doesn't make sense even to those with a minimal math education. "C to the power of 1" is just C. In looking back, one can see how Roddenberry was inspired by these complex studies. One can also see that he did not understand them very well.

Even though no serious scientist or mathematician would ever write an equation like Roddenberry's, it made good copy. In the first-season *Star Trek* episode "Balance of Terror," Dr. McCoy makes reference to Roddenberry's original contrived equation. In this episode, Kirk begins to doubt his abilities as captain, confessing to McCoy, "I look around that bridge, and I see the men waiting for me to make the next move. And Bones, what if I'm wrong?" McCoy replies: "In this galaxy, there's a mathematical probability of three million Earth-type planets. And in all of the universe, three million million galaxies like this. And in all of that, and perhaps more, only one of each of us. Don't destroy the one named Kirk."

Roddenberry made an early effort to publicize the use of these scientific studies while he was creating *Star Trek*. He reasoned that even though television viewers might not fully understand some of the details of what they were watching, they would appreciate the fact that it wasn't just made up but was based on real science done by people such as Drake, Dole, and the RAND Corporation. Roddenberry wanted to impress upon his audience the fact that *Star Trek* was a show that treated both science fact and science fiction seriously.

In early 1965, shortly after principal filming for *Star Trek*'s first pilot, "The Cage," had ended, Jeffrey Hunter, who played Captain Christopher Pike in that pilot, was interviewed by the *Los Angeles Citizen News*. In this interview, he acknowledged Dole's original RAND report, saying, "The thing that intrigues me most about the show is that it is actually based on the Rand Corp projection of things to come. Except for the fictional characters, it will almost be like getting a look into the future and some of the predictions will surely come true in our lifetime."[14]

CHAPTER ONE

In addition to Hunter's direct reference to Dole's RAND report, Whitfield calls out the same study in *The Making of Star Trek* noting "The Rand Corporation recently speculated that there could be 640,000,000 planets in this galaxy alone where you could open the door of a spaceship, step out on the planet surface, and breathe fresh air."[15]

Roddenberry's use of current scientific studies and literature such as the Drake equation and Dole's RAND report shows that he wanted to embed in the production from the very beginning a level of believability and technical accuracy. His premise for the show from the start was that *Star Trek* is: "Real adventure in tomorrow's space, based on the best scientific knowledge and estimates of what future astronauts might face."[16] This meant that he would have to not just call upon such scientific and technical expertise when he first developed the idea for the show and its two pilots, but he would also have to hire a pool of experts to direct and influence *Star Trek* throughout the series' production.

Harvey P. Lynn Jr.

Donald Irwin Prickett was one of Roddenberry's closest friends. He and Roddenberry grew up together and went to the same school in Los Angeles. Eleven days after the bombing of Pearl Harbor, they both joined the army and graduated from training at Kelly Field where both he and Prickett received their officer's commissions as second lieutenants. In September 1942, Roddenberry joined Prickett on his first assignment at Bellows Field, Hawaii, with the 394th Squadron, 5th Bombardment Group, flying B-17 bombers.[17]

Because Roddenberry knew Prickett, it came as no surprise that he would later call upon his old friend in 1964 when he was developing *Star Trek*. By 1964, Prickett, now a full colonel, was deputy chief of staff of the Air Force's Weapons Effects and Tests Group Headquarters, Field Command Defense Atomic Support Agency based at Kirkland Air Force Base in Albuquerque, New Mexico. Here he served as a program director for nuclear weapon tests, witnessing more than 100 atmospheric tests of nuclear bombs between 1951 and 1963 that were conducted in Nevada and the South Pacific.

In May of that year, Roddenberry sent Prickett a letter asking if he knew of anyone who could offer technical advice for the new series that he was developing. Prickett was honest in his reply. After reading Roddenberry's story treatment he knew he was not the person to ask. "To answer your query instead of trying to be an expert in your business, let me assure you that I will be only too happy to put you in touch with personnel from RAND and/or the Space Technology Labs or the AF Space System Division."[18]

Prickett sent a letter to Jack Whitener who worked in the Physics Division of the RAND Corporation in Santa Monica. In that letter he included a copy of Roddenberry's prospectus, explaining that "the writer is interested in the reaction of people in the scientific community such as yourself. He may also be interested in getting some advice on this fairly soon."[19]

Whitener's daughter Theresa recalled, "My father was a nuclear physicist. A colleague of his who worked in the nuclear program headquarters in New Mexico sent him a letter with the pitch and asked him to take a look at it and said that he thought that Gene Roddenberry was probably looking for technical advisors. So, my father looked at that, and he met Gene Roddenberry."[20]

Even though Whitener was approached, he never ended up consulting for the show. "He was a real stickler for science," recalled Theresa. In addition, "he knew that he would be too picky, and he wouldn't let some things go by . . . he really didn't think it [the show] would go anywhere."[21]

Instead, Whitener referred Roddenberry to an Air Force colleague of his, Colonel David R. Jones. Roddenberry wrote Jones asking for "some technical information on a science fiction series I have created and am developing for Desilu Studios."[22] Jones soon wrote back thanking Gene for a cigar, which "was tremendous" and "created quite a sensation at dinner the following night." In Jones's letter to Roddenberry, we could see one of the earliest mentions of Harvey P. Lynn Jr., the person who would soon become *Star Trek*'s first technical consultant.

On June 26, 1964, Jones wrote,

> I talked with Harvey Lynn this afternoon and he is very definitely interested. I definitely believe Harvey is the type suggested, but you'll have to judge as to how well he fits the bill as a person. What recommended him to me was his technical background, plus a very real awareness of how things work in the real world. What I would be most concerned about would be getting a person to do this job who couldn't forget that he was violating the known physical laws. I think Harvey fits the bill.[23]

Later that summer, Roddenberry met with both Whitener and Jones. During this meeting, Jones indicated to Roddenberry that Lynn "evidenced some interest in the project, or at least seeing that we here keep it at least enough in accord with the laws of physics that scientists can enjoy the program too. Just how much evidence is, of course, something for the future. Depending on our final direction, we may want some experienced physicist on a fairly regular basis, or it may work out that intermittent reading and technical advice would be best."[24]

Harvey Prendergast Lynn Jr. was born in San Angelo, Texas, on August 3, 1921. An only child, his father was a traveling hardware and tool salesman, while his mother, Leonie May, was a housewife. Lynn attended San Angelo

CHAPTER ONE

Central High School, where he was noted in his senior yearbook as having "a great big voice for a little boy."[25] Lynn was in the ROTC, and after graduating from high school in 1937 he worked and attended Angelo College for a few years prior to enrolling at Texas A&M in 1939. "Dad told me that every year, his folks had to pawn all the silverware that they had because they had trouble paying for tuition and books,"[26] recalled Dennis Lynn, Harvey's youngest son.

During his time at A&M, he was an associate editor of the college yearbook, the *Longhorn*, as well as active in the San Angelo Club and Press Club. While there, he trained in the Chemical Warfare Service Company "B." He was also a member of the American Institute of Chemical Engineers and the American Society of Chemical Engineers. He graduated as a second lieutenant in 1942 with a degree in chemical engineering.

After graduating from college, Lynn served in World War II for three years while being stationed in Africa, Sicily, Italy, and Austria. "Even though he was in the Army Air Corps, he was colorblind and never flew," said Dennis. While in Italy he was promoted to major and met his wife, Yvonne Petrotti, in Corsica. "There's this place where the women walk up and down the boulevard, and all the men just kind of stand around, drink coffee, and watch the women," recalled Dennis on how his dad met his mother. "Dad was in one of those places and saw my mom walking by with her friends, and that's how they met."[27]

After the war, Lynn was released with the rank of major and returned to the states with Yvonne in 1945. He and Yvonne got married in San Angelo, Texas, in 1947. Lynn stayed active in the Air Force reserves during the time that he briefly worked in the oil business in Marfa, Texas, as well as Necha Butane Products in Port Arthur. "My mom was new to the States," said Dennis. "She didn't speak any English and was now living in West Texas, the worst place ever. It was like Hell on Earth for her. Then Dad hurt his arm really bad on some oil-drilling equipment, so he moved to Virginia. After that incident, he realized that manual labor was not for him and that he should focus on a profession that would make more use of his brain than brawn."[28] Upon leaving West Texas, Lynn returned to the Air Force.

The Lynns had their firstborn, Harvey P. Lynn III, in Bethesda, Maryland, in 1949. The family then moved to Sandia Base near Kirtland Air Force Base in Albuquerque, New Mexico, where they had their second son, Dennis Sebastien Lynn, in 1952. During their four years at Sandia, Yvonne also applied for US citizenship, and Lynn was promoted to lieutenant colonel in the Air Force in 1956.

From New Mexico, they moved to California, where they settled into an apartment on Sepulveda Blvd. about five miles from Malibu Beach. "Dad saw Malibu and just loved it," said Dennis. "I remember he told me he wanted to open a restaurant there right on the beach. He picked out a name for it and everything. It would have been called 'Lynn's Little Place by the Sea.' He liked the idea of grilled steaks as his speciality."29a-b

Being an active-duty officer, Lynn was often gone for long periods of time. "He was going up to Edwards Air Force Base every six months, and it was rough on us kids," said Dennis. "I never saw my dad until I was around four years old. I was told that when I first saw my mom and dad together, I didn't know who my dad was, so I went up and started hitting him saying, "You let go of my mom!"'30

The family tendency toward physical displays of emotion went both ways. When Lynn was home, he served as the disciplinarian. "Dad always wore this huge college ring and hit us with it whenever we did something wrong. He would turn that ring over and say, 'You want knots on your head?' and whack us with it. Man, that would hurt."31

While living in Malibu, Lynn took up a position as a liaison officer between the Air Force and the RAND Corporation. It was during his time at RAND that he began consulting for *Star Trek*.32

Star Trek's First Technical Consultant

Roddenberry was impressed with Lynn when they first met during the summer of 1964 at Lynn's RAND office in Santa Monica. Lynn, like Roddenberry, was originally from Texas, and they both served in the military even though at the time of their meeting, Lynn had retired as a full colonel but still remained active in the Air Force Reserve.

"On the basis of meeting with a number of personnel there, [Lynn] seems the best qualified to serve as technical consultant and/or coordinator of technical advice once *Star Trek* is rolling," wrote Roddenberry. "He also has access to government and industry libraries, film footage and even some miniaturized models of planet exploration ships and machines."33

Working full time for RAND, Harvey initially viewed his relationship with Roddenberry as casual. "Neither I nor Desilu make any commitments that we will now or at any time enter into any agreement or contract with you," wrote Roddenberry in a 1964 letter to Lynn. However, being a good businessman, Roddenberry left the door open for future opportunities writing in that same letter, "You and I have, however, met and found a certain rapport and enjoyment in working together and also mutually hope this pre-pilot film association will mature into an arrangement for your services as technical advisor or consultant or coordinator during the production of the series." Roddenberry also recognized Lynn's association with his current employer when he stated, "Also, I understand that your primary loyalty and obligation is to Rand Corporation and any arrangement would have to realistically reflect your commitments there."³⁴ Even at this early stage of meeting, Roddenberry hinted at the importance of Lynn's affiliation by mentioning RAND and his desire that Lynn remain working for them.

In their getting-to-know-each-other correspondence during the summer of 1964, Lynn shared with Roddenberry some of his early ideas for the series even before seeing the script for the first pilot. Lynn explained to Roddenberry that the distances between the stars are great, and that interstellar travel is "likely to be of long duration (years), unless we find a new dimension or something." Lynn added that "such a premise would be helpful." Lynn then offered the novel suggestion that the crew of the USS *Enterprise* never returns to Earth and that the ship should include children.

"Assume that the *Enterprise* left the Earth, never to return," wrote Lynn. "Aboard are the crew, scientists, etc., but, in addition, are a number of youngsters who are to grow up as the craft travels through space. These youngsters, ages 10 and up, are to become the colonizers and space travelers of the future. As the ship travels, they receive classroom education, not only in what the future may bring and what might be expected of them, but also in their heritage. They will be told of a world and a way of life that they have left, never to return."³⁵

Lynn's idea that the *Enterprise* would basically serve as a generation ship was intriguing but not outlandish. American rocket pioneer Robert H. Goddard was among the first to write about long-duration interstellar voyages in his work from 1918, "The Ultimate Migration." Here Goddard proposed the use of an "interstellar ark" that would take its suspended-animation human crew to another star system after the death of the sun. The crew would take centuries to arrive, and Goddard speculated that they would have genetic and psychological changes over multiple generations as a result of their long mission. Konstatin Tsiolkovsky, one of the founders of modern rocketry, described the need for multiple generations of passengers in his 1928 essay "The Future of Earth and Mankind." Here the crew changed so much over the millennium of their journey that they lose track of Earth as being their home planet.

Roddenberry wanted the show to focus on the lives of crew members serving the Federation while aboard the *Enterprise* but also their lives while off duty. This was something that Roddenberry emphasized from the very first pilot, when, in one of the ship's corridors, we see a casually dressed couple in sandals and flip-flops passing by a very tense, fully uniformed Captain Pike. Likewise, the first-season episode "Balance of Terror" opens with a wedding ceremony performed by Captain Kirk.

It would have been interesting to see children routinely on board the *Enterprise* in the original series, but this did not happen until *Star Trek: The Next Generation*. By then, Roddenberry updated the *Writers Guide* to reflect the changes as a result of the new *Enterprise*. Now many times larger than the original, the new *Enterprise* could accommodate whole families. In one early episode, Commander Riker is seen shooing kids out of the observation lounge, but this only really happened in the first season. After that, we didn't see children living aboard the ship except only once in a while. Dr. Beverly Crusher's son Wesley, played by Wil Wheaton, while technically a child, also was a prodigy and soon became a full member of the crew. He was in the main cast in the first four seasons of TNG but then left the show, only to occasionally return during the remaining three seasons.

Tractor Beams, Phasers, Starship Designs, and Yeoman Rand

What followed from that early-summer correspondence in 1964 was a series of additional letter exchanges and phone calls in which Lynn offered suggestions and technical advice for the first *Star Trek* pilot, "The Cage." It helped that Lynn was invited onto the production studio to view the sets. "I remember him telling us that when he first went to view the sets with Gene Roddenberry, he toured the bridge of the *Enterprise* that had been built for the first pilot," recalled Dennis of those early days. "He told them to put a bunch of blinking lights on the bridge, and they did."[36]

In reading the first drafts of the script, Lynn wrote, "It is highly exciting and hard to put down when you start to read it."[37] Some of Lynn's contributions during this early period of *Star Trek*'s development have since become recognizable trademarks of the series. Notable *Star Trek* terms that evolved from ideas originally put forth by Lynn include the following:

TRACTOR BEAM: The idea of having some kind of beam for use in the series, visible or otherwise, that would track and grab hold of an object from a distance to either hold it in place or move it was first put forth by Lynn in the following letter from Lynn to Roddenberry: "From what I have read, docking is likely to occur by having the large ship ahead of the space shuttle or taxi. As the *Enterprise* slows down, the taxi nears if from the rear. To accommodate smaller shuttles, taxis, and tugs, I visualize the *Enterprise* having something like a bomb bay. When a ship is to be docked, the doors open, and a ring, two or three feet in diameter, is lowered. Upon this ring, pointed to the rear, is an intense beam of light. As the shuttle nears, a hook is extended above the smaller ship, and upon this hook is something like a photoelectric cell. The cell directs the hook to the ring as it 'rides the beam.' When the hook is automatically locked on the ring, a hoist lifts the shuttle into the 'docking bay,' doors close, valves are opened to let air into the area, and we have our shirt-sleeve atmosphere."[38a–c]

CHAPTER ONE

PHASER: The idea of modular power packs for the iconic personal weapon carried by Star Fleet[39] had its origins in the following comments by Lynn: "Don't forget to have a connection between the guns and the 'powerbelt.' I visualize the belt looking something like a waist-type life preserver, having individual power units, say, three inches by six inches. These units can be replaced, just as bullets in a gun can be replaced." In addition, Lynn first encouraged changing the name from "laser" to what eventually became "phaser." "LASER stands for Light Amplification Stimulation Emission Radiation—the M in MASER stands for Microwave. Now, your guns (and the big gun used later on) are highly advanced pieces of hardware. Don't you think it is likely that they will have a new name? After all, LASER and MASER are only five or six years old." The show's art director, Matt Jefferies, also pointed out that "Lynn's suggestions included MASER for a microwave weapon, which eventually became phaser."[40]

Lynn was also present during the time that Jefferies was trying to finalize the design of the *Enterprise*. With walls covered in sketches, Lynn was called in by Roddenberry to give advice to help narrow down the design to its finished form.[41] Lynn had made use of his aerospace contacts through his work at RAND. Sensing Roddenberry's urgency for design ideas for the ship, Lynn wrote, "I'll send you a package of magazines for your office, as you requested. I still haven't heard from the North American people in Downey. I am checking to see if there is anything at Hughes Aircraft which warrants a tour."[42]

By October 6, 1964, Roddenberry submitted the final draft of his proposed pilot to NBC. Of course, the network did not accept the script as final without their own input, so three more drafts were made before principal photography finally began on November 27, 1964.

For Lynn's work on the first pilot, he was paid $100, equivalent to nearly $1,000 in today's dollars.[43]

Soon a second pilot was ordered, and it was this one that finally convinced NBC that it wanted *Star Trek*. But Lynn wasn't sure if Roddenberry still wanted him.

On March 15, 1966, soon after NBC notified Desilu that *Star Trek* was a go for the fall season, Lynn sent Roddenberry a brief hand-written note that said: "Congratulations on selling *Star Trek* to NBC! But, please don't let them give you a time period opposite *Bat Man*. I want to be able to see both!" Joking aside, Lynn knew that the show would need a regular technical consultant, and he wanted that job.

On June 10, 1966, Lynn wrote Roddenberry, "I still submit that you could use a technical advisor for several reasons . . . one reason is to catch obvious goofs . . . and secondly, a person who makes it a point to read a variety of scientific literature might be able to come up with novel ideas which you could use. I know that you don't want *Star Trek* to turn into a parade of scientific gimmicks."[44]

Eventually, Lynn was retained by Roddenberry as the series' technical consultant. While remaining a full-time RAND employee, he was paid $50 per show (around $450 in today's dollars), for his part-time technical advice to the series. By the summer of 1966, Roddenberry indicated that he appreciated all the help that Lynn provided, but also that it was difficult for him to fund his services. "I've hesitated to ask you to generally read scripts since I could not legitimately take money out of budget to make any reasonable, non-insulting, payment to you."[45]

Lynn worked on *Star Trek* as a technical consultant for about eighteen months before abruptly leaving the show in April 1967, just as the second season was starting production.[46] Lynn had struck up a close friendship with Roddenberry as he advised the series on the side while still working full time for RAND. Documents in the Roddenberry Collection at UCLA's Special Collections Library revealed many exchanges between the two. All the letters from Lynn were either handwritten on Lynn's personal stationery or typed from his home residence. "Dad was a great typist, and he was always on the typewriter at home. He had great handwriting, but he told me that he could type faster 'so his mind could flow,'" recalled Dennis.[47]

Only one letter was found in the Roddenberry Collection written on official RAND stationery. This letter, dated December 21, 1966, was written by Lynn in response to a fan's question about how the *Enterprise* traveled through space.[48] Four months after that letter was sent, a production memo circulated from Ed Perlstein, attorney for Desilu Studios, stating that Lynn "will no longer serve as Technical Advisor to *Star Trek*."[49] When Perlstein's memo appeared, production of the second season of *Star Trek* had barely begun. Lynn's services as a technical consultant would certainly have still been needed, so why the abrupt termination?

We can only speculate, but management at RAND may have grown tired of the attention generated by Roddenberry as a result of Lynn's affiliation with the series. Even though his consulting work was done outside his job with RAND, his affiliation mattered as much as his technical advice, and Roddenberry didn't hesitate to associate his show, through Lynn, with the corporate think tank. Roddenberry was fond of boasting about Lynn's support for *Star Trek* and how he contracted technical advice from RAND. "I did most of my space research with scientists from the California Institute of Technology," Roddenberry said, "and I spoke to the Air Force Research and Development people at RAND Corporation."[50] Roddenberry recognized the value that a RAND affiliation gave to *Star Trek*'s credibility, and he made sure to include mention of that in the promotional material used to help sell the series. After all, as one television reporter observed at the time, "RAND is where all the brilliant researchers and scientists go to work out things for the future."[51]

CHAPTER ONE

A good example of this is a slick promotional brochure issued by the Desilu studios sales office in 1966, just prior to *Star Trek*'s network premiere. In it, Roddenberry used deceptive but persuasive language to imply more than what actually occurred with Lynn's contributions. The brochure failed to point out that Lynn worked as a part-time consultant from his home in no real official capacity with RAND. Instead, the brochure suggested the consultation in question was the work of many individuals, all employed by the same organization. "To maintain the proper blend of fantasy and fact, scientists from the world-famous Rand Corporation, America's 'space think factory,' act as consultants to *Star Trek*," stated the brochure. "As a result, although the series might be described as science fiction, it is so intelligently conceived and realistically presented that the word 'fiction' hardly seems to apply."[52] When *Star Trek* was first aired by the BBC in July 1969, the BBC Press Service issued a similar release that included mention that "scientists from the famous Rand Corporation act as consultants for the series," concluding that "*Star Trek* has drawn high praise particularly for its attention to detail and technical accuracy."[53a]

In an interview given for the June 7, 1966, issue of *Daily Variety*, Roddenberry claimed that "I read about lithium fuel, went to Rand, but they wouldn't talk about it. They refused because we had stumbled on a project now underway."[53b] One can see from such boasts how RAND officials might get a little annoyed for this example and others failed to acknowledge the whole story. Only one scientist, Lynn, was actually employed by *Star Trek* and that was in no official capacity with RAND. As we have found, Lynn worked for the show part-time outside of his job at RAND. In addition, Dole was never directly consulted or employed by *Star Trek* and would probably have been surprised to find himself affiliated with the series through Roddenberry's claims. This in no way minimizes the overall contributions to the series that such individuals as Lynn and Dole made, but at the same time they also serve as examples of how deceptive Roddenberry could be as he set aside factual evidence for the sake of marketing his fictional universe.

To this day, the RAND Corporation is still asked about their connection with the original *Star Trek* television series. This happens so frequently that the privately held corporation eventually felt compelled to clarify their association on their website. Their "frequently asked questions" page includes this statement: "RAND works in subjects of interest to a wide range of people, and the line between fact and fiction can blur (especially with regard to our historical research)." Their website addresses some of the most persistent rumors, including one associated with *Star Trek*: "A RAND researcher designed the bridge of the Starship *Enterprise* on the original *Star Trek* television series." This is then followed by the answer "A RAND researcher, Harvey Lynn, was consulted, but as a private citizen, not as part of a RAND project."[54]

Recall that crazy equation used in the beginning of Roddenberry's sales pitch when he tried to sell the series to the networks? Stephen Whitfield, in a conversation that he had with Roddenberry during the spring of 1968, gave a humorous story in his book *The Making of Star Trek* on how Roddenberry came to use that fabrication:

"One of the early pages of the format was supposed to explain how many stars there are in the Galaxy, mathematically how many 'M' Class planets there are, and so forth. I had read a theoretical study on this but couldn't remember the formula used. I asked a friend of mine at Caltech to look it up for me, but in the meantime, I wanted to see how it might look on paper. So I just made up a complex-looking formula to visually give me an idea how it would look. Before I could get back to my friend at Caltech, MGM began asking me for the formula. My secretary typed it up the way it was, under the pressure of time, and we sent it out that way. From there, the formula made its way to RAND, Caltech, Duke University, and a few other places. I got busy with other details and forgot about the phony formula. Now, over four years later, no one . . . scientists, mathematicians . . . no one has ever questioned that formula, and yet it's been pretty well publicized across the country. Perhaps someone has lectured on it, for all I know!"[55]

In spite of the lack of scientific authenticity, Roddenberry's made-up equation has survived to become part of *Trek* lore. Among fans it is referred to as "the second variation" of Drake's original equation.[56] When *Star Trek: Voyager* aired decades after the original series, Michael Okuda recalled how both the Drake equation and Roddenberry's version appear in the episode "Future's End." That episode involves a twentieth-century astronomer named Rain Robinson, played by Sarah Silverman, who uses a radio telescope to search for extraterrestrial life. Okuda recalled, "On the wall of her office, I put a placard containing the real Drake equation, but because it was *Star Trek*, I also included a copy of Roddenberry's fake version. You can glimpse the placard very faintly in the background of a couple of shots. Afterwards, I faxed a copy of the placard to Professor Drake and spoke with him briefly on the phone. He seemed amused but noted gently that Roddenberry's version included a couple of variables raised to the first power, which, of course, is exactly the same as the variables themselves."[57]

Perhaps in acknowledgment for the numerous contributions that the RAND Corporation played either directly or indirectly in the development of the series, Roddenberry introduced a new character during the first season of *Star Trek*. In a memo dated April 11, 1966, Gene submitted a list of possible names for the character. The typewritten carbon copy found in the Roddenberry Papers at UCLA lists seven first names of "JANICE" neatly typed in a column, followed by a parallel column of suggested different last names. The name "JANICE RAND" is the eighth and final name added to the list, and it clearly had been

CHAPTER ONE

typed directly onto the carbon, possibly as a last-minute addition.[58] Rand, played by actress Grace Lee Whitney, served as the captain's personal yeoman for part of the first season.[59]

The *Original Series Roddenberry Vault* Blu-ray, which was first issued in 2016 for the fiftieth anniversary of the show, contains behind-the-scenes footage from the making of the original series along with alternate takes, deleted scenes, omitted dialogue, outtakes, and original visual FX elements. Included is a photo supposedly of Harvey Lynn Jr. It turns out that this is not Lynn but his oldest son, Harvey Lynn III.

Lynn's oldest son, Harvey P. Lynn III, followed in his father's footsteps and pursued a career at RAND working in their computer sciences division. In an August 2, 2002, email exchange with *Star Trek* fan Greg Tyler, Harvey P. Lynn III had this to say about his father's work on *Star Trek*: "He graduated as an electrical engineer [this is not true, since Lynn's degree was in chemical engineering]. Worked at RAND as a liaison officer between RAND and Project Air Force. Was never starstruck and had little interest in TV, films, or science fiction. Apparently, he met Mr. Roddenberry through a mutual friend and was selected for the technical consultant job more because he hit it off with Mr. Roddenberry than his technical expertise. When offered the job, he boned up on physics, astronomy, etc. He picked up surprisingly quickly on how to express the technical elements simply . . . i.e., not having to explain how a phaser works . . . sort of how most people know that a light switch turns on the lights but don't wonder about the mechanics."[60]

Harvey Lynn's influence on *Star Trek* can be seen even though he never returned as a technical consultant to the original series after the first season. However, during Filmation's *Star Trek: The Animated Series*, which aired from 1973 to 1975, Roddenberry brought back Lynn, reasoning that even though the animated series was designed as a Saturday morning television show, it still needed someone to oversee its scientific and technical accuracy.[61]

On September 8, 1986, Lynn attended a special twentieth-anniversary birthday celebration of *Star Trek* that was held on the lot of Paramount Studios. He brought his younger son, Dennis, with him. "Everyone was there," recalled Dennis of the event. "They had the best food in the world I'd ever tasted. The new *Enterprise* was there, and all the cast from the new series [*Star Trek: The Next Generation*] as well as the original series were present. It was great. I walked up looking for Dad, and a crowd of character actors and recurring cast were complaining that there were no royalties for most of the cast, only the main actors. They were not in a good mood." During this event, guests could have their pictures taken on the set of the transporter room to the new *Enterprise*. Lynn posed with Dennis for a souvenir photo. "I remember driving home from the *Star Trek* anniversary event, and Dad asked me to take care of my mom after he died," said Dennis. "I thought that was a strange question to ask and didn't think much of it at the time."[62] Several months after the Paramount event, Dennis learned that his father was diagnosed with cancer. Harvey Prendergast Lynn Jr. died on December 31, 1986, of pancreatic cancer and was buried at Fairmount Cemetery in his hometown of San Angelo, Texas.

Although Lynn left the series after the first season, he remained an influential force in the show's development. Why did Lynn even want to be involved in the show? Clearly it was not the pay. What is clear is that both Lynn and Roddenberry got along well in part because they both served in the military, and even though Lynn never flew, they both were airmen. Lynn may have also been motivated to contribute after seeing the show's positive look at the future, which he might have appreciated after growing up in a household of limited means. Having made a career in the military, Lynn might have also welcomed the show's quasi-military presence as represented by the Federation—the main governing body for the *Enterprise* and its crew. Whatever the reason, Harvey Lynn liked *Star Trek*, and Roddenberry not only welcomed his technical expertise but also his RAND affiliations and made use of both to inspire the creation of the series and promote its standing to a public that remained skeptical of a television show that treated science fiction seriously.

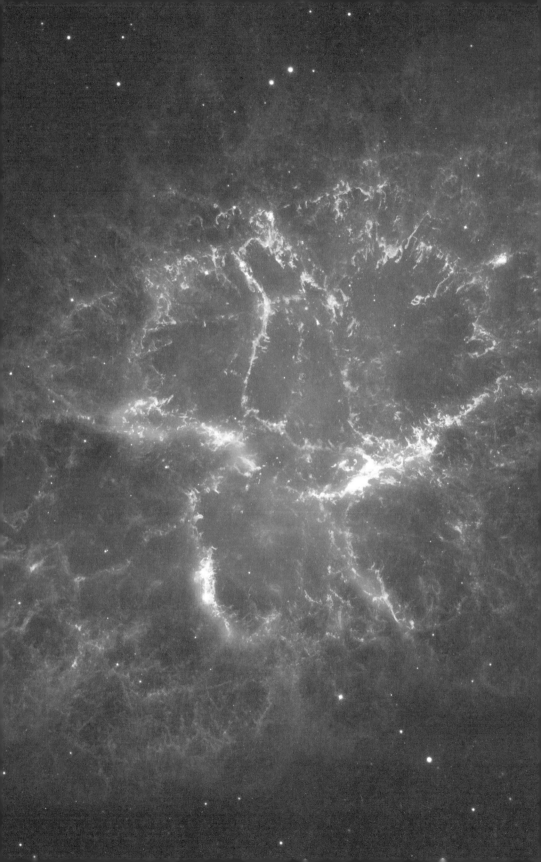

CHAPTER 2

Professional Nitpicker, Kellam de Forest

Kellam de Forest was one of two technical advisors employed during the production of the original *Star Trek* television series. The other, as discussed in the previous chapter, was Harvey Lynn. However, unlike Lynn, who lent his advice to the show on a part-time basis typing memos and reports from his home, de Forest's input came from a full-time staff based in multiple offices located around Hollywood. Kellam's business, called "De Forest Research," provided a vital service of fact-checking that helped ensure that a show's content was accurate, consistent, plausible and most importantly, not susceptible to litigation.

Both de Forest and Lynn served as keepers of continuity and stewards of known facts, collectively seeking to present a believable twenty-third century universe to a skeptical television audience of the 1960s.

De Forest Research Service

Born in Santa Barbara, California, on November 11, 1926, Kellam de Forest grew up in nearby Mission Canyon. A child of the region's cultural and political royalty, de Forest was a distant cousin of radio pioneer Lee de Forest, inventor of the basic radio amplifier tube, and son of famous landscape architects Lockwood de Forest Jr. and Elizabeth Kellam.

Kellam attended the same boarding school in Ojai Valley as his father. After graduating from there in 1944, he tried to enlist in the Army. "He was only in the service for one year because he had some physical issues," explained Kellam's oldest daughter, Ann, during an interview about her father. "He was little as result of being born very premature. He also had some coordination issues, so the Army did not take him. That is why he ended up in the reserves. He was sent to a camp in Pullman, Washington, where he was a supply clerk."[1]

After leaving the reserves, he went to Yale, where he graduated with a degree in history. Returning to the West Coast in 1949, he worked for a year in San Francisco with an aptitude-testing firm. From there he found employment closer to home at San Ysidro Ranch in Montecito. Here he served as a desk clerk and assisted guests, many of whom were Hollywood types. Ann recalled an interesting story about her father from that time: "The story that he told me was that John Huston and James Agee rented cottages at the Inn for several months while they were working on filming *The African Queen* (1951). Humphrey Bogart and Lauren Bacall both showed up at one point. Bogart's middle name was de Forest, so that proved to be a topic of conversation."[2]

It was while he was working at San Ysidro that he met his wife-to-be, Margaret MacCormick. "Everybody called her Peggy," Ann recalled in the story of how they first met. "She was from Massachusetts and went to Middlebury College. She was a schoolteacher that taught French. She had also worked in boarding schools. She got offered a job at the Montecito School for Girls. She went out to California at age twenty-five since her parents told her, 'Well, you're never going get married. You're too old, so just go out and take this job in California.' While out here, her aunt and uncle were visiting, and they came to visit her. They were staying at the San Ysidro Ranch Inn, and there was this handsome young desk clerk there. My mother had gone to the restroom and came back out. At that point my great-aunt introduced them. My father then asked my mother out, so that's how they met." In 1952 they got married.[3]

Kellam de Forest inherited many of the talents of his parents, particularly writing. He soon moved to Hollywood to try his hand as a scribe in the growing field of television. He found the area overrun with individuals like himself, those competing to make a living as a scriptwriter. But de Forest soon discovered that he was able to offer his skills as a researcher to check what others wrote. "I wanted to get into show business but found you can't get anywhere in this town without something to sell," said de Forest during a 1967 interview.[4]

One of his first jobs was researching a film for the National Safety Council. "I went around and knocked on doors to see if I could find a job," said de Forest about his early days in Hollywood. "There was this short-subject producer working on a film about driving safety. He needed a gopher to help him do research, so he hired me. When that was done, I was let go because he didn't have any other projects in the offing."[5]

Even though the National Safety Council job was short term, it taught de Forest that there was a need for the same kind of service that he had done on that first short subject. De Forest found that while most major TV networks had their own research departments, independent television producers and writers did not. He quickly learned that people were willing to pay to have someone take the time to delve more deeply into a subject.

CHAPTER TWO

De Forest found a niche and soon established De Forest Research Service to provide writers and directors with information, along with checking scripts for clearance and accuracy. "With the rise of television, I could see there was going to be a need for this service," said Kellam.[6]

De Forest conducted what is formally referred to in Hollywood as "script clearance and research," two important parts of studio production that are now routinely done both in the film and television industry. By definition, "script clearance is the legal process of ensuring that the proper names of people, products, companies, etc., in film or television do not present a conflict with ones that actually exist," wrote Michael Kmet, a media scholar who extensively researched this aspect of the film and television industry. Kmet described script research as "the process of checking for scientific, historic, geographic, and series-specific accuracy and continuity. It is the job of a researcher to point out every possible legal conflict, inaccuracy, and continuity error and to suggest legally cleared and accurate alternatives."[7]

During the height of the fully integrated studio system, when Hollywood was producing hundreds of pictures a year, every studio had its own research library. But by the late 1940s, most major studios had either downsized or closed their internal departments responsible for script clearance and research. Warner Brothers donated its collection to the Burbank Public Library, while MGM boarded up their library. A few studios kept their libraries intact, such as Universal and Disney.[8] De Forest explained that the main reason for the demise of the studio library was because "the primary function of the studio research departments originally was to provide pictorial material to aid in the building of sets," he said during a 1984 interview with the *New York Times*. "If they wanted to know what a Welsh village looked like for *How Green Was My Valley*, they had files of material for reference. But as soon as the studios began to shoot most movies on location, the vast picture collections they had accumulated were no longer necessary."[9]

As the television industry evolved, de Forest's professional nitpicking business grew. He expanded his service to provide aid to the burgeoning sitcom industry. He and his staff worked to verify facts, find flaws, and check details, using the preinternet tools of his time: newspapers, books, magazines, reports, and word of mouth.

Though de Forest did not have a background in science, he did excel in basic research. A consummate reference librarian with an amazing ability to instantly recall facts, he moved his business out of his home and set up shop in Hollywood, which eventually grew to include an office in the boarded-up libraries at MGM Studios.

De Forest's research staff made extensive use of the MGM library files, but his research service's main office was housed in the research libraries at Paramount

Pictures, which included the old RKO Studio library, which later became the prop department at Desilu. With numerous shelves piled high with magazines, eighty file drawers filled with photos of almost anything imaginable, a 1910 edition of the *Encyclopedia Britannica*, and five thousand reference books, de Forest and his assistants kept busy researching scripts for *The Lucy Show*, *Mission Impossible*, and other studio productions.[10a-b]

The Move to Desilu and *Star Trek*

De Forest's Desilu office was the largest of two locations that made up his research services company in Hollywood. In 1958, Desilu handed de Forest control of Paramount's and RKO's libraries and gave him rent-free office space at the studio's Gower Street Lot.[11] This proved beneficial in that it guaranteed de Forest the business of all of Desilu's television shows, as well as added business from all the other companies that rented studio space on the Desilu lot.[12] It was also while de Forest was working at his Desilu office that Gene Roddenberry found him and hired him to consult on the very first *Star Trek* pilot.

"Since my office was on the Desilu lot, I did most of the Desilu productions. As a result, they approached me about consulting for *Star Trek*," recalled de Forest. "They would send a script over, and we would read it and report on it. Usually this was a couple-page report indicating that we checked all the names. We also checked for continuity and if there were any gross anachronisms."[13]

"It was a great place to work growing up," recalled Ann. "When we were kids, we would often visit the sets, especially when people from out of town or relatives came to visit. We went to the *Star Trek* sets a lot because they were so close to where my dad worked at the old RKO library. The show's soundstages were right around the corner from him. We found that the cast members were always very friendly. James Doohan was especially very cordial. He would always come over and talk with us. He acted like a host, being very friendly and accessible. I remember going to the sets and picking up these large rocks or boulders that weighed nothing. They also were working on the bridge, and I remember seeing the transporter room."[14]

In recalling her dad's interest in *Star Trek*, Ann said that he "was not that familiar with science fiction as a genre other than what he got from reading scripts. But Dad knew a lot of things that could be applied to the series." In addition, Ann said that "our whole family watched *Star Trek* every night that it was on. Dad would never really comment on the episodes, but he would watch them with us growing up. I remember he had a big photo in his office. It showed Captain Kirk buried in Tribbles. I think he liked that scene because it showed how they were at times completely buried if not overwhelmed by their work."[15]

In 1964, Roddenberry copied de Forest with memos asking for input on a variety of subjects concerning the development of his new series. He even sought his advice when it came to helping create the starship *Enterprise*. De Forest was included in a memo that Roddenberry sent out on August 25, 1964, soliciting input for the ship's design. The memo read, "We are dangerously near the time we must settle on a shape and configuration for our spaceship of the future but are running into considerable difficulty of settling on that design. Would much appreciate your checking if there is any 'far out' very futuristic selection of sketches or drawings of spaceships which we could examine. Please understand, we're talking more science fiction here than we are anything available in space tech manuals or on scientific drawing boards today. Perhaps the best direction will be science fiction magazines and books. It would be most helpful to find if there has ever been any collection of such sketches and drawings, or any surveys or articles devoted to the subject."[16] De Forest sent Roddenberry ideas. "I provided him with pictures of science fiction magazine covers and books but little else," said de Forest. "Matt Jefferies was the art director of the show, so I gave those to him."[17] The Gene Roddenberry Papers at UCLA include file photographs of nine pulp magazine covers in the same folder that houses Roddenberry's memo to de Forest petitioning him for help with the ship's design. These covers are from *Science-Fiction Plus*, *Science Wonder Stories*, *Astounding Science Fiction*, *Air Wonder Stories*, *Analog*, *Wonder Stories*, and *Amazing Science Fiction Stories*. Although there is no proof that specific elements of the ship's final design were taken directly from these examples, these specific magazine covers nevertheless were deemed important enough to be saved.

Stardates, the Gorn, and Spican Flame Gems

One of de Forest's most significant contributions to the series was in developing the idea of the "Stardate," which he supplied for *Star Trek*'s second pilot, "Where No Man Has Gone Before." "In *Star Trek*'s first pilot, 'The Cage,' they initially tried some futuristic dating in the script, using the Gregorian calendar," said de Forest. "I felt that surely they would come up with something better in the future as the system they had did not really work in tracking intergalactic movement."[18] De Forest recalled how the idea originated. "So, there was something called the Julian calendar that uses a star, and it was superior to the Gregorian calendar because it didn't have leap years; it was all numerical. For example, December 25, 2023, is Julian date 2460304.1843403. So, I suggested to Gene that if he used this concept of a numerical date and decimal system that it might give a little more futuristic feel to the series. He felt that was a good idea, and bought it."[19]

Peter Sloman was still in high school when he landed one of his first jobs working at de Forest Research. "My father was Ernest Sloman, but he was known as Easy Sloman," said Peter during an interview. "Dad was head of press information at CBS West Coast, and he was doing a piece on Kellam for the *TV Guide*.[20]

Kellam mentioned that they were getting shorthanded, and that's how I started working for him."[21]

Sloman's father had overheard de Forest talking about ways to improve upon the show's antiquated dating system and thought he would offer his own suggestions. "My dad was an amateur astronomer," said Sloman in recalling how the idea of stardates evolved. "He and Kellam were discussing some of the really early stuff before the show was actually written. Dad suggested that they adopt the Julian date used by astronomers. The Julian date used the Julian day plus the fraction of a day since the preceding noon in Universal Time. Say for example you were observing at night, and an event occurred between, let's say, ten at night and three in the morning. You would have two dates ordinarily because the date changes at midnight. With the Julian date, the date changes at noon. Basically, you pick a starting place, and you pick a number and as long as everybody says, 'Okay, this is where we're starting,' then it's just one up from there. The parts of the day are decimal fractions, so a stardate of, say, 1374.5 means that it is day 1,374 from whatever the starting point was, and the "point five" means it is taking place halfway through the current day."[22]

Attempts to portray life in the future could get a bit confusing for a show that also sought to be plausible to viewers of the 1960s. *Star Trek*'s believability was sometimes damaged by its own nettlesome scientific techno-jargon babble. "One time I was talking to Gene Coon, and I said to him, 'By the way, Mr. Coon, we're what? 250 years in the future here? Don't you think we would be using the metric system by now?'" recalled Sloman. "Coon said to me, 'Pete, we don't want it to be authentic. We just want it to *sound* authentic.' So, that's the whole thing. If you dazzle them with footwork, you don't have to have any content. But it was the only game in town as far as science fiction goes at that time. And some aspects of it were excellent, and some aspects were the best we could do in those days."[23]

Sloman noted upon reviewing Gene L. Coon's script for the *Star Trek* episode "Arena" that the story bore a similar resemblance to a copyrighted story of the same name by Frederic Brown. He approached his boss about this, and, as de Forest recalled, "As a result there was some scurrying around thinking what to do. They eventually contacted the author, bought the story rights, and went ahead with the production of the episode."[24] It was Coon's script that originated the Gorn in name, but it was de Forest's research that led producers into quick action to avoid any copyright infringement and bring the episode to the airwaves. Sloman was sixteen when he started working for de Forest. "It was a fascinating job, and I'm still doing it after more than fifty years."[25]

During production of the series, sometimes de Forest had a paradoxical relationship with *Star Trek*. In a July 6, 1966, memo from Roddenberry to

CHAPTER TWO

Herb Solow, the series' executive in charge of production, Roddenberry complained that "none of us are very satisfied with what we get out of the Kellam de Forest research department, weighed against money spent there. While he undoubtedly has some value to the studio Legal Department in checking names, the rest of his advice, to be frank, does not meet our needs."[26] Roddenberry's attitude toward de Forest then seemed to flip four months later, when Roddenberry wrote, "We've all been gratified by the excellence of the comments you've been sending in on *Star Trek*." He noted that his "scientific corrections and common-sense suggestions" were "increasingly helpful and more germane as the series progresses" and "mean a great deal to scripts." Roddenberry was eager to receive his "suggestions on scientific accuracy, logic, common sense, etc., which we can turn to our associate producer and story editor so that these comments can be taken into account during the rewrite."[27]

Dorothy Catherine Fontana (known more commonly as D. C. Fontana), whom Roddenberry first hired as his secretary then as the show's story editor, contributed many ideas that were used in the series. She also wrote the scripts for the episodes "The Enterprise Incident," "That Which Survives," and "The Way to Eden," all under the pseudonym Michael Richards because women at that time had more difficulty getting work as television writers. She wrote of de Forest's work, "I have just gone over DeForest Research comments on your script. I am really pleased at the way he is taking hold and, although some of his suggestions are impractical or too involved in miniscule [*sic*] detail, they get better week by week. I recommend them to you not only in the case of this particular script, but as a helpful guide in all future rewrites. I think we both agree that it is far better to change a word or delete a reference which is wrong than lose the affection of the SF fans and Scientists in our audience."[28]

In working for Roddenberry, de Forest never guaranteed protection from the law, but he advised his clients to err on the side of caution. For example, in the third-season episode "For the World Is Hollow and I Have Touched the Sky," there appears a Starfleet Command flag officer sending orders removing the *Enterprise* from responsibility for the asteroid spaceship *Yonada* after Kirk initially proved unable to divert its course from the inhabited Daran V system. The officer, played by actor Byron Marrow, was initially called Admiral Westervliet. De Forest recommended changing the name because at that time in 1967, there was an existing Captain Westervliet. "We try to protect people who might get ribbed about it," said de Forest.[29] Westervliet's name was never mentioned on-screen during the episode, but it did appear in the end credits.

De Forest also evaluated planet names submitted by scriptwriters. For example, in one memo dated February 23, 1967, de Forest referred to the planet in the episode "Miri" as "Another Earth," which is mentioned in the teaser at the beginning of the episode. In that same memo, the planet in the episode "Shore Leave," which is called "Treblank IV" in the script, was cleared by de Forest; however, the planet's name was never actually given in the final aired episode.[30]

It was also de Forest's staff that pointed out an issue in David Gerrold's script to the famous second-season *Star Trek* episode "The Trouble with Tribbles." The title creature in Gerrold's story is a "tribble," sold by space trader Cyrano Jones and described as a "featureless, fluffy, purring animal, friendly and loving, that reproduces rapidly when fed, and nearly engulfs a space ship."[31] The problem was that Gerrold's tribbles shared a remarkable similarity to the Martian flat cats that play a major role in the plot during the last one-third of Robert A. Heinlein's 1952 science fiction novel *The Rolling Stones*. De Forest's report warned that Gerrold's storyline was too similar to that of Heinlein's and could lead to legal issues. According to Herb Solow, Roddenberry simply made a phone call to a very understanding Heinlein and sorted things out. Solow explained, "Since the episode had not yet been broadcast, a simple 'mea culpa' from Roddenberry was sufficient to remedy the problem."[32]

In addition to rescuing Gerrold's tiny, furry titular creatures from oblivion, de Forest put his stamp on another part of the same episode. He suggested an alternate name for the gems that Cyrano Jones brought to the K-7 space station (along with the aforementioned tribbles). In the script, the jewels were originally going to be called "Tellarite Flame Gems," but de Forest's report explained that "tellurite is an Earth mineral, TeO_2 occurring in tufts of yellowish or white crystals." Even though both are spelled different, they sound the same. As a result, de Forest suggested naming them "Spican flame gems," most likely after the real star Spica.[33]

Roddenberry considered de Forest's reviews of every episode with the same high regard as Harvey Lynn's science critiques from RAND. De Forest was a Yale-educated historian, and his verification of character names, places, customs, and other checks on past and present norms that viewers could relate to were just as important for story development as the Texas A&M–trained chemical engineer Lynn's critiques on the science and technology that placed everything in the future. "What we have is authentic drama that just happens to take place in outer space," said Roddenberry about *Star Trek* in an interview that appeared in the *Boston Herald* shortly before the series premiered.[34]

Even so, the reports that de Forest sent back were often too detailed, to the point where Roddenberry sometimes had to step in to streamline things under the intense deadline pressures of television writing. In a November 3, 1966, memo to de Forest, Roddenberry wrote, "Far too often we find ourselves wading through legal information such as clearing of names in order to get to the meat of your comments, the scientific correctness and common-sense suggestions which mean a great deal to scripts. Therefore, is it possible for you to divide your comments into two memos? The first, clearing of names and all other legal, insurance and similar matters. This part of it we can turn over to our clerical staff for handling. The second memo, then, to contain the suggestions on scientific accuracy, logic, common sense, etc., which we can turn over to our associate producer and story editor so that these comments can be taken into account during the rewrite."[35]

Eventually, everybody found de Forest's reports consistently useful. "I continue to be overwhelmed by the tremendous research job being done for the studio and *Star Trek* in particular by Kellam de Forest," wrote Herb Solow in December 1966. "The more I think about the whole thing, it would be wonderful if more of our writers had the understanding and feeling for our series, *Star Trek* in particular, that Kellam has."[36] Roddenberry concurred, stating that de Forest's research "seems to get better with each episode"[37] and that as the series moved along into its third and final season, "he's become very, very good and increasingly necessary to us in double-checking the scientific content of our show."[38]

When asked if it was common for a televisions series to have more than one technical consultant, de Forest replied, "No. In fact it was unusual for most of them to have any at all. . . . Most crime shows if they had any consultants, they were on the set. I'm sure some of the writers consulted with the LAPD or other law enforcement agencies about procedure, but generally it was on the set. . . . They didn't have history experts on the sets of westerns. With *Star Trek*, Roddenberry wanted to avoid having bug-eyed monsters, which up to that time generally was how science fiction had been depicted on television and motion pictures. That made our work even more important because he would rely more on technology and dialogue than on shock value."[39]

During the first quarter of 1967, Gulf and Western, then the parent company of Paramount Pictures, announced that they planned to buy out Desilu studios. Later that year the deal was finalized, and De Forest Research began conducting script clearance and research for all of Paramount's television shows and big-screen productions. By 1984, the *New York Times* described De Forest Research as "the largest Hollywood research firm operating today."[40]

Even though Harvey Lynn was no longer associated with *Star Trek*, having left the show in April 1967, de Forest stayed on as technical advisor, offering his services for the entire run of the original series as well as *Star Trek: The Motion Picture* and subsequent film spin-offs. Like Harvey Lynn, de Forest never received screen credit for his work and was paid $50 per episode for his services.[41]

After *Star Trek* went off the air, de Forest left Desilu but was soon hired by Paramount to maintain their library. In addition, when MGM held its grand auction in the early 1970s, when they sold thousands of items from films, including props, miniatures, costumes, and photos, their library was spared since it remained under lease by de Forest.[42] The same was true for 20th Century-

Fox's research library, since they also were able to survive the auction block as a result of similar leasing agreement under de Forest.[43] All three places stayed open and allowed de Forest and his team to continue providing a service that helped keep writers on track. Eventually, though, time ran out. Paramount's library closed in 1983, followed by MGM's library in 1986. The library at 20th Century Fox soon followed, shutting down a few years after that.

Upon retiring in 1991, de Forest donated all his papers to UCLA. Today, the Kellam de Forest Collection is a valuable resource on the history of film and television. Publicly accessible in the Charles Young Library, it comprises some 250 linear feet (600 boxes) and includes material from more than 255 television shows and over 500 motion pictures that de Forest researched during his nearly forty-year career.

Kellam died on January 19, 2021, at age ninety-four. He and his staff of fact-checkers and research assistants were among the unsung heroes in the creation of the original *Star Trek* television series. Together, they offered an underacknowledged pool of creative talent that read through every story outline, memo, and script produced for the series. Their influence was great, since they cleared names, provided factual and technical advice, and pointed out issues that helped avoid potential legal problems for the studio and network.

The two main technical consultants to *Star Trek*, Harvey Lynn and Kellam de Forest, came from distinctly different backgrounds. Perhaps because of this, both Lynn and de Forest were able to contribute a wider range of advice to the show than if Roddenberry had to rely on just one individual. Even though Lynn left *Star Trek* after only eighteen months, his contributions made a lasting impact. By comparison, de Forest stayed with the series through its entire three-season run. It didn't hurt that de Forest's degree was from Yale, an association like that of Lynn's RAND connection, which Roddenberry often used in his efforts to legitimize and promote the series' scientific accuracy and credibility. Together, Lynn and de Forest made significant contributions to *Star Trek* by lending their knowledge and experience from their respective fields to help enrich each episode.

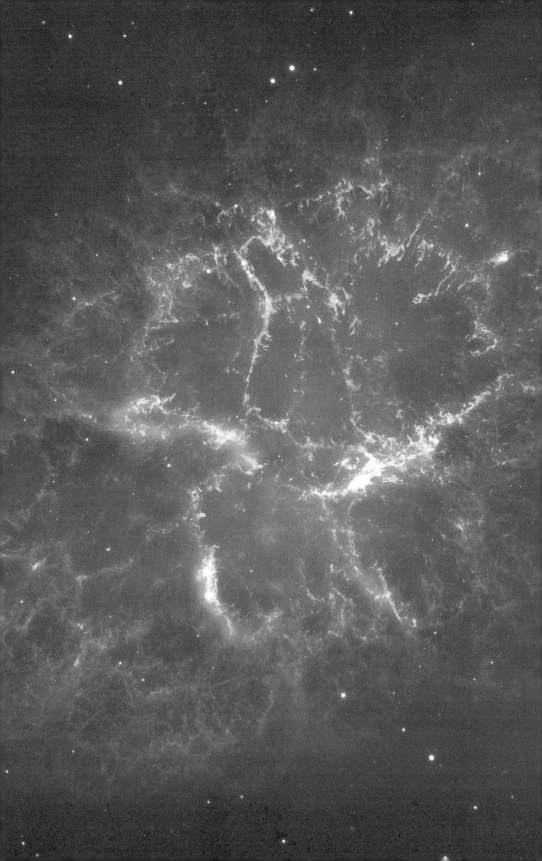

CHAPTER 3

Stephen Whitfield, AMT, and *The Making of Star Trek*

Stephen Whitfield

When *Star Trek* premiered on television, producers of the show licensed products during its original production run in an attempt to cash in on viewer interest in the series. In a memo written by Roddenberry in 1967 he asked "Would much appreciate getting from each of you a list of your ideas on what items seem to you to be good merchandising possibilities for *Star Trek*. Also, any ideas you have for new items we might introduce into shows for merchandising potential."[1]

Such commercial efforts were not new, as other shows during this time period—such as *Lost in Space*, *Bonanza*, and *The Munsters*—all had marketed products and toys to their television viewers. Most of these licensed products were duds, such as rebranded board games and off-the-shelf ray gun toys. However, two success stories emerged from *Star Trek*'s early attempts to replicate what others had done. One was the appearance of several *Star Trek* plastic model kits manufactured by Michigan-based AMT Corporation. The other was the book *The Making of Star Trek*, published by Ballantine Books. Both the models and the book proved to be extremely popular and helped generate interest in the show during the vulnerable second and third seasons of the series.

The summer 1967 release of a scale-model kit of the starship USS *Enterprise*, the first produced by AMT for the new television series, was a phenomenal success. The 18-inch model broke all records in the plastic-modeling industry for volume and sales, facts that did not go unnoticed by the show's producers. *The Making of Star Trek*, first published in September 1968, unfortunately came along too late to have an impact on the series' fate during its original network run on NBC. But large sales of the book encouraged publishers to produce others, such as the popular novelizations of the episodes by James Blish and the later *Star Trek Blueprints* and *Technical Manual* created by Franz Joseph. Sales of these and other books helped promote the show long after its network run had ended, by continuing to breathe new life into a franchise that was further enhanced by the show's popular syndication. Key to the success of both *The Making of Star Trek* and the early AMT model kits was a man named Stephen E. Whitfield.

Stephen Edward Whitfield was born Stephen Edward Poe on March 18, 1936, in Tulare, California. A distant cousin of nineteenth-century writer Edgar Allan Poe,[2] he was the middle child of three born to Reuben Hadley Sylvester Poe and his wife, Mary. He had an older brother, James (Jimmie), who was born in 1933, and a younger sister, Roberta (Bobbi), who was born in 1938.

Reuben Poe, who went by the name of Jack, was a carpenter and died on September 31, 1937, from an industrial accident. Only twenty-nine at the time, he was in Tulare working on a warehouse when something brushed against a tall ladder that he was on, causing him to fall. He ended up with severe cuts, a broken left leg, and a broken right ankle. While recovering in the hospital, he contracted tetanus and died.[3]

"The death struck the family very hard," recalled Bobbi Poe, Stephen's sister. "His wife, Mary, faced with raising three children in the midst of the Depression, never fully recovered."[4] Mary Poe remarried in 1941, to Harold Lee Whitfield, and they added two more children to the family, David Leslie and Susan Kathleen.

"It was hard," recalled Bobbi, of growing up back then. "We lived in a little two-room cabin by the railroad tracks just a few blocks from the grammar school on the west side of Tulare. Our mother worked in the gas station, and we didn't know that we were poor. We went to school barefoot every day, and one day the principal called us into the office and said, 'You can't come to school anymore until you wear shoes.' We didn't own any shoes. That was the day we learned we were poor."

Growing up, Whitfield loved to write. While in high school he, along with his sister, Bobbi, worked for the school newspaper. He was also in the drama club and played the lead in *Harvey*, the senior-class play. He was very active in scouting and wrote a series of articles for his hometown newspaper, the *Tulare Advance-Register*, chronicling the 1950 Boy Scout Jamboree held in Valley Forge, Pennsylvania.

Whitfield was around 6 foot 4 and had very large feet. "He wore a size 16 shoe," remembered Bobbi and recalled how her brother's large feet helped feed the family. "There was a doctor in town who was fascinated by the perfect bone structure in Stephen's feet," said Bobbi. "The doctor would pay us something like thirty-five cents an hour to palpate[5] the bone structure in his feet. For a while, this was the only income that my family had."[6]

After graduating from Mt. Whitney High School in Visalia in 1954, Whitfield enlisted in the Navy at age nineteen. He received his new recruit training in San Diego before being shipped to the Naval Aviation Cadet Program in Pensacola, where he earned his wings as a first lieutenant in the Marines.

CHAPTER THREE

On Sunday, June 10, 1956, at the Community Presbyterian Church of Three Rivers in Tulare County California,[7] Whitfield married Ann Jaren before moving to Hawaii.[8] He flew helicopters while stationed at the Marine Corps Air Station in Kaneohe Bay, where they had a son named Mark. Though active in the military, Whitfield never served in wartime because of a medical condition. "There was a story of him trying to be a pilot, but at some point later in his military career he had a sinus blowout when he was up in the air, and as a result he couldn't fly anymore"[9] explained his stepsister, Susan.

After leaving the service, Whitfield settled in Arizona, where he worked in the public-relations field, serving as the western states director for the Muscular Dystrophy Association.

By the mid-1960s, Whitfield worked for Ptak & Richter Advertising Agency in Phoenix, Arizona. Founded by Roland Daniel Ptak, the agency was one of the largest advertising, public-relations, and marketing-research firms in the West. It was while working for Ptak that Whitfield first became associated with *Star Trek*.

"My first personal contact with *Star Trek* was in August 1966," wrote Whitfield in the introduction to his book *The Making of Star Trek*. In recalling how he got involved with the series from the beginning, he explained, "At the time I was employed by an advertising agency in Phoenix, Arizona. One of the agency's accounts was AMT Corporation, a manufacturer of scale model plastic hobby kits. AMT had acquired the merchandising rights on the USS *Enterprise* and intended to market a scale model plastic kit of the *Star Trek* spaceship. Part of my job became one of working closely with Desilu Studios and NBC-TV in order to generate publicity that would reflect favorably on AMT and, hopefully, future sales of the *Enterprise* model kit."[10] Whitfield's initial involvement at Ptak to help secure and promote the production of plastic model kits for *Star Trek* soon led to a full-time position with AMT as their national advertising and promotion director.

AMT

The link between the aerospace industry and the hobby of scale-model building is strong. Early trade publications such as *Popular Science*, *Popular Mechanics*, and many others devoted pages on how to build scale models. The ability to mold and cast tooling to make scaled-down plastic models drew upon the same set of skills used to fabricate such things as full-sized fighter aircraft and heavy bombers. Likewise, advances in chemistry and tooling technology that produced lightweight and durable plastics, plus the means to mold them, helped engineers build the rockets and spacecraft during the Cold War and the space race. Some of that technology made its way into the toy and hobby industry.[11a–e]

With an ever-increasing number of people watching television, viewers wanted to connect with shows they enjoyed. One way to do that was to purchase model kits that allowed them to more closely associate with their favorite TV characters and vehicles. Models of ancient dinosaurs, Frankenstein's monster and the Wolfman, King Kong and Godzilla, Superman and Batman, the TV hero the Lone Ranger, and spaceships, both factual and fictional, were all mass-produced and sold in scale-model form.[12a-b]

One of the more successful of these scale-model companies, the Aluminum Model Toys Corporation, or AMT, began as a side business in 1948 by West Gallogly Sr. He focused on a very real-world product that millions could relate to—the automobile. He made preassembled plastic promotional car models for the big-three automotive companies in and around Detroit. These were first made out of metal such as aluminum, and then later, they were injection molded plastic. The company, based in Troy, Michigan, later expanded to make model kits.

The 18-inch plastic model kit of the USS *Enterprise* that began to hit store shelves in the middle of 1967 was AMT's first spaceship model. Prior to its release, the company had never produced a science fiction model. AMT was the world's largest manufacturer of model cars and trucks—not rockets and spaceships. Its *Star Trek* model was the first time that AMT sold a kit having anything to do with space.

AMT had previously established itself as the producer of "Star Cars"— model kits that featured popular movie and television vehicles and their celebrities. For example, shortly before the release of the *Enterprise* kit, the company produced the highly successful "Munster Koach" and "Drag-U-LA" coffin car, two vehicles that were seen on the hit 1960s CBS television series *The Munsters*.

As a result, industrial-trade insiders were surprised when AMT announced in the fall of 1966 that it would begin tooling to produce a model kit relating to *Star Trek*—a television show that most people had not even seen yet. Desilu's attorneys wrote the Licensing Corporation of America, the legal arm of the studio where *Star Trek* was made, that AMT clearly had great confidence in the show, since "no manufacturing company would normally incur $50,000 in tooling for heavy goods before the show went into its second year."[13]

Memos and letters between Desilu and AMT issued during the time when both were negotiating terms for the first *Star Trek* model kit reveal that the deal originally included a second model in addition to the *Enterprise*. This second model stemmed from *Star Trek*'s creator, Gene Roddenberry, who wanted another means other than the innovative ship's transporter for getting crew members to and from the *Enterprise*. The idea for the transporter was brilliant. Rather than land the entire ship of 400-plus crew members each time somebody wanted to

CHAPTER THREE

get on or off, *Star Trek* had the transporter—a device that allowed individuals to dematerialize and then rematerialize to get them from one place to another within seconds. This was a huge cost-savings measure for the production.

Fans liked the idea but were quick to ask, "What if the transporters broke down or the ship was out of range?" After airing "The Enemy Within," the fourth episode of the first season, viewers could see that something was clearly needed after watching Sulu and other crew members nearly freeze to death while stranded on the surface of a planet as the *Enterprise* orbited above them.

Roddenberry's answer was to create a separate fleet of smaller spacecraft. Those smaller ships, or "shuttlecraft," would be stored in a large hangar bay located in the aft end of the secondary hull of the *Enterprise*. Such an addition seemed natural, since a hangar bay had been built into the ship from the very beginning.

Walter Matthew "Matt" Jefferies Jr., *Star Trek*'s art director and production designer, and creator of the original starship *Enterprise*, designed such a shuttlecraft, christening it the Galileo Seven. The shuttlecraft could take off and land, much like fighter aircraft do on present-day aircraft carriers. The problem, however, was that the studio could barely afford to build the models and sets required to show the shuttlecraft in the series.

In letters and memos issued during August and September 1966, there emerged a deal that solved that problem. In exchange for the rights to produce both the *Enterprise* and shuttlecraft model kits, AMT would construct "at its cost, an exterior and interior of Galileo Seven" for an estimated cost of $24,000. In addition, AMT would also construct a filming miniature of the Galileo for an additional cost of $650.[14] Desilu agreed to fund the building of a scaled hangar deck model so that the miniature shuttlecraft could be filmed coming and going from the *Enterprise*. By the end of September 1966, AMT intended to originally produce not only a model of the *Enterprise* but also one of the Galileo shuttlecraft. In an October 28, 1966, letter, Desilu Studios told AMT, "For your information, we have used the Galileo Seven on two episodes thus far and we hope in the future to be able to use it quite often so that you will be encouraged to do the tooling on the Galileo Seven for sale."[15]

AMT was uniquely qualified to do the construction of the shuttlecraft model and its full-sized sets. At the time, it was the only model company that also built "full-sized" cars. This was done at their Speed and Custom Division Shop in Phoenix, Arizona. Established in 1966 as an AMT subsidiary, the shop built scale models and full-scaled mockups from which it pulled templates or masters used to make dies for model-kit production. The full-scale mockup would then be used at auto shows and other popular hot-rod events to promote not only the custom car work done at AMT's Arizona shop but also the model kits that the company was making and selling. The shop was headed by legendary hot-rod builder Gene Winfield, a well-known automotive designer, who was tasked with bringing to life the shuttlecraft designs submitted by Jefferies. Winfield sought to make Jefferies's original shuttlecraft designs less curvy to help reduce the cost of constructing the full-size prop and filming miniature. Winfield tapped the talents of industrial designer Thomas "Tom" W. Kellogg to help with the redesign effort. Kellogg simplified Jefferies's original renderings to produce the final shuttlecraft design that appeared in the show.

Less than a month after the first episode of *Star Trek* premiered on NBC on September 8, 1966, AMT produced a catalog to pass out at the fall model kit toy fairs. On the cover was a new model labeled "No. 921 Space Ship," with a suggested retail price of $1.70.[16] AMT did not associate the odd-looking model with the new television series. Only folks fortunate enough to watch the new show on TV, or who had read about it in the media, would know what they were looking at.

AMT may have been intentionally vague about the identity of this new model for a reason. By the time the dies were cast to produce the new kit, there were no guarantees the show would succeed. AMT may have been more than just a little uncertain that the model would fly, since the show could have been canceled after only a few episodes had aired.

The *Enterprise* model kits were originally planned for release during Christmas 1966,[17] but production was delayed.[18]

Kits finally began hitting store shelves in late spring of 1967.[19] By the end of July, retailers reported strong sales across the country. The trade publication *Toys and Novelties* noted, "The surprising leader, for this time of year, of the top sellers last month was the *Star Trek* plastic kit. One wholesale-dealer put results as 'fantastic' and most others agreed it was better than good."[20]

Early sales of the kit were in fact so strong that AMT requested that "the license of our company be extended for a minimum of one year," noting that "this extension is stimulated by the very excellent acceptance of our kits in the market." At the same time, AMT sought "to extend the license to the entire world," stating that it was "considering going into double tooling and an association with a toy company in England" that would allow the kit's release in the United Kingdom and Canada.[21] AMT eventually made an agreement with the Aurora model company to license its *Star Trek* kits overseas.

As AMT's director of national advertising and promotions, Stephen Whitfield kept busy working behind the scenes to help promote the new *Star Trek* kit. Letters showed that he was actively putting complimentary copies of the model into the hands of influencers that would do the most good not only for sales of the model but also for promotion of the show.

CHAPTER THREE

Shortly before the twenty-fifth annual World Science Fiction Convention (WorldCon), also known as NyCon 3, held in New York City during Labor Day weekend, August 31–September 4, 1967, Whitfield sent several dozen AMT *Enterprise* kits to the Americana Hotel where Roddenberry and his wife were staying. Roddenberry knew the kits would be highly desirable among influential convention-goers, so he made sure to have a supply on hand to pass around.[22]

In an October 19, 1967, memo, *Star Trek* associate producer Bob Justman could barely contain his enthusiasm when he told Roddenberry about the record-breaking numbers of kits being sold: "The machine which turns out the plastic parts for the kit goes continually 24 hours a day, 7 days a week[,] and AMT is rushing another machine into production, so that they can keep up with the demand."[23] Less than one year after its initial release, AMT's *Enterprise* kit had sold over one million copies—outstripping AMT'S Munster Koach, which had previously been "the most successful plastic model kit in the business."[24] In the same memo, Justman lamented that "pretty soon AMT will be sorry they did not undertake a Galileo Model Kit."[25]

Given the phenomenal sales success of the *Enterprise* kits, it is not known why AMT failed to immediately issue a model kit of the shuttlecraft Galileo. Perhaps the financial terms were less favorable than those AMT received for the *Enterprise*.[26] Even though a shuttlecraft model was not in the immediate future, AMT wanted to make another *Star Trek* kit while the series was still on the air.

With strong sales of AMT's *Enterprise* kit showing no signs of stopping, the company sought to produce another *Star Trek* model that would be equally if not more successful than the first. As a result, by early August 1967, AMT had negotiated a deal with Desilu Studios to produce a model of the Klingon ship. Bryce W. Russell, AMT's vice president of finance and planning, wrote Roddenberry on August 11, 1967, "I have now had several discussions regarding the 'Klingon' spaceship with our people here and we are most excited about the prospects. We would very much like to add another *Star Trek* ship to our line as soon as possible."[27]

Even though the Klingons were depicted as the enemy of the Federation since their appearance in the first season, nobody knew what their ship looked like until the start of the third season. Until then, AMT could only offer ideas: "We should consider making the 'Klingon' ship a three-way type of kit—possibly a freighter, transport and fighter."[28]

It's not clear exactly when the formal contract was signed to make the Klingon model kit, but AMT ironed out a deal sometime in late 1967. Memory Alpha, a Wikipedia equivalent on all matters concerning *Star Trek*, claims that the creation of the Klingon ship was exclusively done by AMT.[29] Though the model company did push to have it made, Desilu Studios was not left out of the loop during its development. They still had final say on the design and held all the rights to the finished product.

In a deal like the one that secured the rights to produce the *Enterprise* model kit, AMT again traded something that the cash-starved series could use in exchange for the rights to produce another kit. In this case, AMT provided the studio with a filming model of the Klingon ship. *Star Trek*'s art director, Matt Jefferies, was in charge of coming up with the design.

"We already established the essential character of the Klingons, so we really had more to draw on in background than we originally had on the *Enterprise*," explained Jefferies in an interview done by Dorothy Fontana in October 1968. "The Klingon character was different and clearly defined in several scripts. We tried to keep some of that character in the design of the ship—cold and, in a sense, vicious. We tried to get into it some of the qualities of a manta ray, shark, or bird of prey, because the Klingons follow that general feeling. Another requirement was that we had to get a feeling their ships were on par with the starships in equipment, power, size, etc. After many, many sketches and many evenings, it finally evolved. Everyone liked it, and that's what we built."[30] Jefferies then created a full engineering drawing, stating, "I established what size it would be in relation to the *Enterprise* and did a scale drawing, which is what we handed to AMT."[31]

By January 1968, while the cast and crew of *Star Trek* were wrapping up the filming of the last episode of the second season, Jefferies continued to work on his own in developing his design for the Klingon ship. "We had no need for a Klingon ship, nor did we have a budget to do one, or the time to design or build it," explained Jefferies during a 2002 interview for *Star Trek: The Magazine*. "But AMT wanted a follow-up to the USS *Enterprise* NCC-1701 kit because it had sold over a million in the first year. So, although the Klingon ship was something new that would fit the show, it was primarily done for AMT."[32]

Whitfield was again actively working behind the scenes to help AMT and *Star Trek*. This time he focused on helping the Klingon model kit get off the ground. In a January 2, 1968, letter to John Reynolds, vice president of Paramount Pictures Television, Whitfield boasted of the success of AMT's first kit, stating that "extraordinary heavy demand for this kit has pushed sales towards the 800,000 mark, and gives every indication of reaching an all-time record of one million." Whitfield went on to explain that AMT had begun "developing a second kit, with the assistance of Gene Roddenberry and Matt Jefferies. This second spaceship is in the 'clay model' stage of production now and will be ready for the final surface detail to be added sometime around January 15." AMT wanted Jefferies to be available to oversee the final details of the Klingon model, and Whitfield encouraged Paramount Television to allow Jefferies "to guide our production staff, at our plant in Troy, Michigan, in this last vital stage." Whitfield concluded his letter stating that "we are quite convinced that a properly detailed model of the second ship will result in sales records paralleling those of the *Enterprise* kit. We are also convinced that, poorly done, sales will be disappointing, at best."[33]

The studio balked at footing the bill to bring Jefferies to Michigan. Marvin Katz, director of business affairs for Paramount Studios, wrote AMT, "We have reviewed your letters of January 2 and January 11, 1968, regarding Matt Jefferies' visit to your plant during the week commencing January 22, 1968. While we would very much like to assist you in this matter, it would be unfeasible to carry Matt on our payroll for an additional week as he is scheduled to conclude his services for us on Friday, January 19, 1968. We have realized that you have spent much time, effort and money on the Klingon project, but the substantial production costs of the series dictates that we must draw the line somewhere."[34]

Whitfield's persistence paid off in getting Jefferies to come to AMT. In a follow-up letter to Whitfield from Marvin Katz, Paramount Television's director of business affairs, Katz was impressed by Whitfield's confidence regarding the projected success of a second *Star Trek* model. Katz told Whitfield in a January 17, 1968, letter, "If you are as positive about the economic success of the Klingon model ship kit as your letters indicate, then consider this suggestion for resolving the matter." Thanks to Whitfield, the matter was resolved. Jefferies completed his drawings for the Klingon model and was able to travel to AMT's plant to oversee the final detail work.[35]

Two 29-inch-long master models of the ship were built at AMT's Speed and Custom Division Shop in Phoenix.[36] The studio used one for filming effects shots, while the other was used by AMT to create molds for their model and to photograph for the box art.

The AMT model kit has some minor differences from that of the television model, mostly in details on the engine nacelles. "They used what they call a pantograph; at one end there was a stylus that traced its way over the master model, and at the other end of it there was a tool that carved out the same shape in tooling steel, which became the mold they built the kit from," explained Jefferies during an interview about the model-making process that he observed at AMT. "I was there at about 2 o'clock in the morning when they ran the first two or three through the machine. They weren't perfect, so they said, 'We'll take out a fraction here, and a fraction here.' Then they'd run two or three more. If I remember correctly, it was about 10 o'clock when the first one came out that they said was perfect. They ran maybe another half a dozen, and checked those out. . . . Then the machine was put in operation, and after that, one came out every 20 seconds."[37]

AMT knew it had another hit model on its hands as trade publications and stores began heavily promoting the new 14.5-inch model. In the August 1968 issue of the fan-produced *Star Trek* newsletter *Inside Star Trek*, D. C. Fontana reported details about the new Klingon ship model:

> The Klingon ship was designed by Walter (Matt) Jefferies, *Star Trek*'s art director, who also designed the *Enterprise*. The new vessel is the equal to the mighty Star Fleet cruiser and is comparable in many ways. It is approximately 747 feet long from the tip of its bubble nose to the tail end of its power pods. It is decorated with Klingon symbols, and its numbers are also in "Klingonese."

Fontana continued:

> In shape, the vessel appears much like a large bird of prey in flight and is distinctly different from the saucer-and-nacelle configuration of the *Enterprise*. The control room or bridge is in the top-front of the bird-like "head." This forward area also contains quarters, labs and armament control. On the underside of this area facing forward is a combined sensor-weapon device. Heavy weapons are also mounted in the housings in front of the dual power pods. The vast wing-like section is used for storage, fuel, power source, environmental control units, etc.
>
> Matt Jefferies informs us it took him about a month to get his ideas "squared away" on the design and another month to get the drawings out for AMT. All this was done on his own time, beyond his normal work on *Star Trek*, but the resulting unique and graceful design of the battle cruiser was well worth the extra effort. To date, Klingon ships have been used in the upcoming third[-]season episode, "The Enterprise Incident[,]" and may be seen in another titled "Elaan of Troyius."

Fontana concluded:

> The AMT kit will be on the same scale as the *Enterprise* model and has control deck and crew's quarters illuminated by operating lights. The Klingon symbols and numbers are included on decals, and there is a display option of "space base" stand or "skyhook" ceiling mount. Also included will be several sheets of sketches which show this vessel in comparison to the *Enterprise* and containing other new information. AMT has announced the kits will be in stores and hobby shops sometime in late August and will retail for $2.50.[38]

Gene Roddenberry speaking on Saturday, September 3, 1966, before showing the first *Star Trek* pilot, "The Cage," to attendees of Tricon, the twenty-fourth annual World Science Fiction Convention, held in Cleveland, Ohio, that Labor Day weekend. Three days later, the series premiered on television in Canada, followed by its first US NBC network airing on September 8. *Photo of courtesy the Jay Kay Klein Collection, Special Collections & University Archives, University of California, Riverside.*

Roddenberry while attending Tricon, the twenty-fourth annual World Science Fiction Convention in Cleveland, Ohio, in 1966, is shown participating in a local radio panel called "The Monster Craze" on September 5, 1966. Shown here is Harlan Ellison (*face behind microphone*), Gene Roddenberry, and L. Sprague de Camp. In the foreground to the right is Forrest J. Ackerman. Not shown off to the left foreground is Cleveland attorney Nicholas A. Bucur.

The main registration table at Tricon. Shown are Harlan Ellison (*in sunglasses*), and to the right of him is Tricon president Ben Jason. Gene Roddenberry is shown seated next to Jason.

This photo was taken at a dinner held by the Science Fiction Writers of America (SFWA) during Tricon on September 3, 1966. The dinner was held at the Tavern Chop House on 1027 Chester Avenue, about a ten-minute walk from where the convention was being held in Cleveland, Ohio. Gene Roddenberry and Harlan Ellison are shown seated directly opposite each other. Gene's first wife, Eileen-Anita Rexrot, is seated between her husband and Elizabeth Pickering, a professional model who wore the famous "Andrea" costume designed by Bill Theiss that was first worn by actress Sherry Jackson, who played Andrea in the *Star Trek* episode "What Are Little Girls Made Of?" That costume along with several others from the show were shown during Tricon's costume party. Next to her left is Jerry Sohl. With back to the camera and sitting to Ellison's left is actress Sherri Townsend. Shown to her left is Norman Spinrad.

Tricon president Ben Jason presenting Roddenberry with a special "Distinguished Achievement Award" for *Star Trek* given to him by the World Science Fiction Convention Committee during the Hugo Awards banquet held on Sunday, September 4, 1966, at Tricon.

Roddenberry at NyCon 3, the twenty-fifth annual World Science Fiction Convention, held in New York in 1967. *All photos courtesy of the Jay Kay Klein collection and used by permission of Special Collections & University Archives, University of California, Riverside*

Harvey Prendergast Lynn, Jr., age eight

Harvey Lynn, age thirteen

Lt. Harvey Lynn, age twenty-two

Col. Harvey Lynn, age forty-two

Harvey Lynn and his two sons Harvey P. Lynn III (*center*) and Dennis Sebastien Lynn, 1965

Dennis Lynn with his father at Paramount Studios during the twentieth-anniversary birthday celebration of *Star Trek* held on September 8, 1986. *All photos courtesy of Dennis Lynn*

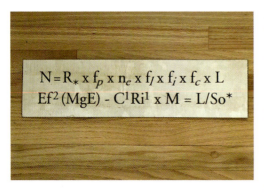

Okuda Drake equation. In the *Star Trek: Voyager* episode "Future's End," viewers meet a twentieth-century astronomer who used a radio telescope to search for extraterrestrial life. Taped to the astronomer's computer monitor in her lab is a placard designed by Mike Okuda that shows both the real Drake equation and Roddenberry's fake one. Viewers of the episode can glimpse the placard very faintly in the background of a couple of shots. Okuda later faxed a copy of the placard to Professor Frank Drake and spoke with him briefly on the phone. Okuda recalled of the conversation that he seemed amused but noted that Roddenberry's version included a couple of variables raised to the first power, which, of course, is exactly the same as the variables themselves. *Photo courtesy of Michael Okuda*

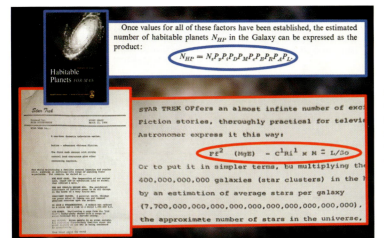

Habitable Planets for Man book cover superimposed with equations. *Photo courtesy of the author*

Stephen H. Dole. *Photo courtesy of RAND*

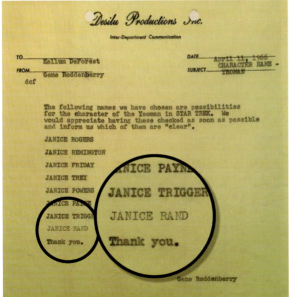

An early 1966 memo showing what may have been a last-minute typed addition of "JANICE RAND" to the list of suggested names for the Yeoman character. *Photo courtesy of the children of Kellam de Forest and the Gene Roddenberry Papers (Collection PASC 62), UCLA Library Special Collections, Charles E. Young Research Library, University of California, Los Angeles*

Kellam de Forest is shown with two of his research assistants, Rona Kornblum (*right*) and Charlotte Worth. *Photo courtesy of CBS Films and the children of Kellam de Forest*

Individual photos of Kellam de Forest. *Photos courtesy of the children of Kellam de Forest*

Kellam and his wife, Peggy, at Paramount Studios during a special twentieth anniversary of *Star Trek* birthday celebration held on September 8, 1986. The event took place on Stages 9 and 10, where the series was originally filmed. *Photo courtesy of the children of Kellam de Forest*

A Gorn mask made from a casting off the original studio prop that was used in the *Star Trek* episode "Arena." The hollow mask sculpture was cast and finished by Tony Hardy. *Photo courtesy of Karl Tate*

Kellam de Forest with the author, December 22, 2019. *Photo courtesy of the author*

During a December 22, 2019, author visit with Kellam de Forest, he pulled out this photo to show me where his office was located at Paramount. During the interview, he explained that during the time of *Star Trek*, his company's main office was housed in the research libraries at Paramount Pictures, which included the old RKO Studio library that later became the prop department at Desilu. The solid line indicates the border of the Paramount lot, with Desilu studios bordered by the broken line. *Photo courtesy of the author, and Kellam de Forest*

Stephen Whitfield (*far right*) is shown with his brother, Jimmie, and sister, Bobbi. *Photo courtesy of Bobbi Moore*

Stephen and his mother are shown with his brother, Jimmie (*far left*), and sister, Bobbi (*far right*). *Photo courtesy of Bobbi Moore*

Stephen Whitfield, age ten. *Photo courtesy of Bobbi Moore*

Stephen's high school photo, March 18, 1954. *Photo courtesy of Bobbi Moore*

Naval aviation cadet Stephen Whitfield. *Photo courtesy of Bobbi Moore*

Naval aviation cadet Stephen Whitfield standing next to a North American SNJ "Texan." "That's an SNJ I'm leaning on," wrote Stephen on the back of this photo. "As you can see, it's not too big. I can just barely touch the wing of the T-28, the plane I'm flying now." On the basis of the BA tail code on the aircraft in the background, the photo was probably taken at Naval Air Station Pensacola, where the Navy's intermediate training (using the SNJ) was done. *Photo courtesy of Bobbi Moore*

Marine first lieutenant Stephen Whitfield getting his wings. *Photo courtesy of Bobbi Moore*

Far left: Stephen Whitfield, 1968. *Photo courtesy of Bobbi Moore*

Photo of author with Bobbi Moore taken at her California home on July 27, 2021. *Photo courtesy of the author*

One of the very first announcements of the new AMT *Star Trek* model kit appeared in an October 1966 AMT dealer promotional flyer. Not knowing if the new *Star Trek* television series would last, AMT simply listed the model as "No. 921 Space Ship." *Photo courtesy of the author*

Advance dealer insert issued in 1968 for the new AMT Klingon model kit. *Photo courtesy of the author and Christopher Beamish*

Advance dealer insert issued in October 1966 for the new AMT *Enterprise* model kit. *Photo courtesy of the author and Christopher Beamish*

The 1:650 AMT *Star Trek Enterprise* model kit reissued by Round 2. *Photo courtesy of the author*

Left: Peter Carsillo shown with an AMT *Enterprise* model kit and a scratch-built Klingon D7 model. "Since about six, I had been making my own paper and cardboard versions of things. Superheroes, spaceships, monsters," said Peter. "Oddly enough, I still do today." Peter grew up and became a senior creative director at Walt Disney Imagineering. He now works for Universal Creative, where he designs theme parks for a living. *Photo courtesy of Peter Carsillo. Photo taken by Edmond Deraedt*

Right: Young *Star Trek* fan Blair Stanridge is shown holding an AMT model in a photo taken in 1968. *Photo courtesy of Kipp Teague*

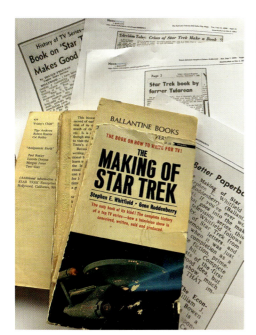 

This AMT kit was built and modified by Richard G. Van Treuren while serving aboard the naval aircraft carrier USS *Kitty Hawk* (CV-63) during Vietnam. Van Treuren donated the model to the Smithsonian in 1973, where it is now on display. *Photo courtesy of the author*

My well-worn tenth printing of Whitfield's *The Making of Star Trek* that I purchased in 1972 when I was nine years old. Now worse for wear, this copy traveled with me to NASA and was used to help write *Inspired Enterprise*. *Photo courtesy of the author*

In his book *The Making of Star Trek*, author Stephen Whitfield (a.k.a. Stephen Poe) knew not only what words to write but also what pictures to show. The photographs in his book gave insights into the inner workings of *Star Trek*. On this spread are photos belonging to Whitfield that never made it into the book. Shown are the hand phaser, tricorder, Klingon disruptor, Type 3 computer, data card reader, high-intensity goggles, logbook and hand-held medical and engineering scanners. All of these photos are stamped on the back "Please Return" followed by Stephen's Reno, Nevada mailing address. The man in the plaid shirt is Jim Rugg (1919-2004), a special effects artist that worked on all three seasons of *Star Trek*. He is pictured overseeing the wiring and mechanical effects of the *Enterprise's* Bridge set. The stage lighting photo is from Westheimer, one of the visual effects companies that worked on the series and depicts the filming of one of the many vari-colored planets seen in the show. The set of Gary Seven's office from the second season episode "Assignment: Earth" is also shown along with a rare news release photo of Walter Koenig who played Ensign Chekov. All photos courtesy of the author.

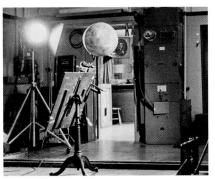

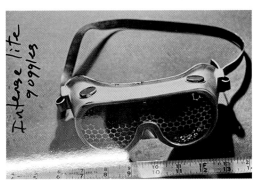

Intense lite 9/6/66

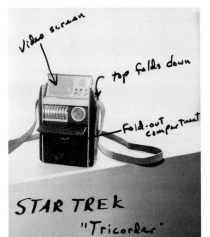

Video screen
top folds down
Fold-out compartment

STAR TREK
"Tricorder"

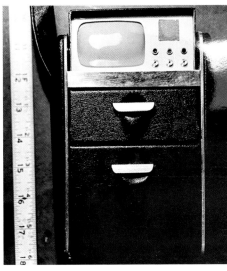

Early prototype photos of AMT's bridge kit. Both the bridge kit and a kit of the Romulan ship were issued in 1975 as the first of AMT's *Star Trek* kits to be packaged in shorter boxes. Prior to this time, all *Star Trek* kits were issued in larger long boxes. This decision saved AMT money while, at the same time, allowed stores to display more kits on their shelves. Note the shorter-backed bridge chairs depicted in this early prototype configuration. *Photos courtesy of John Mueller and Round2/AMT; collage courtesy of Karl Tate*

Early prototype photos of AMT's Leif Ericson model kit, which was first issued in 1969. The kit was later reissued as the Interplanetary U.F.O. Mystery Ship in 1974, which was the same year that AMT first issued their Galileo shuttlecraft model kit. *Photos courtesy of John Mueller and Round2/AMT; collage courtesy of Karl Tate*

Former NASA astronaut José Moreno Hernández, who flew aboard the Space Shuttle, was heavily influenced by the AMT *Enterprise* model kit when he was a kid. While meeting with Hernández in 2023, he confirmed that the series played a highly influential role in his life. He told me that "both that show, and the model really made a difference in my life," while posing with a finished AMT *Enterprise* model that I gave him. "When I became frustrated or depressed, that model gave me hope and reminded me that everything was going to work itself out and be just fine." *Photo courtesy of the author.*

USAF Lieutenant Colonel Jack L. Hartley, displaying his 1-pound, 4-ounce, dental repair kit that he developed to prevent astronaut toothaches. Photos were taken on March 2, 1966.

Lt. Col. Hartley is shown during a January 10, 1966, episode of the television show *To Tell the Truth*. As honorary chairman of the American Dental Association's National Children's Dental Health Week, Hartley appeared on television and radio shows to help promote awareness of the event.

Dr. McCoy's screen-used medical satchel or bag and hypospray pouch can be seen along with various other medical instruments as displayed at the Greg Jein Collection Auction held October 14–15, 2023, at Heritage Auctions in Dallas, Texas. In reality, McCoy's medical bag was a repurposed Dopp Kit. *Photo courtesy of the author*

A cut scene from the end of the episode "Balance of Terror" shows Kirk saluting his defeated opponent during a video exchange between the two just before the Romulan commander blows up his own ship. This scene is from a Lincoln Enterprises unaired trim. *Photo courtesy of Christopher Beamish*

CHAPTER THREE

On the cover of that issue of *Inside Star Trek* there appeared a line drawing of the Klingon ship, a first view of the new ship for most, done by Greg Jein, a model maker who created miniatures used in many science fiction films and television series, including *Close Encounters of the Third Kind*. He also worked on the *Star Trek* films as well as the various television spin-offs.

Kits of the Klingon ship first began arriving in stores by early September 1968. AMT packaged them as the "Klingon Alien Battle Cruiser," proclaiming on the box "As Seen on *Star Trek*" and "As Seen on NBC." The only problem was that by this time, nobody had yet seen it on the show.

At the end of September 1968, "The Enterprise Incident" beamed into living rooms, finally showing the Klingon ship, but a Klingon ship entirely occupied by Romulans, another enemy in the show's universe. In the episode, Mr. Spock explains that the Klingons shared their technology with the Romulans, which was why the *Enterprise* crew encountered a Klingon ship in Romulan space. In reality, the original filming-production model of the Romulan "Bird of Prey" ship could not be found after the earlier first-season shooting of the episode "Balance of Terror," in which the new spaceship was featured. Roddenberry was forced to improvise by substituting a Klingon ship rather than build a new Romulan one. Meanwhile, by December of that same year, "Elaan of Troyius" finally aired, showing Klingons aboard a Klingon ship.

The resulting confusion became a topic of discussion among fans. In a letter that was published in issue 9 (March 1969) of *Inside Star Trek*, C. John Fitzsimmons asked, "Why did the Romulans stop using the boomerang-shaped ships and start using the Klingon models?" D. C. Fontana replied in her column, "The Romulans and Klingons formed a limited alliance to meet the threat of the Federation. Their treaty included help from the Klingons for the Romulans to change over to the faster, more powerful type of ship used by the Klingons. Also, the production staff of *Star Trek* felt that the Klingon model was visually more interesting and should be stressed to help sales of the model, besides."[39] Despite the controversy and confusion, the new AMT Klingon ship model was a big seller.

Since AMT model kits of both the *Enterprise* and Klingon ship came out while *Star Trek* was still in production, it is not at all a surprise that Desilu considered using them in the production of the original series. Even though assembled model kits of the Klingon ship never made it onto the show, built-up AMT kits of the *Enterprise* were featured in several episodes.

AMT *Enterprise* model kits appeared in three second-season episodes. In "The Doomsday Machine," an AMT kit was assembled and shown as the heavily damaged USS *Constellation*, which Commodore Matt Decker commanded while unsuccessfully trying to fend off a giant planet-eating machine. If you look closely, the *Constellation*'s registry number, "1017," is a scrambled version of "1701," so that they could use the same decal set that came with the original model kit. "USS Constellation" seems to have been set in regular dry transfer lettering. AMT did not release decal sheets that allowed builders to create other ship registries until later.

Reused footage of the damaged *Constellation* from "The Doomsday Machine" can also be seen again in the episode "The Ultimate Computer," but this time it is portrayed as the damaged USS *Excalibur*.

How ironic that the first *Star Trek* models to appear in actual episodes depict the AMT kit in a less-than-flattering form. Perhaps this helped comfort those who suffered trauma from the kit's persistent sagging engine nacelles. Those who assembled the AMT *Enterprise* kits experienced frustration in not being able to properly align the two warp engine struts. This was no fault of the model builder, but rather a design flaw by AMT, one that has somewhat been corrected over time but never fully perfected.

Finally, in the episode "The Trouble with Tribbles," a fully assembled AMT model kit of the *Enterprise* is shown moving outside K-7 space station manager Lurry's office window. Beginning in 2019, a traveling exhibit called *Star Trek: Exploring New Worlds* displayed a host of props from the original television series that included this AMT model as well as the original D7 Klingon ship and *Enterprise* bridge helm from the series.

The exact color of the Klingon ship has long been a topic of debate among fans, ever since the model was first seen on television. In a rush to get the model filmed, the first two episodes that show the ship revealed it to be all gray with a green light illuminating the model from below. In some scenes, you can even see the green lights reflecting off of the front bulb-shaped forward hull section. According to Greg Jein, who worked for Lincoln Enterprises, Roddenberry's merchandising company that started to sell *Star Trek* paraphernalia in 1968, he remembered seeing one of the Klingon ship models arrive back then uncrated and painted all gray. The Jein story about seeing the all-gray color of the model has been confirmed by others.

Footage of the Klingon ship was seen primarily in three episodes from the third season: "Elaan of Troyius," "The Enterprise Incident," and "Day of the Dove." It was first glimpsed on television by the public in the episode "The Enterprise Incident." It is a challenge to find any good screen captures of the ship that clearly show it painted anything other than gray. What we do know is that at some point, both models were painted a two-tone greenish gray. Jefferies spoke about the color of the ship in a 2002 *Star Trek: The Magazine* article: "The coloration came directly from a shark; it's a grayish green on top and a lighter gray underneath."[40]

Once they were done filming the first studio production model at Howard Anderson's optical-effects company in Hollywood, and after AMT finished making its molds and photographing the second studio production model for their model box art and advertising, both studio models of the Klingon ship were returned to Paramount. In 1973, Jefferies gave the filming model of the Klingon ship to the Smithsonian's National Air and Space Museum.

"I got a hold of one of those tooling masters," said Jefferies in a 2002 *Star Trek: The Magazine* interview about the Klingon ship model, "and I gave it to the National Air and Space Museum." Because of its unique size and shape, safely getting it to the Smithsonian proved to be a challenge. "I was trying to figure out how to pack it up to ship it," explained Jefferies. "Dorothy Fontana was headed back to DC and agreed to take it for me; we put it in a plastic garbage bag, which was not deep enough to take the whole thing, so the head of it stuck out. Somebody on the airliner recognized it, so they unwrapped it and it toured the airliner!"[41] Frederick C. Durant, the assistant director of astronautics for the National Air and Space Museum, was extremely pleased when the model arrived, and he indicated as much in a November 2, 1973, thank-you letter that he sent to Roddenberry: "We were excited when she [D. C. Fontana] marched in with the studio model of the Klingon Battle Cruiser."[42]

Few photographs are known to exist of the actual filming model. In a 1976 article that appeared in the fan-produced publication *Trek: The Magazine for Star Trek Fans*, there is a black and white photo of the model taken shortly after it arrived at the Smithsonian. Even though the photo was not of the best quality, a close study of the image reveals subtle shading differences in the two-tone paint scheme, along with bead-like objects scattered along various points on the perimeter surface. During filming, the model was hung by piano wire, and these beads served as attachment points. Oddly enough, the photo also reveals a small shadow cast by a piece of piano wire that remained attached to one of the beads atop the cobra head.[43] Several other photos unfortunately, all in black and white, showed the filming model mounted on a stand, and missing the beaded wire-hanging attachment points. This version of the model may have been used in those filming sequences that didn't require direct underside views.

In an effort to chronicle the scant known details of the original Klingon filming model, Doug Drexler, an Emmy and Academy Award–winning visual-effects artist, was allowed to take a closer look at the model. A mega *Star Trek* fan who, during the 1970s, helped found a popular New York retail store in Long Island that catered to *Star Trek* fans called "The Federation Trading Post," Drexler also was editor along with Ron Barlow of the *Star Trek Giant Poster Books*. These poster books proved popular with fans, since each eight-page issue consisted of a 34-by-22.5-inch poster, with additional photos and articles about *Star Trek* printed on the back. Seventeen regular issues plus one special issue were published from 1976 to 1979.

In 1977, Drexler visited the National Air and Space Museum to photograph the filming model of the Klingon ship for a feature article that appeared in one of the poster books. During this time, he confirmed that the model was painted in a two-tone greenish gray. "For years I had been led to believe that the battlecruiser was a fragile, almost make-shift sort of conglomeration. Nothing could be farther from the truth," wrote Drexler while examining the model. "Solidly handcrafted, it is indicative of Mr. Jefferies' great talent, Dimensions: 31"L × 20"W × 7½"H. Painted grey and pale green."[44] Drexler described other details on the model, including the fact that its chrome-like sections were adhesive Mylar. He also spotted the bead-like hanging points as seen in the 1976 *Trek* magazine article photo.[45] Drexler's article proved to be among the most-popular in that series.[46]

Effects shots aren't always done in the same order as live action. The first-season episode "The Corbomite Maneuver" was the first episode shot in the series, but it did not air until months later because the effects shots took so long to finish. It is possible that during post production, painters added the two-tone color for additional filming sequences of the Klingon model. Perhaps Jefferies stepped in to do the two-tone painting himself after seeing how good it looked with the green lighting projected onto it.

Influence of the Early *Star Trek* Model Kits

Today, the original *Enterprise* AMT kit still remains challenging to assemble, just like it was when it first came out. When Roddenberry received some of the first kits back in April 1967, he was not very pleased. In an April 24, 1967, letter to AMT, Roddenberry wrote, "I must be quite frank in telling you that I am most disappointed in the *Star Trek* model kit. I have professional model builders working for me on the show, experts with a considerable knowledge of the model kit market, and all of them tell me the *Star Trek* kit is below par for your company and for the quality kit market in general."[47] Even Bob Justman wrote of the model in an October 19, 1967, memo in which he told Roddenberry, "You and I know that our Model Kit is not that great a kit."[48]

Over the years, changes in tooling to the original kit have been made (most for the better), but the model pretty much remains the same as was when it first appeared in 1967.

Unlike today, there simply was not that much available in the way of high-quality products to support one's fandom back when the original *Star Trek* television series first aired. That changed in 1967, when AMT began producing a plastic model kit of the starship *Enterprise*. That 18-inch hunk of plastic was not perfect, but many fell in love with it, sagging warp engines and all.

Thomas Valmassel, national sales manager from the AMT Corporation, said in a 1978 interview that "the typical model builder is a 10-to-16-year-old boy apparently trying to kill time through the fall and winter months, the biggest sales season."[49] During the 1970s, the show gained greater popularity and won over legions of new fans through syndication. More people were exposed to *Star Trek*, and, as a result, more model kits from the series were produced. But the original *Enterprise* remained the biggest seller. In the ensuing years, millions of copies of that kit were produced and sold. Since that first model, AMT and its successors went on to release at least twenty-three unique *Star Trek* kits, making it one of the longest-running and most-successful licensees in the history of the franchise.

Remarkably, the *Enterprise* model kit is still being manufactured and sold today through Round 2 LLC. When asked about the enduring popularity of the original kit, Round 2 senior designer James Hood said, "The original AMT *Enterprise* has been touted as the best-selling model kit of all time because it has essentially been in perpetual release since it was originally brought out in 1967."[50] Indeed, the *Star Trek* models were, in fact, "credited with boosting the [AMT] company to its best year ever in sales and profits," during the recession of the mid-1970s. "Nostalgia is in," Valmassel explained in commenting about his company's association with *Star Trek* and the enduring success of AMT's models. "Our all-time best-selling models were not cars or trucks, but models associated with the popular television series *Star Trek*."[51]

Today you can see an original AMT model of the *Enterprise* proudly displayed in the newly refurbished National Air and Space Museum. As part of these renovations, a built-up model kit has been added to a brand-new gallery called *Nation of Speed: America's Pursuit of Going*. Sponsored by Rolex, this new exhibit is a collaboration of the Smithsonian's National Air and Space Museum and the National Museum of American History. According to Smithsonian media representative Amy Stamm, "This new exhibit explores the intersections among technology, business, culture, and America's popular quest for speed by showcasing various vehicles known for not just going fast but for going the fastest. Whether it be Mario Andretti's Indy 500–winning race car, a motorcycle such as the one used by 'Evel' Knievel or a spaceship like the starship *Enterprise* from *Star Trek*, all of these have served to inform us how speed has impacted popular culture and the public imagination."[52] The *Enterprise* model is one that was originally manufactured in 1967 by AMT and is part of the Smithsonian's permanent collections. The model was built and modified by Richard G. Van Treuren while serving aboard the naval aircraft carrier USS *Kitty Hawk* (CV-63). He donated the finished 18-inch model to the Smithsonian in 1973, but until now it has not been displayed in public.

Whitfield's contributions while associated with AMT are important in the making of *Star Trek*. Whitfield helped with the creation of several of the first commercial products connected with the new series. Both the *Enterprise* and

Klingon model kits built by AMT went on to be highly successful, and their sales helped bring *Star Trek* products into the hands and homes of television viewers. Whitfield knew that making these two model kits was important. In the January 2, 1968, letter to the vice president of Paramount Television, Whitfield wrote of the forthcoming Klingon kit that "Paramount therefore has a potentially large sum of money at stake."[53] The stakes were high, but the phenomenal sales of *Enterprise* kits in 1967–68 sent a message to network executives that *Star Trek* was worth supporting. Though the Klingon model kits came out after *Star Trek* was ensured a third season, continual record sales of both model kits long after the show ended its original network run helped prove that *Star Trek* remained popular even if NBC did not think so at the time.

Roger Mitchell, a fan of the original television series who lived in Ireland when the first AMT *Star Trek* model kits began to appear on store shelves overseas, remembers the early kits. The UK did not show *Star Trek* until the BBC began airing it in the summer of 1969. By then, the show had already finished its network run in the United States. During an interview that I conducted in May 2022, Mitchell shared his thoughts about the AMT *Enterprise* model kit that he built while growing up in Ireland:

> That model was built about 1972 when I was 12, and it was the second one I built. This time I'd learned to stand the nacelles up against a pile of books to stop them tipping backwards and sideways while the glue was drying! I bought my first-ever can of car spray paint: "Brilliant White" and sprayed it in the garage. The decals fell to pieces when I put them in the water, but my father sold Letraset rub-down lettering in his shop, so I got my letters and numbers from him—and put them on very badly . . . but you know, in those days it kind of didn't matter—there were NO other *Star Trek* models about, or not in my part of Ireland, anyway (apart from the Dinky Toys version, which was so dreadful I couldn't bring myself to buy it). I hung the completed *Enterprise* over the stairs and every time I went up to my bedroom, I would look at the ship and watch the angles change on it. I was so impressed that AMT had made such an accurate and beautiful-looking kit. *Star Trek* was very important to me—it was the only view of an exciting future that I had, and my model of the *Enterprise* epitomized many of my hopes for life ahead of me. By the early 1980s, the saucer section was covered with dust and the red electrical-tape lines I'd stuck on the nacelles had fallen off. Soon after that the thread snapped, the *Enterprise* went into free fall and hit the floor, reducing itself to kit form again. So, I bought another one! Long live the AMT *Enterprise*!

CHAPTER THREE

As of this writing, Mitchell's father still lives in the house where he grew up in Ireland. "I was there a few months ago," said Mitchell in December 2022, "and noticed that the knotted loops of the 1970s thread that held my original *Enterprise* model in endless flight in midair are still there, tied to a post at the top of the staircase. I do have an *Enterprise* here (of course), though it's not an AMT. . . . My home could never be complete without that image of an adventurous, optimistic future."[54]

The Making of Star Trek

The first extensive history of *Star Trek* came from the same individual that helped produce the first two AMT model kits. In May 1967, Stephen E. Whitfield obtained a copy of the *Star Trek Guide* (also known as the *Star Trek Writers Guide*) from Matt Jefferies.[55] This document was produced by Roddenberry to help writers become more familiar with the series and to serve as a guide for creating episodes. It contained a wealth of information about the show, and after Whitfield read it, he was inspired to write a book about *Star Trek*. Whitfield admitted "I was not the first person to approach Roddenberry with the idea [of writing a book]."[56] For instance, Roddenberry mentioned in a June 19, 1967, letter that other publishers were interested in doing a making of book about *Star Trek*, but he wanted Whitfield to write it because of "his enormous interest and enthusiasm in *Star Trek*, backed by his considerable knowledge of the show which the ordinary writer selected by a publisher would not have."[57] The fact that Roddenberry already knew about Whitfield, having worked with him as a result of the AMT model deal, may have also influenced his decision. What's more, Roddenberry seemed to like Whitfield, stating, "I suspect he is a young man who is going to go places in this or some related field."[58]

According to Herb Solow and Robert Justman in their book *Inside Star Trek: The Real Story*, "Roddenberry procrastinated and finally read the book after it was typeset, and in galleys, and spent 'one long night' with Whitfield 'making changes.' Owing to the book's printing deadline, very few changes were incorporated, and the book was published pretty much as Whitfield had written it." Roddenberry was given cowriter credit on the book's cover, but when you look at the inside title page, it clearly states "Written By Stephen E. Whitfield." By including Roddenberry's name on the cover, Solow and Justman claimed that Ballantine Books considered this to be a "marketing ploy."[59]

The Making of Star Trek uses material that was previously written by Roddenberry himself just for the show. This includes sections from the *Star Trek Guide* (i.e., *Star Trek Writers Guide*), along with script excerpts, story treatments, internal memos and letters. Even though this material was not written specifically for the book, what Whitfield chose to include offers insight into the author's ability to know what would be of interest to readers while, at the same time, assuring that the voice of the show's creator remained heard. Indeed, readers can easily distinguish when Roddenberry was talking as Whitfield chose to print his words in CAPS. While the percentage of text that Roddenberry wrote for the book may be relatively small, Whitfield knew that Gene's words would greatly enhanced the book's credibility. Besides quoting Roddenberry, Whitfield also conducted many new interviews himself noting, "I interviewed everyone from the Vice President of Television Production all the way down to the man who sweeps the floor on the set,"[60] said Whitfield in his book. Even though Whitfield had unique access to the show's creators, he still sought "to tell a story, from an 'outsider's' point of view, of the making of an hour-long weekly television series, the history of how it began and what makes it go."[61]

Whitfield truly was an outsider. He had never written for Hollywood or television, and it was through pure luck that he managed to find himself working for a company that held an advertising account with *Star Trek*. Whitfield offered a fresh perspective. He wrote in his book, "In many cases I have used quotes from individuals and excerpts from printed material in order to underscore a particular point or activity. In such cases, I have acknowledged the source. In all other respects, however, the observations and the words expressing them are mine."[62]

Since the two master production models of the Klingon ship were made at AMT's Speed and Custom Equipment Division in Phoenix, and the advertising agency that Whitfield worked for that held the AMT account was also located there, Whitfield spent much of his time in Arizona. However, working on his book necessitated that he be closer to Hollywood.

While working on *The Making of Star Trek* in California, Whitfield met his soon-to-be second wife, Judith Creifelds, who worked as a special assistant to a Hollywood television producer at the time. Whitfield's sister, Bobbi, said that he and Judith eventually got married at a ceremony at Lake Arrowhead in San Bernardino County on November 22, 1969. Roddenberry served as best man.

"The Hollywood crowd was there, and then there was me and my husband," recalled Bobbi about the wedding. "My husband had a coat and tie on. I had my gloves and my little hat, and then the justice of the peace came in to do the ceremony. He had a turtleneck on. We were the most formally dressed people there." The wedding was held at a large house that Roddenberry rented for the weekend, and everyone who worked on *Star Trek* was there. "It was at this great big house that had a balcony that was attached to a large bedroom that Gene Roddenberry and his wife, Majel Barrett, occupied. She was trying to get pregnant at the time, so every half an hour or so, Majel would rush out onto the balcony and announce to the crowd below that she hadn't had her period yet so she still could be pregnant. And they all cheered. We were then all informed that if we wanted our own rooms, we had better grab one now. We had to run through

CHAPTER THREE

the house and find a room to lay claim to . . . so we'd have a place to sleep, because they were all over the place. It was a real Hollywood scene."[63]

Because of their close personal relationship, Roddenberry gave Whitfield 50 percent of the book's royalties. "He never regretted the fifty-fifty deal he made with Roddenberry," wrote Solow of the deal. "It was a substantial portion to give away, but otherwise he never would have had the opportunity to become the first chronicler of television's most successful unsuccessful series."[64]

It is clear that Whitfield knew not only what to write about that would appeal to *Star Trek* fans but also what to show them. First issued at a price of ninety-five cents, the 415-page book included generous black-and-white photo inserts that showed the cast, crew, and hardware from the series. In addition, there were also detailed line drawings that depicted sets from the show, as well orthographic views with dimensions of the *Enterprise* and Klingon ships.[65]

There was once a time when, if you were a fan of a particular television series, it was difficult, if not impossible, to learn much about that show beyond what you saw each week on television. If you missed an episode, it was challenging to try to watch it again, especially before the series was syndicated. As one fan wrote, "Being a *Star Trek* nut takes a lot of work. It means hours of article cutting, perusing bookstores, stalking stores and building models."[66]

Whitfield's book was a treat not only for fans of *Star Trek*, but also for those interested in the making of television. As the cover declared, it told "What it is—how it happened—how it works!" Marketed by the publishers as "the only book of its kind!," it offered readers "the complete history of a top TV series—how a television show is conceived, written, sold and produced."[67] Whitfield's work was pioneering, since it not only presented the first detailed book about *Star Trek* while the series was still being made, but also presented the first book-length review of television in general during the 1960s. "This is the first such attempt to relate the history of a television series, which also happens to be one of the most trying and ambitious shows ever produced in any medium," wrote one reviewer of the book in 1968.[68] *Publisher's Weekly* wrote at the time, "For would[-]be TV writers, directors and producers, this will be an education in itself, a polished but non-varnished look at how TV really works."[69]

Prior to the first publication in September 1968 of *The Making of Star Trek*, there appeared several article-length stories that offered a behind-the-scenes look at *Star Trek* beyond the material contained in the studio-issued press releases. The December 1967 issue of *Popular Science* for example, contained an article that not only acknowledged NASA's contributions to the series but also that of "a research firm [i.e., Kellam de Forest's Research Service] to go through each script" and "contacts with scientists across the US." [i.e. Harvey Lynn at the RAND Corporation]."[70] The February 1968 issue of *Analog* published an article by G. Harry Stine titled "To Make a 'Star Trek,'" which was basically a reprint of material from the *Star Trek Writers Guide*. The article included drawings of the *Enterprise* done by Matt Jefferies.[71] In the October 1967 issue of *American Cinematographer* magazine, there was also an article that gave details about the optical effects used in the series.[72]

The Making of Star Trek became the "bible" for fans obsessed about every detail of the show. It also served as a template for other "making-of" reference works. Bettelou Peterson, a television columnist based with the *Detroit Free Press*, wrote at the time, "'Star Trek' fans are 'nuts' in the best and worst sense of the word. They can tell if a rivet is out-of-place from one episode to the next. Producer Roddenberry loved the loyalty but was driven right out of his tree by the letter writers who wanted minute details about the show—That's why the book [*The Making of Star Trek*] was written."[73]

Whitfield's book was used in public school classrooms as well as college curricula. According to a 1971 newspaper article, the book was used as a science textbook in California high schools.[74] The book also was the subject for advice columnists. "I borrowed a paperback book entitled 'The Making of Star Trek' from a friend—but alas, I left it in a phone booth in the Nashville airport. Where can I get another copy?" wrote Peggy L. Johnson of West Palm Beach, Florida, to a newspaper advice columnist. The columnist replied to the reader with the publisher's mailing address, along with directing them to their local bookseller.[75] In another column, a request offered a glimpse into the life of an adolescent *Star Trek* fan: "I'm only 16, but I'm rapidly becoming disillusioned with the system. Over the summer, I ordered and paid for three paperbacks from a New York book company—*The Making of Star Trek*, *The Teachings of Don Juan*, and *The Sensuous Couple*. I was happy when the books arrived so quickly—less than one month after my order—but annoyed when I ripped open the package to find *Star Trek* and something called *The Lady of the House*. I sent back *Lady* and asked for *Sensuous Couple* and *Don Juan*. When I didn't hear from the company, I wrote again and said forget the books, I'll take a refund. Still no answer. Now I've decided I'd really rather have the books. Can you help?"[76]

Whitfield's Work Inspired Others

After the initial release of the AMT *Enterprise* and Klingon model kits, Whitfield came up with the idea for a new television show that was not related to *Star Trek*, to be called *Space Cadet*. Matt Jefferies designed the ship for it, and AMT then made a model kit. The idea would be that viewers would follow cadets on a mission aboard a different type of spacecraft each season. The model would have served as part of an overall proposal for this new television series, and AMT would design and sell a new model kit each season to support the show. While Matt Jefferies's ship never managed to make an appearance on TV, it did make it as far as a few storyboards for the animated *Star Trek* series that Filmation produced in the early 1970s. Sadly, it did not appear in any of the actual episodes.[77]

AMT first produced a model kit of this new ship in 1969, the year after the original Klingon ship was released. Called the "Leif Ericson," the model consisted of a two-piece molded hull with chrome-plated parts, clear engines, and a small scout ship that could be stored inside a rectangular hangar with hinged doors built into the center of the ship's hull. Like the first *Enterprise* and Klingon kits, the model also included lights. It was packaged with a paper record that played "Sounds of Outer Space" and came with a two-page short story, complete with typos, that explained the history and story behind the ship, including a statement that the model was "the first in a new series of AMT-designed spaceship model kits."[78]

Beginning in 1974, the kit was reissued and renamed "U.F.O. Mystery Ship." AMT molded the ship in light-green glow-in-the-dark plastic and left out the lights along with all the chrome and clear parts. The paper record also was not included. Still later, in AMT's move to repackage all their *Star Trek* kits into smaller-sized boxes, the company split the ship's neck in two so that parts would more easily fit into the smaller-sized packaging. As a result, the kit was shorter.

Round 2 reissued the kit in 2009 and unfortunately retained the split neck. The newer kit does include all the parts, even the ones that were originally chromed. An extensive decal sheet is included, as well as a new short story. Like the original U.F.O. Mystery Ship, it's molded in glow-in-the-dark plastic.

AMT was eventually acquired by the Lesney Corporation, a company known more for its Matchbox brand of small metal toy cars than plastic model kits. Under Lesney's ownership, AMT came out with model kits supporting the various big-screen film versions of the franchise, which began with the December 1979 release of *Star Trek: The Motion Picture*. Since that time, other companies have been licensed to make *Star Trek* model kits.

As mentioned previously, Round 2 LLC is the current licensee of the original AMT *Star Trek* models. The company also has produced new kits from the franchise, including a larger-version model kit of the original television series *Enterprise*.

During the summer of 2020, Round 2 issued a completely new model kit of the Galileo shuttlecraft that is larger and more detailed than the original 1974 AMT kit. In 2022, they issued a version with a fully detailed interior.

Upon leaving AMT, Whitfield continued to write. In the 1970s, he started a videocassette magazine called *VideoPlayer*. He was also tapped by Roddenberry to write *The Making of Star Trek: The Motion Picture* but was unavailable at the time.

Whitfield eventually returned to the *Star Trek* franchise to write another book called *A Vision of the Future—Star Trek: Voyager*, a behind-the-scenes look at the making of the *Star Trek: Voyager* television series. The book, published in 1998 under his birth name, Stephen Edward Poe, would be his last. While preparing for a minor surgical procedure, doctors discovered he had leukemia. Stephen died on January 6, 2000.

Because of their close relationship, Roddenberry gave the second Klingon production model to Whitfield. "Stephen had a model of the Klingon ship in his living room. I remember seeing it when I was about 18 and thought it was very cool,"[79a-b] said his stepsister Susan Whitfield during a 2020 interview. Whitfield sold the model to a private collector in 1998 to help pay for medical expenses as he fought leukemia. It subsequently went up for auction multiple times after that before the model was purchased in 2006 by the late Paul Allen, one of the founders of Microsoft.

In 1998, only two years before his death, Whitfield said that *The Making of Star Trek* was in its twenty-ninth printing and had sold more than four million copies.[80a-c] Such a number is not unrealistic, since newspaper accounts going back to the early 1970s reveal that each printing sold, on average, around half a million copies.[81]

Whitfield's book and the *Star Trek* models that he helped create made an indelible impression on fans of the television series. As one young fan wrote, "The most valuable things in our house are the TV which I can watch *Star Trek* on and my four *Star Trek* books, along with some models. About a month or two ago I had a real bad accident. I accidentally stepped on my *Enterprise*, and there went the best model in the world."[82]

Even astronauts have gone on record to describe how AMT's first *Star Trek* model kit made an impression on their lives and influenced their careers. Former NASA astronaut José Moreno Hernández, who flew aboard the space shuttle during the STS-128 mission, explained in his 2012 autobiography *Reaching for the Stars* how he applied eleven times before he was finally selected as an astronaut in 2004. In his book, Hernández wrote about how he had an AMT model of the *Enterprise* that he played with as a kid in the 1960s. "*Star Trek* was my favorite show growing up," explained Hernández. "My brother Chava had a toy model of the USS *Enterprise* spaceship from the show, which was my favorite toy to borrow and play with for hours and hours." He also described how later, when he applied to become an astronaut with NASA, astronaut Franklin Chang Diaz interviewed him as part of the selection process and asked, "Why do you want to be an astronaut?" Hernández replied, "I simply told him that my desire to become an astronaut originated in 1969 as a seven-year-old, when I watched the very first Apollo mission on our old black-and-white TV. This event, coupled with watching the *Star Trek* original TV series, fueled my desire to become an astronaut."[83]

CHAPTER THREE

While interviewing Hernández in 2023, he confirmed that the series played a highly influential role in his life. "Both that show and the model really made a difference in my life," Hernández told me while posing with a finished AMT *Enterprise* model. "When I became frustrated or depressed, that model gave me hope and reminded me that everything was going to work itself out and be just fine."[84]

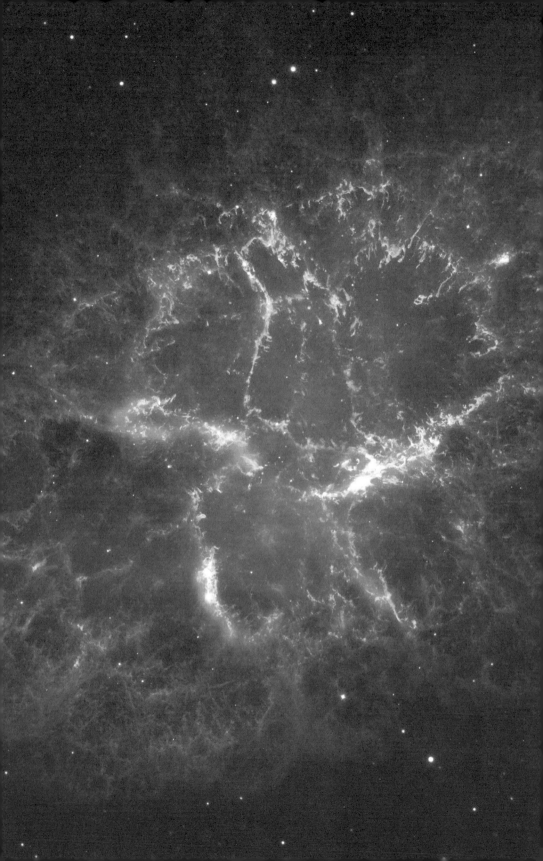

CHAPTER 4

The Military and Nationalism in *Star Trek*

During the show's production, many of the details that went into the making of *Star Trek* were still being ironed out even while the cameras were rolling. This included what exactly to call the entity that commissioned the *Enterprise* and governed its crew. During the introduction of every episode, television viewers were told that the mission of the starship *Enterprise* was "to explore strange news worlds" and "to seek out new life and new civilizations." Parallels between the military and *Star Trek* have been evident throughout the series since the show's inception. But under whose authority was all this exploring and seeking being done?

The Military and *Star Trek*

Starting with the first season, terms such as "Star Service," "Spacefleet Command," "United Earth Space Probe Agency," and "Space Central" were used to refer to the operating authority that owned the *Enterprise* and governed her crew. In the series' second pilot, "Where No Man Has Gone Before," an "academy" is mentioned (though not "Starfleet") along with reference to "The Service." In the episode "Mudd's Women," the show established that captains, like their modern-day counterparts, have authority to convene a board of inquiry. It wasn't until the fourteenth episode, "Court Martial," that audiences first heard the term "Starfleet." The third edition of the *Star Trek Writers Guide*, which came out near the end of the first season, stated that "the USS *Enterprise* is a spaceship, official designation 'starship class'; somewhat larger than a present-day naval cruiser, it is the largest and most modern type vessel in the Starfleet Service."[1] "Starship Class" is the designation given on the dedication plaque on the bridge. The term "Service" is still used throughout the second and third seasons, long after Starfleet was first heard.

If Starfleet is *Star Trek*'s made-up quasi-military organization responsible for the ship and its crew, then the histories and traditions of the Royal and

United States Navies are the closest real-world entities that served as its reference.[2] Naval influence is obvious in *Star Trek*, since Roddenberry himself admitted as much when he said, "We do keep a flavor of naval usage and terminology to help encourage believability and identification by the audience."[3] Roddenberry went on to say that this "helps link our own 'today' with *Star Trek*'s 'tomorrow,'"[4] adding that the show's character of James T. Kirk is seen as "a space-age Captain Horatio Hornblower . . . in which nothing or no one moves without his command."[5]

Roddenberry's fixation with the fictional Nelson-era British sea captain, brought to life through a series of novels written by C. S. Forrester, is apparent in his show. Like in the Forrester stories, Roddenberry sought to portray the same excitement and adventure in *Star Trek*'s future universe as that experienced by Hornblower's past enlightened era of exploration that characterized the age of discovery during the eighteenth and nineteenth centuries. Naval historian John E. Fahey noted that even *Star Trek*'s famous soundtrack begins with the same three notes as Robert Farnon's opening score from the 1951 Warner Brother's film *Captain Horatio Hornblower*.[6]

Rather than being powered by wind or steam, *Star Trek* used antimatter to propel its starships through the sea of space.[7] Even with the high tech of the twenty-third century, Kirk still longed for earlier days when navy captains were dependent upon the wind, the waves, and their wits to ply through the sea. In the second-season *Star Trek* episode "The Ultimate Computer," Kirk is faced with the possibility of being replaced by a computer. As he wistfully recalls those days of sail, he says, "'All I ask is a tall ship and a star to steer her by.' You could feel the wind at your back in those days. The sounds of the sea beneath you, and even if you take away the wind and the water it's still the same. The ship is yours. You can feel her. And the stars are still there."

Herb Solow, the executive in charge of production for *Star Trek*, liked the fact that Roddenberry patterned the show after the Navy. "The one thing I thought was very, very good that Gene had done was to make science fiction familiar to an audience by handling it like the United States Navy. . . . It invited the audience into something they knew about," said Solow. "They knew terms like 'Captain' and 'Admiral,' 'starboard,' 'port,' and the USS *Yorktown*, which later became the *Enterprise*."[8]

The Starfleet of the twenty-third century has many similarities to the naval organizations of today. The huge starships used historic military names such as *Constellation*, *Yorktown*, *Lexington*, *Defiant*, and, of course, *Enterprise*. They all were considered "starship class," which were "somewhat larger than a present-day naval cruiser" and built in the shipping yards of San Francisco.[9] Stephen Whitfield in *The Making of Star Trek* gave a naval analogy for the purpose of the starships: "The *Enterprise* is doing the same kind of job naval vessels used to do

CHAPTER FOUR

several hundred years ago. In those days ships of the major powers were assigned to patrol specific areas of the world's oceans. They represented their governments in those areas and protected the national interests of their respective countries. Out of contact with the admiralty office back home for long periods of time, the captains of these ships had very broad discretionary powers. These included regulating trade, fighting bush wars, putting down slave traders, lending aid to scientific expeditions, conducting exploration on a broad scale, engaging in diplomatic exchanges and affairs, and even becoming involved in such minor matters as searching for lost explorers or helping down-and-out travelers return to their homes."[10]

Writers in talking about space often refer to nautical terms such as a "sea of space" and "exploring the shores of distant new worlds." Many of the show's episodes begin with Captain Kirk saying, "Captain's log," and the *Enterprise* is piloted by a helmsman. Officers attend Starfleet Academy, and the chain of command, training, ranks, and military themes addressed in the show all point toward the influence that the military had on the series' development.

The first-season episode "Balance of Terror" opens with Kirk conducting a marriage ceremony before the episode launches into a story that was deeply influenced by postwar films like *The Enemy Below* (1957) and *Run Silent, Run Deep* (1958). The episode is filled with imagery and military terminology that comes from the Navy, as the *Enterprise*, cast as the defender, tries to protect the United Federation of Planets and its territories against the Romulans, who have a cloaked ship that, like a submarine, can't be seen. The submarine analogy is apparent in a scene that unfolds when, in desperation, the Romulan commander orders that the ship's "launch tubes" be filled with debris and the body of the ship commander's friend, along with a bomb set with a proximity fuse. The bomb manages to strike a severe blow to the *Enterprise*, but, in the end, after Kirk and his crew deliver a critical hit to the Romulan vessel, its commander decides to destroy his crippled ship rather than have it taken into the hands of the enemy. In a scene that was filmed but edited out of the final episode, Kirk is shown saluting the Romulan commander. The cut filmed scene as originally described in the script is as follows: "Kirk is looking at the commander, who stands among the rubble and ruin of this half-crushed bridge; he is burnt and bleeding. . . . Kirk realizes that the other can see him in his own view-screen—and his hand comes up in a salute. The commander responds with a small, stiffly precise bow. Human and alien, studying each other."[11]

If that scene survived to the final cut, it would have been the only time in the series when a military edge-of-hand-to-the-head salute was shown. The scene as described in the James Blish novelizations of the series had no final exchange of messages between the two, and the *Enterprise* destroyed the Romulan ship. But Blish's novelizations were often the results of early script drafts and not from as-aired episodes. For example, in Blish's telling of the episode "The Naked Time," he had a crewman salute Scotty after he ran up and interrupted the chief engineer while he was trying to cut into the engineering-room door with a phaser.

In another first-season episode, "Tomorrow Is Yesterday," viewers saw more indications of a military influence in the *Star Trek* universe when captured Air Force pilot John Christopher asks if the *Enterprise* was a ship of the Navy. Kirk responds that Starfleet is a "combined service; our authority is the United Earth Space Probe Agency."

"Rules of engagement," which define the circumstances, conditions, degree, and manner in which force or actions that might be construed as provocation may be applied is another example of a military idea that influenced *Star Trek*. In this case, Roddenberry created something called the "Prime Directive." Also known as Starfleet General Order #1, the Prime Directive dictates there can be no interference with the natural internal development of alien civilizations.[12a-b]

Gene Roddenberry's service in the Army Air Force and later as a Los Angeles police officer influenced the series.[13] Roddenberry said, "I'd been an army bomber pilot in World War II. I'd been fascinated by the navy and particularly fascinated by the story of the *Enterprise* in World War II, which at Midway really turned the tide in the whole war in our favor. I'd always been proud of that ship and wanted to use the name."[14] During *Star Trek*, Roddenberry also surrounded himself with key production staff who had military experience in branches of the service, which included the Navy. Associate producer Robert Justman was a Navy veteran who served on destroyers and escort ships.[15] Art director Matt Jefferies, who designed the *Enterprise*, was a bomber pilot. Like Roddenberry, both Justman and Jefferies were veterans of World War II .

With such a military presence evident in the series, it was not a surprise that the Pentagon itself would take interest in the show.

One of the more popular stories in *Star Trek* fandom involved the US Navy sending a group to study the layout of the *Enterprise* bridge. According to Roddenberry, the group was looking for ideas to model a new communications center.[16a-b] The Navy also invited Roddenberry as their guest aboard one of its aircraft carriers. "I was always rather envious of the Navy and rather wished I had joined that service instead," wrote Roddenberry just before embarking in 1968 on a weeklong tour of the nuclear aircraft carrier USS *Enterprise* while it cruised off the California coast doing air-training maneuvers.[17]

The Navy liked the new series enough that it took steps to get *Star Trek* into the hands of its servicemen. During the series' original network run, 16 mm prints of *Star Trek* were made available through Consolidated Film Industries (CFI). The Hollywood-based firm was contacted by the Bureau of Naval Personnel to manufacture prints of *Star Trek* to be used for "direct projection solely for Navy personnel aboard ship."[18] This was a service done by CFI for the military at minimum cost and as a cooperative effort for the Navy morale program. As described in a letter from CFI to Roddenberry, "Today we received a call from Washington, requesting us to contact you for permission to manufacture prints of your series *Star Trek*."[19] The letter went on to explain that these prints would not be used for "any land-based stations or through TV facilities which might interfere with your syndicated program."[20] The request for *Star Trek* was not unique, since the Navy sought similar permissions for shows from other networks. Even though the networks waived their standard licensing fees associated with these requests, the government still found it expensive to produce the films, so they remained selective in what shows they sought to use. To help prevent their use outside the military, after the films were circulated for a period of two years, the prints were gathered up and destroyed.

Star Trek fan Richard Van Treuren, who served aboard the carrier USS *Kitty Hawk* (CV-63) in the early 1970s, explained, "We never had *Star Trek* when I was on board, but I knew it existed because I saw it listed in the catalog to order films. We would get television shows and present them during different shifts and on Saturday nights. Any ship could have a 16 mm projector. They were showing *Star Trek* episodes in the Navy; I just didn't have them to show aboard the *Kitty Hawk*."[21]

The Navy even appreciated having a copy of the infamous *Star Trek* "Blooper Reel" being made available to them. This 16 mm film contained a selection of outtakes, gags, and bloopers from the series that later proved to be one of the most popular items requested on Roddenberry's lecture and convention circuit. After loaning the Department of the Navy a copy, the commander of the Navy's Public Affairs West Coast Office wrote Roddenberry, saying, "Thanks so much for loaning me the print of the out-takes. The film certainly broke up what could have been a terribly sticky situation. Please let me know when and how I can return the favor." The details as to what the "terribly sticky situation" was weren't included in the letter.[22]

For those who served in other branches of the military during the time of *Star Trek*, similar arrangements were made with the studios to show 16 mm films of television shows to servicemen. In a social media exchange with Terry Rhodes, who grew up with the original series, he explained that his brother "was in the Marines and was in country [i.e., in Vietnam] from '67 to '69." Rhodes went on to say that "they had *Star Trek* night in the rec tent on a regular basis. They also saw *Laugh-In* and *Get Smart*. There was a noncom who evidently had the sole job of escorting the film around the country. Wish I had that job."[23]

In addition to the Navy, the Air Force also took interest in *Star Trek*. In 1966, Dr. Jack L. Hartley, Aerospace Medical Division dental scientist, and a lieutenant colonel in the Air Force, wrote a book on dental standards for the selection and examination of space crewmen. He also was an avid *Star Trek* fan.[24] He saw the very first episode, which aired on September 8, and sent a telegram to Gene Roddenberry the very next day that said, "TREMENDOUS THOUGHT PROVOKING EXCELLENT CAST AND SCENERY SUSPENSEFUL OUTSTANDING PLEASE ACCEPT CONGRATULATION AND PLEA TO CONTINUE."[25]

Hartley served at Brooks Air Force Base in San Antonio, Texas, where he studied the problem of what to do if an astronaut got a toothache while in space. His solution caught the attention of Roddenberry and may have influenced some of the design decisions for medical instruments that we see used in *Star Trek*.

Faced with the time-consuming problem of halting a simulated spaceflight in a vacuum chamber because an astronaut had a bad tooth, Hartley began to think about what would happen if the victim had the same problem in outer space. An article on the subject that appeared in a 1966 issue of the *Dental Times* noted that "easing these pains would not be as simple as shutting down a chamber and trotting the victim off to the neighborhood dental emporium."[26]

As a result, Hartley developed the science of astrostomatology (space dentistry) and created a portable dental kit for the Aerospace Medical Division of the Air Force Systems Command. Called the "buddy-care" kit, it weighed just a few pounds and contained twenty-six instruments. It was specifically designed for use by NASA's astronauts and was the result of two years of work at the Air Force's School of Aerospace Medicine at Brooks Air Force Base. The kit contained everything needed by astronauts to perform dental work on each other while in space, including an electric toothbrush that would not interfere with spacecraft radio communication and a digestible toothpaste that does not need water. A hypodermic syringe for administering local anesthetic was included in the kit, along with a lighted mirror, equipment to replace lost fillings, and forceps to pull a tooth.[27]

When Hartley's oldest daughter, Patricia, was asked about her father's role in *Star Trek*, she replied, "He collaborated a fair amount with Roddenberry. He [Roddenberry] was really very good about trying to get things right, in spite of the fact that he had a very limited budget. And he didn't have a lot of tools to show. But he created a whole lot of things. And Dad was just one of the people that he was talking to that helped him design this stuff, such as Dr. McCoy's surgical instruments and medical bag."[28]

In addition to the congratulatory telegram that Hartley sent to Roddenberry, Patricia indicated that her father offered his new kit for use in *Star Trek*. After all, Dr. Leonard "Bones" McCoy, the ship's doctor, used a portable medical kit in the series. *Star Trek* fans sometimes saw the good doctor carrying it along

CHAPTER FOUR

with a small pouch that contained a hypospray, the twenty-third century's equivalent of today's syringe.[29] "Dad always wanted to watch *Star Trek*," said Jackie, Hartley's younger daughter. "He just loved the way the show taught us that good was good and how good overcomes evil. That basic concept is what he loved about it and what I still love about the series, because so many TV shows of that era were really dumb. *Star Trek* showed us a positive future and one that we are not all being killed by alien robots or that everyone is running around in some dystopian postapocalyptic world massacring everybody."[30]

As a result of numerous technical papers that he published, along with newspaper articles that were written about him, Hartley received considerable attention for his work. In addition, he promoted his portable buddy-care astronaut dental-hygiene kit through appearances on television shows such as *To Tell the Truth* and *The Tonight Show Starring Johnny Carson*.[31] According to Patricia, their father's original astronaut buddy-care dental kit was donated to the Smithsonian in the late 1960s.[32]

Roddenberry's efforts to seek advice from other professionals with military affiliations who could help infuse believability in the show was demonstrated by his association with Hartley. But Hartley was not the only person. Roddenberry also turned to his own production crew to find creative ways to help depict such things as a twenty-third-century operating room. In a memo addressed to Matt Jefferies, Roddenberry sought to instill realistic and believable medical instruments into the show. Roddenberry wrote, "I'm sure you've already thought of this, but I think we should be medically accurate on which instruments we decide to show on the bed, and then very carefully label each of them so the audience can easily read it and know exactly what these gizmos are doing."[33]

The cast members themselves were also encouraged to engage in events sponsored by the military to help further develop their character in the show. DeForest Kelley, who played Dr. McCoy, was encouraged by Roddenberry to tour the Naval Air Development Center, stating, "DeForest needs this not only for background information on his role but also has long deserved some travel and tours and publicity in this area."[34]

In 1968, the Air Force's prestigious Aerospace Pilot Research School, located at Edwards Air Force Base, California, where NASA obtained some of its first astronauts, invited Shatner and Nimoy to attend the graduation exercises of its class of 1967. The invitation read "The staff and students will be wearing their Air Force winter mess dress uniforms," as written by Colonel Eugene P. Deatrick, commandant of the school, "and we would be pleased if Messrs. Shatner and Nimoy would wear their Star Ship dress uniforms."[35] Shatner and Nimoy could not attend.

Both the military's influence and its admiration of *Star Trek* lasted well beyond the show's original network run. In 1975, members of the US Air Force

Academy Cadet Squadron 19 contacted Gene Roddenberry to obtain permission to use the starship *Enterprise* and the phrase "Where No Man Has Gone Before." Roddenberry was impressed by the request and wrote back: "You not only have my permission to use our starship on your squadron patch, but also my very best wishes to the entire group and its officers. May it convey good luck in carrying you all to places in both inner and outer space 'where no man has gone before.'" Roddenberry added that "it is a particular pleasure for me to grant this permission, since I once flew for our country in what was then known as the Army Air Corps, graduating in Class 42G from Kelly Field, and serving through the war in both combat and Stateside assignment. I still get a sentimental and warm feeling when I hear the music and words 'Off We Go . . . ,' and I'm sure that background had much to do with the creation of *Star Trek*."[36]

Seventeen years later, in 1992, Squadron 19 changed its name from Starship 19 to Wolverines 19, a change that remained in place for thirty-eight years. In 2023, however, members of the squadron decided to change its name back to Starship 19. The design of the squadron patch returned to its *Star Trek* origins, complete with an updated rendering of the original *Enterprise*. The patch, designed by Cadet 2nd Class Cassidy Rosa, now reads "STARSHIP 19." In place of the original patch phrase, "Where No Man Has Gone Before," the squadron inserted its new motto, "Vive Diu Et Prospera," which is Latin and means "Live Long and Prosper." The new Starship 19 emblem also features four stars on the top left of the patch that represent the different class colors. The arrangement of the stars with equal spacing signifies the squadron's unity among the classes. Above the stars is a large Polaris with the earth behind it and the *Enterprise* flying in the direction opposite that of the original Starship 19 patch.

Cadet 1st Class Mark Lema, who led the redesign campaign, said they made the changes to pay homage to their history and alumni. He also noted that the changes reflect the 2019 creation of the US Space Force and acknowledge the reinvigoration of civilian and military space activities.

Brian Laslie, the Air Force Academy's command historian, praised the cadets for considering the US Space Command in the squadron name change. "The cadets from Starship 19 have put a lot of time and effort into making this happen," said Laslie. "It's obvious that it was very important for them to make this change. As a cadet joins the US Space Force, it is important that representation from that service is found at an early point in the cadet's career. It's important for at least some of these squadrons to have some type of space flair to them."[37] Although the Space Force didn't yet exist in 1975, "space flair" seems to have already been important.

CHAPTER FOUR

Nationalism and *Star Trek*

"Space . . . the final frontier. These are the voyages of the Starship *Enterprise*, its five-year mission . . . to explore strange new words . . . to seek out new life and new civilizations . . . to boldly go where no man has gone before."[38]

With this introductory monologue, Captain James T. Kirk (William Shatner) invited television viewers to keep their dial tuned to NBC, where they witnessed how Roddenberry's vision of life in the twenty-third century would unfold each week.

The words that launched each opening episode were the result of nationalism that emerged from a postwar popularization of space that was very much present during the time of *Star Trek*'s creation. The viability of human exploration of space during the middle of the last century helped revitalize a belief in the American frontier. This was especially appealing to those who sought to use the growing popularity of space rhetoric that emerged at the time to spread their message of exploration and conquest. This cultural effort in space exploration, fueled by the Cold War, reawakened a sense of manifest destiny in postwar America by reviving notions of freedom, courage, and Western exceptionalism—the same ideals that originally drove expansionist boosters to America's West. After all, how could a country that had tamed the great North American frontier by the end of Davy Crockett's nineteenth century not succeed in conquering the frontier of space by the time of Roddenberry's twenty-third century?

With the resulting technological, industrial, and economic expansion that occurred in America after World War II, especially in the burgeoning field of space, writers sought to understand and portray these changes, which were rapidly occurring. To help address the increasing popular interest in space exploration, new pulp magazines such as *Space World* emerged to fill a need for layman interest in the subject. In addition, within the nascent aerospace industry, there emerged trade publications such as *Missiles and Rockets* and *Aviation Week* (later changed to *Aviation Week and Space Technology*) that either adapted their original editorial content or were founded anew to cover the broad spectrum of space-related, scientific, and engineering cultures that began to emerge in this new industrial field.[39]

In the October 1956 premiere issue of *Missile and Rockets*, the publisher wrote, "This is the age of astronautics. This is the beginning of the unfolding of the era of space flight. This is to be the most revealing and the most fascinating age since man first inhabited the earth."[40]

The lines between fact and fiction became more blurred as writers wrote about the very real possibility of human spaceflight. The idea that outer space and even the moon as being a real place that humans could travel to became a growing topic of interest in popular culture. Nonfiction books that once romanticized humanity's future in the new frontier of space started to borrow the look and feel of the popular pulps to convince their readers that the idea of traveling through space could be real.

Another medium that helped bring to life these visions from the printed page was television. *Star Trek* stood out among a group of network shows that emerged in the mid-1960s to address the topic of human spaceflight because it was the first to treat the subject seriously. By the time *Star Trek* premiered, associating space with frontier had firmly been planted into Roddenberry's mind. The fact that every episode began by using the words "space . . . the final frontier" tells us something about how he was influenced by the rhetoric of the time.

Frederick Jackson Turner claimed that the American frontier had shaped America and defined the characteristics of being American. Turner was a historian who espoused a view on how the idea of the frontier shaped American identity and culture. This was known as the "frontier thesis," and Turner posited that it was the idea of the western frontier that drove American history and, as a result, explained why America is what it is. The frontier facilitated a certain rugged individualism in those who explored it. In this manner, Turner argued, the story of this continual westward push, "with its new opportunities, its continuous touch with the simplicity of primitive society, furnished the forces dominating American character."[41]

Turner's seminal essay outlining his thesis was first presented at a special meeting of the American Historical Association at the World's Columbian Exposition in Chicago in 1893 and was published later that same year. Though it was well regarded for decades after, Turner's framing of American history began to be challenged in the early 1940s and has been largely rejected by historians of today.

But in the 1960s, the western frontier and the myths associated with it remained in the American consciousness. Books, movies, and television shows about the West were wildly popular in the mid-twentieth century, mainly because the settling of the American West was, according to Turner, the story of how Americans came into their own unique character. "The frontier," Turner explained, "is the line of most rapid and effective Americanization."[42]

While some people lamented what they considered the closing of the American western frontier, other Americans at the same time searched for a "new" or "final" frontier. *Star Trek* was heavily influenced by these discussions of a new frontier, particularly the language that politicians and government officials applied to spaceflight.

As television became more popular by the late 1950s, the main genre was westerns. By 1959, thirty westerns aired on prime-time television each week.

CHAPTER FOUR

One of the hundreds of writers for TV westerns openly wondered, "I don't get it. Why do people want to spend so much time staring at the wrong end of a horse?"[43]

The western genre held a prominent place in America's culture because it provided a reliable formula for filmmakers to explore American history and character. Westerns remained popular as a television mainstay well into the 1960s, but another frontier, a *new* frontier, was beginning to attract people's attention.

John F. Kennedy first used the term "new frontier" during the 1960 Democratic National Convention. In giving his acceptance speech as his party's nominee at the Los Angeles Memorial Coliseum on July 15, Kennedy said, "We stand today on the edge of a *New Frontier*—the frontier of the 1960s, the frontier of unknown opportunities and perils, the frontier of unfilled hopes and unfilled threats. . . . Beyond that frontier are uncharted areas of science and space."[44] During his presidency, the phrase became a label for his administration's domestic and foreign policies both on Earth and off it.

During the 1960 presidential campaign, Kennedy exploited a growing public concern about the space race that was fueled, in part, by an insistence that there was a "missile gap," an illusion that his party put forth at the expense of the Eisenhower administration. This feeling of technological inadequacy was further enhanced by a series of successive space spectaculars performed by the Soviet Union that, when compared to early American launch failures, created a public fear that the United States had indeed fallen behind their Russian counterparts in the production of ballistic missiles. The reasons behind the missile gap were more complex than this, but for the sake of this narrative, these were the main issues.

In October 1960, Kennedy issued a campaign statement on space: "We are in a strategic space race with the Russians, and we are losing. . . . Control of space will be decided in the next decade. If the Soviets control space they can control Earth, as in past centuries the nation that controlled the seas has dominated the continents. This does not mean that the United States desires more rights in space than any other nation. But we cannot run second in this vital race. To ensure peace and freedom, we must be first. . . . This is the new age of exploration; space is our great New Frontier."[45]

For those who lamented the end of the frontier, there now emerged the new frontier of space, which allowed fearlessness, rugged individualism, and other American qualities to reemerge in the face of imminent danger posed by the Soviet Union.

On May 25, 1961, less than three weeks after astronaut Alan Shepard safely splashed down in his Mercury spacecraft, marking the first time an American had flown into space, President Kennedy addressed a joint session of Congress where he outlined an audacious plan to raise America's eyes to the stars and land men on the moon. This was done, in part, as a response to the Soviet Union's ambitious space efforts, which not only launched the first artificial satellite, Sputnik 1, into space on October 4, 1957, but then, less than four years later on April 12, 1961, launched the first human to have flown in space, Yuri Gagarin.

Throughout Kennedy's twenty-two-page speech, he used the word "space" seventeen times, and all but three of those occurrences appeared on the last six pages, where he described space as the "new frontier" before launching into his manned lunar-landing goal.

"Finally, if we are to win the battle for men's minds," said Kennedy, "the dramatic achievements in space which occurred in recent weeks should have made clear to us all the impact of this new frontier of human adventure. . . . I believe that this nation should commit itself to achieving the goal, before this decade is out, of landing a man on the Moon and returning him safely to the Earth. No single space project in this period will be more exciting, or more impressive to mankind, or more important for the long-range exploration of space; and none will be so difficult or expensive to accomplish."[46]

Even though Kennedy first used the phrase "new frontier" in his 1960 presidential campaign speech and was quick to associate it with his presidency, NASA used it before him in one of the first publications issued for the general public by the new space agency. In 1959, the year after NASA was formed, NASA produced *SPACE: The New Frontier*, a heavily illustrated fifty-page publication that described what the new space agency would be doing. This proved to be so popular among the public that it was reprinted multiple times.

Gene Roddenberry recognized that *Star Trek* could be used to shed light on social problems that might be too sensitive when addressed in other types of television programs. He also sought to employ some of the public enthusiasm generated by Kennedy's support for human spaceflight to help launch *Star Trek*. Indeed, Roddenberry used "Wagon Train to the stars"—a reference to a popular episodic television western—to help sell his idea of *Star Trek* to network executives. He reasoned that in doing so, it offered a natural extension to President Kennedy's "New Frontier" rhetoric while also capitalizing on the way Americans were then talking about space.

Kirk's memorable opening monologue to every *Star Trek* episode may have been influenced by President Kennedy's New Frontier rhetoric. If true, this would not be the first time the origins of *Star Trek*'s famous opening narration may have been inspired by the US government.

In 1958, the Senate formed a Special Committee on Space Technology and issued a document titled *Recommendations to the NASA Regarding a National Civil Space Program*. In this report, the nation's civil space research program's mission is "to explore, study and conquer the newly accessible realm beyond the atmosphere," a somewhat similar goal to that of the starship *Enterprise*'s mission to "explore strange news worlds, to seek out new life and new civilizations."[47]

In the wake of the national concern brought on by the Soviet Union's launch of Sputnik 1 in late 1957, Dr. James R. Killian, chairman of the newly created Presidential Science Advisory Committee, produced *Introduction to Outer Space*. The document was created to help explain the new concept of spaceflight "for the nontechnical reader" and to help generate support for Eisenhower's new national space program.[48]

The document proved so well liked by Eisenhower that he ordered it made available to everybody. As a result, in 1958 hundreds of thousands of copies of *Introduction to Outer Space* were produced in pamphlet form and sold by the superintendent of documents of the US Government Printing Office. Gene Roddenberry was most likely among those that read a copy.

In the preface to the document, Eisenhower explained that the study of space "is not science fiction. This is a sober, realistic presentation prepared by leading scientists. I have found this statement so informative and interesting that I wish to share it with all the people of America and indeed with all the people of the Earth. I hope that it can be widely disseminated by all news media for it clarifies many aspects of space and space technology in a way which can be helpful to all people as the United States proceeds with its peaceful program in space science and exploration."[49]

Eisenhower closed by saying, "This statement of the Science Advisory Committee makes it clear the opportunities which a developing space technology can provide to extend man's knowledge of the Earth, the solar system, and the universe. These opportunities reinforce my conviction that we and other nations have a great responsibility to promote the peaceful use of space and to utilize the new knowledge obtainable from space science and technology for the benefit of all mankind."[50]

Noticeably absent in Killian's document is an emphasis on the role of humans in space. The pamphlet stresses the "remotely controlled scientific expedition to the moon and nearby planets," arguing that such efforts "could absorb the energies of scientists for many decades."[51] Eisenhower's vision for space relied heavily upon robotic satellite technology and downplayed the role of human spaceflight.

In a 2005 *Space Review* article, spaceflight historian Dwayne Day suggested that the final line of *Star Trek*'s beginning narration, "to boldly go where no man has gone before," may have been inspired by Killian's 1958 document. On the very first page we read:

> . . . the compelling urge of man to explore and to discover, the thrust of curiosity that leads men to try to go where no one has gone before.[52]

"It should be no surprise that Gene Roddenberry, *Star Trek*'s legendary creator, borrowed language from a White House publication," said Day. "Roddenberry did his research, and it is not a stretch to believe that he read 'Introduction to Outer Space' in his preparation for the show."[53]

Though not exactly the same as Kirk's "to boldly go where no man has gone before," President Eisenhower's original wording was coincidentally echoed years later as Roddenberry edited the line to be a more gender-neutral "where no one has gone before" for the opening narrative by Captain Jean-Luc Picard (played by Patrick Stewart) to *Star Trek: The Next Generation*.

Similar occurrences of lines from *Star Trek*'s opening narration appear in earlier literature. Famed eighteenth-century British explorer Captain James Cook, during an expedition aboard the *Resolution* in search of "the Great Southern Land" of Australia, declared on January 30, 1774, that "ambition leads me not only farther than any man has been before me, but as far as I think it is possible for a man to go." Whether Roddenberry directly borrowed language from any of these sources to form *Star Trek*'s famous opening narration is not known for certain.

Upon reading *Star Trek*'s opening lines of narration, some may interpret them as being less noble in their origins. Scholar Valerie Fulton suggests that Roddenberry's wording is nothing other than classic imperialism. In her paper "Another Frontier: Voyaging West with Mark Twain and *Star Trek*'s Imperialist Subject," she notes that the show's governing body, known as the "Federation," has as its goals both to "seek out new life and new civilizations" and "to boldly go where no man has gone before." Fulton argues that these two missions clearly contradict each other unless read through the lens of frontier ideology, which grants new civilizations existence only to the extent that the original culture has "found" them.[54]

Other scholars argue that *Star Trek* appeals to Western democracies by its ability to maintain a western theme through the show's reinvention of the frontier spirit as it celebrates the space cowboy. Prior to creating the series, Roddenberry had written a number of episodes for such television series as *Boots and Saddles*, *Whiplash*, and *Have Gun Will Travel* and even produced *Wrangler*, a short-lived western that aired on NBC. As a result, it is not at all surprising that such frontier themes would eventually find their way into the stars through *Star Trek*.

In a little over twenty years since the end of World War II, the dream of space exploration emerged from the pages of science fiction to become first a government policy by the Eisenhower administration, which formed NASA, which got America into space, then as political rhetoric used by John F. Kennedy to help propel him to the presidency and take us to the moon.

CHAPTER FOUR

When *Star Trek* first premiered in 1966, it was during a time when many people believed that America was falling apart. The war in Vietnam began to further escalate while civil unrest continued at home. But interest in space exploration was still popular in American culture as Roddenberry's new "Wagon Train to the stars" sought to invoke the same frontier nostalgia as seen in the many popular westerns that competed for television viewers every week. Like the American frontier that allowed a belief in rugged individualism during the nineteenth century, *Star Trek* allowed Americans to believe that the new frontier of space exploration represented a better tomorrow, an optimism that proved to be one of the key factors in the show's enduring popularity.

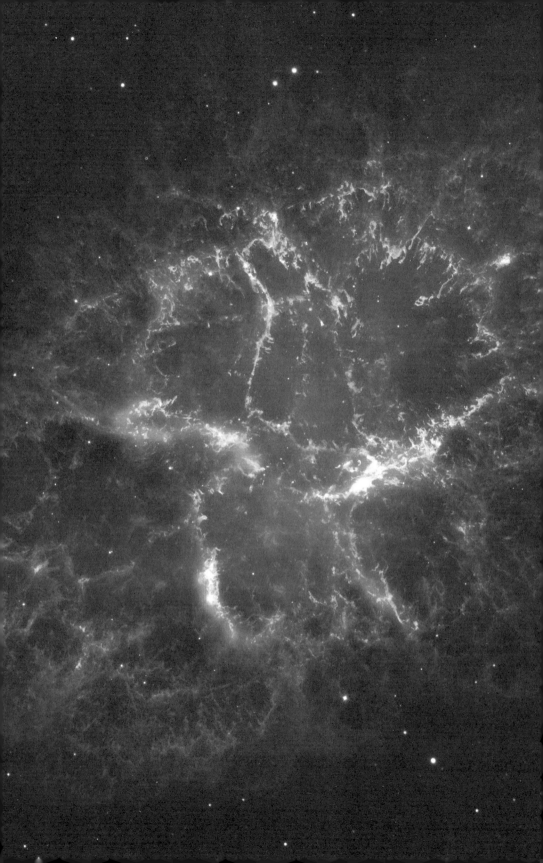

CHAPTER 5

The Aerospace Industry and *Star Trek*

By the end of World War II, 60–70 percent of America's aerospace industry was based in Southern California.[1] While helping win the war, the geography in and around Los Angeles was forever altered by communities that sprung up where giant new plants with names such as Northrop, Douglas, and Lockheed were located.

"Where Hollywood is so expansive and loves to talk about itself and loves publicity, aviation and aerospace is the opposite because you have a culture of secrecy and confidentiality," said Peter Jones, a filmmaker who produced the 2019 documentary *Blue Sky Metropolis*, which chronicles the rise of the Southern California aerospace industry. Jones's father was president of Northrup for thirty years. "People who were born and raised here are still sometimes shocked when they realize that aviation and aerospace was a bigger industry than entertainment for most of the twentieth century."[2]

The good climate and open land that helped draw aviation to the region also lured the motion picture industry. When Gene Roddenberry began developing *Star Trek*, fifteen of the twenty-five largest aerospace companies were in the greater Los Angeles area.[3]

Hollywood and the Aerospace Industry

If you went back in time to the mid-1960s and visited Paramount Pictures, the home of Desilu Studios, where *Star Trek* was made, you would not have had to travel very far from the lot before coming across evidence of a thriving aerospace industry. Along Interstate 405 you would find the Aerospace Corporation and the Los Angeles Air Force Base. Driving still farther, you would pass Long Beach, Downey, and Huntington Beach, where Hughes, Northrop, and North American built aircraft both for military and civilian use, as well as the command and service modules for the Apollo spacecraft. Heading north, you would encounter Caltech and the Jet Propulsion

Laboratory. In the nearby television city of Burbank, you would find the home of Lockheed. Toward the beach along the Pacific Ocean, you would find Douglas Aircraft, builders of the upper stage of the mighty Saturn V. Here, too, was the home of the RAND Corporation, which, as we have already learned, played a pivotal role in helping ensure the technical accuracy and scientific validity of *Star Trek*.

While military spending during World War II helped establish Southern California as a major manufacturing hub of aircraft, it was the space race that transformed Los Angeles into an aerospace giant. As NASA began gearing up with the Apollo program, it drew upon thousands of commercial and private companies that made up the contracts that built the hardware that eventually landed us on the moon. By the time *Star Trek* first appeared in 1966, nearly half a million highly skilled engineers, technicians, and workers were already employed across the country building aviation and space hardware. California was the state that had the greatest percentage of defense spending. It had 50 percent of NASA's business, with the Golden State playing a dominant role in securing nearly half of all of NASA's prime and subprime contracts.[4]

Roddenberry quickly began tapping into this local aerospace market to help with his show. In October 1964 he compiled a form letter that he sent out to area industries to introduce them to what he was doing and solicit their help:

> This is to inquire if your company has live and/or animated film footage about space travel, vehicles and systems which could be made available to us for use in a television show currently being developed for us by NBC.
>
> We are doing an hour television series titled STAR TREK, the time a century or so from now when interplanetary travel has become possible, featuring stories which revolve about a cast of continuing characters who are crewmen of an Earth vehicle patrolling a segment of our galaxy. Although science fiction, it is our intention to be as accurate as the format and dramatics permit.
>
> At this stage of development, it is possible for us to plan our series and the individual episodes to take advantage of film footage which is available. If it appears you have anything which might be useful and are willing to make available, some catalogue or list would be most helpful and appreciated.

The above letter was mailed by Roddenberry to area industries including aerospace companies in Los Angeles such as North American Aviation, Northrop, General Electric, TRW Space, and Westinghouse.[5]

CHAPTER FIVE

Even at this time, when *Star Trek* was not yet on the air and nobody really knew what the show was like, some still wrote back. Northrop replied, "All of our motion picture footage is of a highly technical nature in the form of engineering reports, which would be unsuitable for your requirements."[6] Others were more affirmative in their response. Westinghouse wrote, "Most of our film that deals with space travel, vehicles and systems is classified and could not be used without government approval. However, we do have a number of models of space vehicles available for use."[7]

Roddenberry sought out not only existing space-related motion picture footage to use for his new series, but still imagery as well. In a memo dated August 24, 1964, Roddenberry inquired about the availability of photos from the Griffith Park Observatory and Planetarium in Los Angeles. "There is some question whether the Griffith Park Planetarium authentic slides of interstellar space will be as effective as drawings," wrote Roddenberry. "The enormous telescopic magnification required in these slides may not give us the undistorted dimensionalism our ship would encounter in true outer space."[8]

In Roddenberry's quest to include as much real space as possible, he tapped into a variety of resources to help decorate his off-world starship of the twenty-third century with very real-world set dressings from the twentieth. This included using astronomical photos of the Andromeda galaxy on the bridge as seen in the show's first two pilots and other early episodes.[9]

In the starship *Enterprise*'s transporter room for example, where the crew did their beaming up and down, there was a very large photographic print hanging on the wall, showing the constellation Orion. Visible in the skies of both the Northern and Southern Hemispheres, the star formation as presented in the show looks a little less obvious since it is rotated 90 degrees, presumably to better hide its familiarity from television viewers. The same photo shows up again with black and white stripes added, perhaps to indicate spacecraft trajectories or in an attempt to somewhat disguise the constellation's appearance. During the filming of the first two pilots, a large photo depicting the pinwheel galaxy M-101 appeared on the wall near the transporter console. In addition, several advance publicity shots were taken showing Kirk and Spock standing in front of this photo.

A photo of the Earth's own moon also made an appearance in the show's first two pilots. A large print of the moon was shown mounted in the center console of the transporter controls, as well as on one of the bridge console viewing screens.

As a result of the letter that Roddenberry sent to RCA's Astro-Electronics Division, the aerospace company sent several pictures of the moon's Copernicus crater for use in the show.[10] These photos can be seen in several episodes on overhead screens above various workstations on the bridge. In addition, perceptive viewers will notice that images of Talos IV, the home of the Talosian Keepers depicted in "The Cage," are nothing more than flipped views of the Earth's moon that have been colorized and altered with clouds. Because of RCA's contributions to the show, their program manager was invited to tour the *Star Trek* sets. The visit impressed the representative so much that he wrote Roddenberry a letter thanking him "for a most exciting morning going through Star Trek," adding, "Keep up the good work of motivating people because engineers are hard to come by."[11]

Additional images of constellations, galaxies, nebulae, and other astronomical phenomena obtained from ground-based observatories, NASA, and other aerospace companies appear on *Star Trek*. A star map showing the night sky during late fall / early winter in the Northern Hemisphere or late spring / early summer in the Southern Hemisphere can be seen projecting from the center briefing-room view screen during the episode "The Corbomite Maneuver." Views of the Trifid Nebula can be seen superimposed over scenes of a battling Lazarus and Kirk from the episode "The Alternative Factor." Other astronomical photos can be seen on the bridge in "The Cage" that show the Pleiades, the Dumbell Nebula (M27), and several views of the North American Nebula in the constellation Cygnus.

Surrounded by aerospace companies, it seemed relatively easy for the production staff of *Star Trek* to acquire pieces of present-day hardware that could pass off to television viewers as gizmos from the future, especially if they were used only as set dressing that would be seen on screen for a few seconds. Examples of such off-the-shelf hardware included Geiger counters. These hand-held instruments used to detect radiation served a variety of purposes on the show. Several commercial models of these can be seen in *Star Trek*. Beginning with "The Cage," viewers will notice a Precision Radiation Model 111 Scintillator and its large metal battery case carried by one of the landing-party members when they beam down to the surface of Talos IV.[12] Built out of solid cast aluminum, it really looks like a futuristic piece of hardware.

A Nuclear-Chicago Model 2586 "Cutie Pie" Radiation Survey Instrument can be seen in another episode. This model was introduced in 1954 by Nuclear-Chicago, makers of Geiger counters and other radiation detection equipment.[13] When seen in several episodes of *Star Trek*, it is evident that Desilu's prop department simply scratched off the logo and painted the handle, the barrel, and one knob orange.[14]

Roddenberry tried to optimize the use of the unfamiliar when it came to choosing hardware, instruments, and other tools that would be seen in *Star Trek*. He knew that besides the ever-present Velcro that was used by NASA and employed in *Star Trek* to hold phasers and communicators on the belts of the show's cast, there would still be nuts, bolts, screws, and other fasteners to deal with using the familiar tools that have been around for centuries simply because

CHAPTER FIVE

they've universally proven to have worked so well. The trick was to make such things still seem utilitarian by the twenty-third century.

In the days before handheld digital calculators, engineers and scientists used slide rules. Similar analog devices were used by pilots. Called flight computers, these handheld computational tools were a form of circular slide rule used in the aviation industry and remain one of the few analog computers still in use today. They are mostly used during flight training, but many professional pilots still use them. Once in the air, pilots can use them to calculate ground speed and fuel burn as well as update their estimated time of arrival. Since many of the production crew who worked on *Star Trek* were pilots themselves, including Roddenberry, it is not surprising to find that these devices made appearances in the actual series.

A number of aviation supply manufacturers produced these over the years under various names, including "Slide Graphic Flight Computer" or "CSG." They appeared on-screen in multiple *Star Trek* episodes. The brand most commonly seen were those made by Jeppesen, specifically their CSG-1A (or E6-B) and B-1 models.

The first appearance of one of these devices in *Star Trek* occurred in the episode "The Corbomite Maneuver." Here, one can find both CSG-1A and B-1 models used as well as a mysterious round disk device that not even the most diehard fans have been easily able to identify.

Both the CSG-1A and B-1 flight computers are slide-rule-type devices with measurements and scales on both the front and back. When Mr. Spock is shown using the CSG-1A model in "The Naked Time," pilots may notice that the rectangular slide part has been completely removed from the round wheel section, flipped over face down, and then reinserted. The "calculator side" of the wheel is now trying to perform measurements by using the "wind side" scales, which is incorrect. This may have been intentionally done to make the device appear to be something different than what it normally is. Of course, none of this would have likely ever been noticed during the brief time the episode was originally shown on television. Viewers are, however, treated to a good close-up of Mr. Spock using the B-1 model in the episode "Who Mourns for Adonais?" This particular model is larger and has a bright-yellow ring with a red-and-black wind vector "spinner" that made it more noticeable to television viewers, especially those with color TVs. In the close-up shot, you can even make out the big "J" in Jeppesen's original corporate logo printed across the bottom.[15]

All these devices show that handheld calculating tools are still an essential part of navigating the universe during the twenty-third century. Some of these objects were used in the spirit of most space-themed television shows and motion pictures of this era, as show runners sought to make use of whatever they could find to dress a set quickly and cheaply. The result is that military surplus or laboratory hardware was often used. *Star Trek* was not immune to this temptation, but the show did use less of it than most others, since the production crew focused more on designing new and futuristic equipment.

Polar grids, which depict a two-dimensional coordinate system with each point on a plane determined by a distance from a reference point and an angle from a reference direction, were seen in press images released by various aerospace contractors depicting space hardware. Besides serving a practical engineering purpose, such as being used for alignment and calibration, these bull's-eye-looking devices had a futuristic technical look about them, a fact that Roddenberry did not fail to miss. Polar grids can be seen in *Star Trek* not only in advance publicity shots, but also on the show's sets. A small polar-grid disk was used on the phaser cannon shown in "The Cage." That same disk was repurposed for use on the shuttlecraft's helm as well as at the transporter controls. The Astrogator, as seen on the helm in the original series, also incorporated a polar grid fabricated from a World War II Navy plotting map. These grids gave *Star Trek* viewers the appearance of believable navigational instruments.[16]

Real-World Space and *Star Trek*

Roddenberry's interest in connecting real-world space to his new television series was something that he tried to encourage and promote with the show's production team, even among those who were hired to help sell the series. When *Star Trek* first began, publicity was handled by both Desilu Studios and NBC. While the network created and distributed their own material, Desilu hired McFadden, Strauss, Eddy, and Irwin (MSEI), an outside promotion to do theirs. In the US, MSEI had offices on both coasts, as well as in London. Representing more than two hundred businesses and celebrities, such as the Quinn Martin Production Company and Ford Motors as well as Lucille Ball, they had a proven track record of effectively selling new shows and talent to the media.[17] One of the more unusual PR campaigns that MSEI conducted to promote *Star Trek* involved mailing out actual pieces of material associated from outer space, specifically tektites, gravel-sized bodies composed of black, green, brown, or gray natural glass formed from terrestrial debris ejected during meteorite impacts. During the meteor's impact, some of the material from the meteorite can become embedded within the forming tektite, thus giving it, in part, an extraterrestrial origin. Accompanying each tektite was the following printed media item from MSEI:

> Dear Friend:
> Since "Star Trek" is about outer space, we thought an appropriate memento to you from the crew of the Enterprise would be this genuine antique from outer space.

CHAPTER FIVE

Should you by chance not be an outer space antique buff, this is a TEKTITE. Tektites are found in strewn fields in North America, the antipodes, Europe, and Africa. Before they came to Earth they were melted by high-speed passage through the atmosphere; they solidified, many of them in the shape of drops, and some were curiously sculptured by the air stream as they reach our planet.

Geologists and geophysicists are divided as to the origin of tektites. However, the most popular consensus seems to be that the intriguing objects came from the impact of meteorites on the Moon.

Many geophysicists believe the earth constantly receives debris from the Moon as a result of the impact of meteorites when the latter hit our satellite with immense velocity. The impact produces a crater and hurls into space a considerable amount of matter.

The portions of this matter reaching earth are tektites. Geology of the strewn fields of tektites indicates there have been few tektite falls in the past 30 million years. That's why your tektite may well be considered a genuine outer space antique.

Enjoy it.

Captain Kirk and the crew of the USS *Enterprise*[18]

Brother Guy Consolmagno, the director of the Vatican Observatory and an astronomer and physicist who specializes in the study of meteorites and their connection between asteroids and the origin and evolution of small bodies in the solar system, when asked to comment about the content of this original series promotional item, replied, "What they [MSEI] wrote about tektites was very much the common view before the Apollo samples came back from the Moon and showed some subtle but real differences between tektites and lunar rocks. The generally accepted view today is that they are in fact terrestrial rock melted by the impact that launched them off Earth and briefly into space. Among other new evidence, virtually all tektite strewn fields today have known impact craters that have been identified as the source of the material." Brother Guy agreed that mailing out "Trektites" was a novel promotional campaign.[19]

To keep up with the technical literature produced by the aerospace industry of the time, Roddenberry requested that he be placed on mailing lists to receive whatever information he could.[20] He wrote the US Government Printing Office, asking that any development and research reports be sent to him. He also contacted NASA, asking for specific publications such as the *Summary Report of Future Programs Task Group 1965*, done by NASA, and a copy of *Space Research Direction for the Future 1965*, done by the National Research

Council in Washington.²¹ As discussed in the previous chapter, many of his requests were honored. Companies and organizations such as NASA sent him material, and Roddenberry used it. Matt Jefferies, a fellow pilot and art director for the series who designed the *Enterprise*, recalled the time he first came to work for Roddenberry. "For years I had been collecting material on space exploration and activities from friends of mine in companies like North American, Douglas, NASA, and TRW," said Jefferies. "After Roddenberry's visit I brought all that stuff to the studio, along with everything else we could get our hands on."²²

Star Trek's Image Influenced by Others

During a 1970s interview in which Roddenberry was asked if other films such as MGM's 1958 classic *Forbidden Planet* influenced the creation of *Star Trek*, Roddenberry replied, "Definitely not . . . the only time I ever thought of *Forbidden Planet* specifically when I was laying *Star Trek* out was when I said to myself that here were some mistakes they made in the film that I did not want to repeat. . . . But, no, I cannot remember a single time during the planning of *Star Trek* that I looked at another show and said, 'I will borrow this.'"²³ It is clear that the work of others did influence Roddenberry in his efforts to make his show visually interesting while at the same time guide him with subject matter and themes to help give *Star Trek* a distinctive look not seen in other television shows.

Roddenberry reached out to aerospace companies who had shared with him an interest in the series. Perhaps stemming from the belief that a new television series that took science fiction seriously could be useful for hiring and promoting new workers, as well as inspiring existing employees, representatives from General Electric's Aerospace Division and Caltech's Jet Propulsion Laboratory volunteered to help the show in any way that they could.²⁴ᵃ⁻ᵇ Roddenberry even extended invitations to these companies to tour the show's sets, and, as mentioned earlier, some did. Others invited Roddenberry to come and tour their facilities and he did.

Roddenberry was so impressed by some of the hardware that he saw during these tours that he sought to use some of it in his show. In particular, he wrote to inquire about "an instrument I saw yesterday at North American Advanced Space Research Center," asking, "Is there some way to construct a plain revolving globe on which flicker on and off various small lights, lighted path progressions, projected course lines, etc.?"²⁵ Less than two weeks prior to Roddenberry's visit to the North American Center, Herb Solow sent a memo to Roddenberry stating that he would be getting "pictures of spaceships and spaceship interiors" from the 1956 MGM film *Forbidden Planet*.²⁶ The request for these photos followed an earlier screening of the film done by Roddenberry. In watching the first *Star Trek* pilot, it is clear that Roddenberry was influenced by the film. An August 10, 1964, memo states: "You may recall we saw MGM's

CHAPTER FIVE

'FORBIDDEN PLANET' with Oscar Katz some weeks ago," said Roddenberry. "I think it would be interesting for Pato Guzman to take another very hard look at the spaceship, its configurations, controls, instrumentations, etc. while we are still sketching and planning our own. Can you suggest the best way? Run the film again, or would it be ethical to get a print of the film and have our people make stills from some of the appropriate frames? This latter would be most helpful. Please understand, we have no intention of copying either interior or exterior of that ship. But a detailed look at it again would do much to stimulate our own thinking."[27]

When he toured North American, Roddenberry had scenes from *Forbidden Planet* still on his mind, such as the large, nested navigation globe that was seen on the bridge of the C-57D, the protagonist spaceship. Scenes from *Forbidden Planet* clearly influenced the look and feel of *Star Trek*, since viewers can see evidence beginning with the series' very first pilot, "The Cage."

Another big screen production that influenced Roddenberry was *Robinson Crusoe on Mars*. This 1964 film was the brainchild of the talented and multifaceted Danish-born producer-director-screenwriter and novelist Ib Melchior, who was responsible for other science fiction films, including *Angry Red Planet* (1959) and *The Time Travelers* (1964). *Robinson Crusoe on Mars*, as the title implies, was based on the classic Daniel Defoe story, but instead of centering on a character marooned on an island in the South Pacific during the eighteenth century, Melchior's story follows the exploits of Commander Christopher "Kit" Draper, an astronaut who survives a crash landing on Mars after narrowly escaping death from a flaming meteor that nearly wrecked his spaceship.

In a July 21, 1964, memo to Oscar Katz, executive vice president in charge of production at Desilu Studios, Roddenberry said, "I would like to bring to your attention a science fiction film titled *Robinson Crusoe on Mars*. As yet it is unreleased in this area, but it has been given excellent reviews in *Variety* and *The Reporter* and is regarded as a sleeper. Since it is unlike many of the pictures we have been seeing, dealing directly with planetary exploration and survival, it might be a good idea to screen this one if it is possible to obtain a print."[28] Roddenberry soon saw the film for himself and sent off another memo. "I saw the above motion picture and considered it extraordinarily good, better than anything we have run here. Suggest we get a print when possible so that Oscar [Katz] can run it for himself. Also, would like appropriate department heads and personnel here to see it. Would appreciate your office obtaining a complete credit list for this film."[29]

Robinson Crusoe on Mars has many connections to *Star Trek*. The portable Omnicom console seen in the film and carried to the surface of Mars by Draper incorporated a view screen, tape recorder, and radio, very similar to a *Star Trek* tricorder. The film's promotional teaser boasts, "This film is scientifically authentic.

It is only one step ahead of present reality." As we first saw in chapter 1, science sells and Roddenberry, like Melchior, often invoked phraseology laden with such words as "science" and "scientific" when describing *Star Trek*.

Inspired by what he saw in *Robinson Crusoe on Mars*, Roddenberry sought out Byron Haskin, who directed Melchior's film, to serve as associate producer on the first *Star Trek* pilot, "The Cage." Though the two often were at odds, which resulted in Haskin not returning to the show after his work on the first pilot, Haskin's influence can be seen and heard in *Star Trek*. Roddenberry was impressed by the fast flyby shots of the film's main ship, MGP-1, that appeared in *Robinson Crusoe on Mars*. Those shots may have inspired the "whoosh" footage of the *Enterprise* that appears during the introduction of every *Star Trek* episode. In addition, the sound of the *Enterprise* firing its photon torpedoes was taken directly from the Martian War Machine wing-tip energy weapons as seen in George Pal's 1953 classic film *The War of the Worlds*, which Haskin also directed. In addition to Haskin's connection as director, another production member from *Robinson Crusoe on Mars* was matte painter Albert J. Whitlock, who contributed matte paintings for "The Menagerie" as well as several other *Star Trek* episodes.

Melchior's film also had an influence on Roddenberry in terms of *Star Trek*'s casting. Victor Lundin, who played Friday in Melchior's film, was the first screen-seen Klingon in the *Star Trek* episode "Errand of Mercy." Paul Mantee, who portrayed Draper, was also the first on Roddenberry's short list of actors considered to play Captain Robert April (later changed to Captain Christopher Pike, then finally to Captain James T. Kirk) in "The Cage."[30]

TRW Space Park

On-location filming was almost always more complex and expensive than filming in a studio, but exceptions were sometime made in exchange for the opportunity to establish a required look. For a television series placed in the future such as *Star Trek*, it was difficult to find locations that conveyed a sense of being not in the present. *Star Trek* could have used the Space Needle in Seattle or the TWA Flight Center in New York or the Theme Building in Los Angeles. These places certainly looked futuristic, even today, but they all were built by 1962, so television audiences were familiar with them by the time *Star Trek* first aired in 1966. They were also located much farther away from Hollywood and would have added considerable cost to an episode's budget.

During and after World War II, as the center of the aerospace industry shifted from the US East Coast to the West, companies altered not only the geography of the greater Los Angeles area but also its architecture, with new styles such as Space Age and Googie.[31] Space Age and Googie refer to the type of futurist architecture influenced by car culture, jets, the atomic age, and the space age. Originating in Southern California from the streamline moderne architecture of the 1930s, it was popular in the United States beginning after World War II and into the early 1970s. Features of this architectural style include bold use of glass, steel, and neon, along with geometric shapes and curvilinear features such as upswept roofs. Futuristic designs symbolic of the space age that expressed motion, such as sweeping parabolas, diagrammatic atoms, and boomerangs, all were part of the stylistic conventions that defined this aesthetic. The aerospace companies that populated Southern California sought to project this style of image not only to showcase the dramatic vistas, mild temperatures, and bright sunshine of the region, but also to provide a sense of the future to their employees and the general public.

Examples of this new style of architecture came from Simon Ramo and Dean Woolridge of Thompson, Ramo and Woolridge, founders of the aerospace company TRW. In the late 1950s, they bought land from the Sante Fe Railroad to build their new corporate headquarters in Redondo Beach, California. They hired Albert C. Martin & Associates to design a campus that captured this new style of architecture. The result is a 110-acre aerospace research campus known as Space Park, which was developed between 1960 and 1967 and employed some 11,000 people in its sprawling laboratories, offices, engineering and manufacturing buildings. The futuristic complex sported a different look from the more traditional blue-collar factories that preceded it.

TRW's Space Park campus is an example of the type of self-contained office and manufacturing community that grew popular during this time. To compete with other companies seeking to lure the brightest and finest, companies published advertisements designed to promote "California living at its finest" in various trade publications. Since these companies invented not only the era of the space age but also the architectural type that helped define the period, they sought at the same time an opportunity to reinvent themselves by providing workers with an environment that promoted an atmosphere of intellectual freedom rather than a hardheaded business.[32]

The Los Angeles Conservancy, a nonprofit organization that seeks to recognize, preserve, and revitalize the historic architectural and cultural resources of Los Angeles County, wrote of TRW's Space Park, "It was the country's first group of space science laboratories and manufacturing facilities built solely for the purpose of designing, building, and testing spacecraft—described in the jargon of space technologists as 'blueprint to black sky capability.'"[33] The complex won a "Grand Prix Award" from the American Institute of Architects in 1967. Today, the Space Park campus serves as an example of how the proximity of the Southern California aerospace industry to Hollywood inspired and influenced not only the surrounding area landscape but also the look and feel of television.

On Wednesday, February 15, 1967, during the second day of principal filming for *Star Trek*'s first-season episode "Operation: Annihilate!," cast and crew moved on location to TRW's Defense and Space Group Campus in Redondo Beach. From the director's point of view, it was the perfect choice to depict buildings from the twenty-third century. More importantly, the site was only 26 miles from Desilu Studios, which made it economical to move cast and equipment for filming. Scenes for the episode were shot under the walkway near TRW's cafeteria, as well as around other parts of the campus.

In the episode, the *Enterprise* is investigating what appears to be a case of mass insanity that has wiped out Federation colonies on several planets. The *Enterprise* arrives at the planet Deneva, where they soon discover that a parasitic creature has infected the inhabitants of the main Federation colony living there.[34]

Herschel Daugherty, the director of "Operation: Annihilate!," made creative use of the TRW property by using camera angles that effectively maximized the filming of the futuristic-looking grounds to convey to viewers a real place from the *Star Trek* universe. In one scene, Daughtery used a distant shot to show a long outdoor stairway with TRW's headquarters in the background. Daughtery effectively blotted out the towering "TRW" letters mounted across the top of the headquarters building by positioning his camera so that an overhang from a building in the foreground hid the massive sign from being seen in the shot.

In another scene, a distant shot is shown after the landing party beams down to the surface. This shot was taken from the rooftop of one of the nearby TRW buildings, and as the camera pans below to follow the landing party, it reveals a collection of geometrical shapes in the foreground. These shapes appear to be sculptured pieces, and since TRW had a large metal sculptured fountain on the main campus square, it made sense that similar pieces of sculpture might be present in other locations.

It turns out that these structures were part of an assembly of fixed instruments that TRW put in place for radio frequency (RF) testing. In a different scene, the landing party can be seen on the ground with another geometric object in the distance. Upon closer examination, the object looks like a removable support mast with an early nuclear-detonation-detection satellite known as Vela attached to it.[35] This would help explain the presence of the RF antennas and instruments mounted atop the roof. It is possible that the object mounted to the testing structure is not a Vela satellite at all, but, rather, something else such as an antenna enclosure. Whatever the case, the geometric structures assembled in that brief scene from the show were most likely radio hardware, and they may very well have been testing some type of satellite or other instrument while the show was being filmed.

One final architectural item worth noting in the episode involves views of the exterior of Captain Kirk's brother's laboratory. In this scene, the landing party, rushes in the direction of where screams were heard, which was at his brother's lab. The scene, which lasts only a few seconds, switches to an established shot showing another futuristic-looking building. This building is not part of the TRW Space Park but instead belongs to the University of California's Los Angeles campus, located 17 miles north. The photo of the building seen in the episode is reversed to make it appear somewhat less familiar. The building is Schoenberg Music Hall, which is located on the campus of UCLA.[36]

Numerous behind-the-scenes photos captured during the filming of "Operation: Annihilate!" give more insight into how the location was used for filming, including shots of the cast and crew talking with TRW employees and touring the campus facilities. Several shots show principal cast members reviewing hardware located inside the TRW buildings. One shot shows them being briefed by company officials on the large space chamber used by TRW to test the performance of spacecraft in a simulated outer-space environment.[37]

Today, TRW is owned by Northrop Grumman, but the original Space Park campus is still there, with the area now designated as a historic aerospace site by the American Institute of Aeronautics and Astronautics. On the last Saturday of every month, the local amateur radio club holds a public electronics swap meet from 7:00 a.m. to 11:30 am. This a great opportunity for the public to visit the location to see where they filmed the original *Star Trek* episode as well as other television shows and films that made use of this location's unique architecture.

The K-7 Space Station and Botany Bay

One of the most direct examples of how the aerospace industry inspired the look and feel of *Star Trek* appears in the episode "The Trouble with Tribbles." In this episode, furry little creatures that "coo" and have a tremendous desire for eating and breeding overrun the Federation's deep-space station "K-7."

The principal design elements that make up the K-7, which include the four podlike sections shown on screen, first appeared in a 1959 study done by Douglas Aircraft.[38] The study featured original artwork supposedly done by George Akimoto (1922–2010) that depicts a donut-shaped pod centered around a conical structure. The study describes how the station was launched folded, but once in space it opened like a handheld fan, with the conical center structure serving as a reentry vehicle. Akimoto's artwork depicted a design that was certainly unique and modern, a style that served the young artist well throughout his time working at Douglas and other aerospace firms.

Born in Stockton, California, in 1922 to a family of Japanese origin, Akimoto spent time in an internment camp. While being interned at the Rohwer War Relocation Center, Akimoto created a cartoon character called "Lil Dan'l" that was featured in the camp's journal and became the camp's mascot. The character was also featured in a book that Akimoto later published called *Lil Dan'l: One Year in a Relocation Center* (1943). After the war, Akimoto attended Stockton Junior College and the Art Students League of New York City. He worked at Beacon Studios and later was a partner in Paul George Studios. He then moved to Miami to serve as art director with the August Dorr Advertising Agency, where he was elected vice president of the Advertising Artists Guild of Miami. He returned to California to work as a senior art director for the Edwards Agency in Los Angeles, where his art appeared in many brochures and ads for aerospace companies such as Ramo-Wooldridge and Pacific Semiconductors, a subsidiary of TRW based in Culver City, California.[39]

It was while working for Douglas Aircraft that Akimoto illustrated the space station elements for the 1959 Douglas study that eventually became the K-7 model as seen in *Star Trek*. Physical models were constructed at Douglas that served to help visualize the space station's form as well as show how each of the station's components worked. The 1959 study outlined the operational requirements of an orbiting space station that would be launched into Earth orbit by using a "Saturn-type missile." In addition, "a much[-]larger space station could be boosted by the Nova-type missile." Nova was a more powerful rocket than NASA's Saturn V but was never built.

The station would operate as an "experimental and training facility for periods of at least 30 days at a time." The station would remain in orbit, and the crew would return to Earth via "an escape-return vehicle (capsule) or aboard the resupply vehicle." The crew would ride into space in the combined escape-return vehicle. In the event of an aborted mission, the escape reentry vehicle would separate from the space station and booster section.

The study went on to say that "the station would be constructed in one piece on the Earth and automatically assemble in orbit in its final form with little or no human effort. It folded into a pie-shaped wedge which is placed on the booster vehicle. During the launch phase, the crew members ride in the escape-Earth return capsule. In orbit the space station unfolds and forms a pressurized disc-shaped vehicle 36 feet in diameter with an internal volume of over 5,300 cubic feet. The telescoping and toroidal egg shape was chosen because of deriving the smallest folded shape and the largest unfolded shape. The station has three large compartments and an airlock. The airlock in the center of the station is connected to the main compartment, the escape-return capsule, and to a hatch which permits extra vehicular exploration from the bottom. The main compartment has a hatch and acts as an airlock for the attachment of the resupply vehicle."

Power for the space station would be provided by a solar turboelectric power system. The closed-cycle solar system is composed of a boiler heated by an external parabolic dish that would concentrate solar energy to drive a turbine, a condenser, and a pump. As a backup, a nearby external solar cell could be

CHAPTER FIVE

used. The space station would be manned by a minimum of three men but have a capability of housing up to fifteen people.[40]

At the time, the 1959 report detailing the space station design was well received. Akimoto's artwork from that study was used by Douglas not only in the report but also on the cover of the aerospace company's fortieth annual report, which was published later that same year.[41]

How Roddenberry first became aware of the Douglas design, either by seeing an actual model or reading the published reports, is not clear. A year later, the original 1959 Douglas study was included in the published proceedings from a Manned Space Stations Symposium that was cosponsored by NASA and RAND and held in Los Angeles from April 20 to 22, 1960.[42] Roddenberry may have seen the report in these proceedings along with Akimoto's drawings.

In the 2016 book *The Enterprise NCC 1701 and the Model Maker*, Noel Datin McDonald, daughter of Richard Datin, the model maker who helped build *Star Trek*'s original *Enterprise* model (both the 3-foot and 11-foot versions), explained how her father's K-7 design materialized. In quoting from her father's journals that he kept while working on *Star Trek*, Richard Datin said this of the K-7 model's design and construction:

> I was told upon viewing the original model, and maybe by Roddenberry, that he obtained it [the Douglas space station model] from Douglas Aircraft whose main office was in nearby Santa Monica. Apparently, Gene had a following from people in the space industry, particularly Caltech in Pasadena. At first glance, its original appearance is not much different than what remodeling I performed on it. Among the specific revisions that occurred were the installation of tiny sensor lamps to the upper, or main section; replacement of the connecting passageways or struts that held the station together; and a new paint color.[43]

Those small sensor lamps that Datin refers to appear in the original paintings of the station that Akimoto did for Douglas Aircraft. Datin more than likely saw Akimoto's paintings and included design elements from them in his model of the K-7. If you look at the top of the main center module of the original 1959 station design, you will see a tall mast. Looking more closely, you will see a replica of Akimoto's "sensor lamps" on the K-7 model which look like a pair of rabbit-ear antennas.

Datin worked on the K-7 in his shop, which was in back of the family's house on Klump Avenue in North Hollywood, just around the corner from the house where the exterior shots of the home seen in *The Brady Bunch* were filmed. Datin began building the K-7 from the Douglas model on August 19, 1967, and finished it nine days later. The episode aired on December 29.[44]

Datin described how he built the 20-inch-diameter model.[45] "The K-7 model was similar in coloring to the *Enterprise* models, either a dull off white, or very similar to the light gray/green *Enterprise* color. The model was composed of a material very unfamiliar to me. In perspective, more distinct than other substances from those days, it was a strange material. It was soft, yet resilient, but a rubbery substance, more foam-like but dense enough to hold its shape and be worked satisfactorily by machines. Each of the components, or satellite sections, were cast in this material and the alterations consisted of replacing the existing rods that held them together with Plexiglas tubing to allow electrical wiring to pass through to each other. Looking at the parts purchased for the alteration job, it looks as though I added some dome-like Plexiglas pieces to the satellites, most likely to alter their appearance sufficiently from the original Douglas design."[46]

"During the filming, the model was supported by a pipe from the bottom that served as the center support for the entire model, and also to hide the electrical wiring. The model was labeled with a 'United Federation of Planets' decal on the main section, with 'KA, KB, KC' on the cone-shaped tops of each of the three circular pods."[47]

The origins of the K-7 model's design have been the topic of discussion among fans. In a 2005 article that appeared in the online publication the *Space Review*, spaceflight historian Dwayne Day explained: "A colleague of mine who used to work in the Southern California aerospace industry says that he suspects that the models were more likely fished out of a dumpster by an enterprising collector."[48] Since the publication of that article, early photos have surfaced showing parts of the original Douglas model on display along with other models. These models may have been the ones that Datin was referring to. Somebody at Douglas perhaps no longer needed them and, rather than tossing them out, gave them to someone at the studio. After all, *Star Trek* had been on the air for almost a year by the time that Datin received the models. With the Douglas plant located in nearby Santa Monica, it would not have been difficult for a follower of the series to make these available to the show's production designers.

In the 1960 *Manned Space Stations Symposium* proceedings, which features the Douglas space station design that became the K-7, there appears a paper that focuses on the thermal control of another spacecraft whose design was very similar to one seen in another episode of *Star Trek*. In the proceedings, there is shown an illustration depicting a thermal-protection system for a proposed spacecraft. In comparing this illustration to several made by Matt Jefferies, the artist responsible for designing the *Enterprise*, there is an uncanny resemblance to the sleeper ship *Botany Bay* that Jefferies designed for the *Star Trek* episode "Space Seed." In studying the conceptual sketches created by Jefferies that defined the finished *Botany Bay*, one can see a submarine-like form. One drawing in

CHAPTER FIVE

particular made by Jefferies to illustrate an early configuration of the *Enterprise* is remarkable in its similarity to the *Botany Bay*. The spacecraft exterior grid pattern as seen in these drawings, combined with the angled vertical dorsal tower, raises the possibility that Jefferies may indeed have been influenced by what was shown in these 1960 symposium proceeding drawings. In *Inside Star Trek*, issue 4 from 1968, Dorothy Fontana interviewed Jefferies and asked him about the origins of the design. "The Botany Bay was actually designed before the Enterprise. It was a little idea that popped up and was labeled 'antique space freighter'" said Jefferies. "Later on, we made it look like something else—a vehicle out of the early 2000s."[49a–c]

The examination of *Star Trek*'s early connections to the aerospace industry serves as a reminder of the innovative ways that the series employed then-current space hardware to help inspire and shape its vision of the future. The geographic proximity of both Hollywood and the aerospace industry helped make Southern California a unique source for television writers and designers such as Roddenberry who sought to craft a weekly TV show that looked like the future while at the same time delivering each episode on time and within budget.

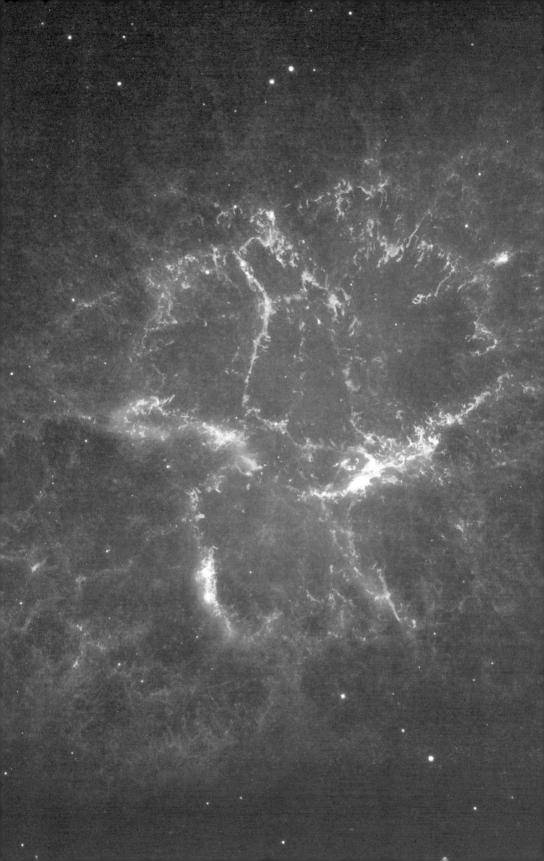

CHAPTER 6

NASA and *Star Trek*

When the first episode of *Star Trek* aired on Thursday evening, September 8, 1966, it was the most-watched program during its 8:30 p.m. time slot.[1] As reviews came out the next morning, NASA prepared to launch two astronauts into Earth orbit aboard Gemini 11, the ninth mission with a crew on board as part of Project Gemini and a precursor to the Apollo program, which nearly three years later would land Neil Armstrong and Buzz Aldrin on the surface of the moon.

Gemini 11 finally launched on Monday, September 12, and the networks were there covering the mission for their viewers. Television broadcasts of the early human space program was still relatively new during the time of *Star Trek*, with ABC, CBS, and NBC competing for ratings.[2] NBC led in its coverage of Gemini 11, with 46 percent of television viewers watching the launch on their network compared to 31 percent on CBS and 23 percent on ABC.[3]

NASA documented the spaceflight of the present while *Star Trek* offered a glimpse into the future. The connection between the two was obvious to Gene Roddenberry, who made every effort to reach out to NASA for help with his new series.[4] In an article that appeared in the December 1967 issue of *Popular Science*, James W. Wright wrote that *Star Trek* was "the only science-fiction series in history that has the cooperation and advice of the National Astronautics and Space Administration."[5] During the early years of *Star Trek*, NASA opened its doors to a show that many within the agency firmly believed was worth supporting. This is a relationship that has continued with the franchise to this day.

NASA Support for *Star Trek*

Between 1962 and 1973, NASA's Jet Propulsion Laboratory in Pasadena, California, designed and built a series of ten robotic spacecraft named Mariner that explored the inner solar system. Mariner 5 (also referred to as Mariner V or Mariner Venus 1967) carried a complement of experiments to explore Venus, Earth's nearest neighboring planet.[6] As the spacecraft flew by the cloud-covered world on October 19, 1967, an official NASA/JPL public-affairs photo was taken showing spacecraft managers Ted Parker and Dan Schneiderman wearing a pair of Spock-looking ears, as colleague Dave Shaw looked over their shoulders.

Though the NASA/JPL engineers could have been portraying the well-known stereotype of a pointed-eared little green Martian, or even Venusian, since the potential for life on both planets was a common theme in films and television shows from that period, they also may have been depicting Mr. Spock. After all, Roddenberry hinted early on in his initial pitch for *Star Trek* that Mr. Spock had a Martian lineage. "His name is 'Mr. Spock.' And the first view of him can be almost frightening—a face so heavy-lidded and satanic you might almost expect him to have a forked tail," wrote Roddenberry in describing the ship's first lieutenant and captain's right-hand man. "Probably half Martian, he has a slightly reddish complexion and semi-pointed ears."[7a-b] According to David Alexander, Roddenberry's authorized biographer, "Gene decided that if the show were a success, explorers might actually land on Mars during its run, so Spock's origin was moved to another, unnamed planet."[8] In the show as aired, Spock was initially called a "Vulcanian."

As writers, directors, and actors worked on a television series designed to depict fictional life and hardware from the future, it made sense for them to use whatever real space technology was available to them at the time. During the production of *Star Trek*, Roddenberry sought to actively engage the actors who played characters from the future by having them take part in space-related events of the present. NASA sent the cast and crew invitations to attend conferences and events that were either cosponsored by the space agency or held directly by them. It was a symbiotic relationship, since, at the time, NASA was one of only two agencies with the ability to put people into space—the other being those employed by the Soviet Union. It was clear from the airing of the very first episode that NASA and *Star Trek* would benefit from the interaction.

On March 31, 1967, DeForest Kelley, who played Dr. Leonard "Bones" McCoy, the ship's doctor, was invited by NASA's Manned Spacecraft Center (now Johnson Space Center) in Houston, Texas, to be the guest of Dr. Charles Berry, chief surgeon for all the astronauts. NASA asked Kelley to "tour facilities and watch experiments and tests," reasoning that "this would be good for *Star Trek* and helpful in his portraying of the chief surgeon of the USS *Enterprise*."[9]

Later, NASA employed the good doctor to give comfort to the first manned Apollo flight into space. Shortly after launch on October 11, 1968, Wally Schirra, Walter Cunningham, and Donn Eisele, the crew aboard Apollo 7, developed head colds. This prompted Kelley, who playfully wrote as Dr. McCoy, to try to provide some relief. During their ten-day mission, he sent them the following telegram: "I do not make house calls but under the circumstances would be pleased to beam aboard and take care of the common cold. [signed] Bones."[10]

CHAPTER SIX

Interactions between NASA and the *Star Trek* cast continued during the spring of 1967, when NASA and the US Air Force extended an invitation to have the cast and crew of *Star Trek* come and see some real space hardware at NASA's Dryden (now called Armstrong) Flight Research Center at Edwards Air Force Base in California. Col. Robert S. Buchanan, the commandant of the US Air Force Aerospace Research Pilot School at Edwards, wrote, "The students and staff of the USAF Aerospace Research Pilot School would like to invite you and members of your organization to visit our facilities at your convenience. Our mission here is to train experimental test pilots for the military services, NASA and NATO countries and to train astronauts for this Nation's space program." Both Buchanan and NASA were impressed by what they saw on *Star Trek*, noting, "You apparently have some very loyal supporters in my organization, and they feel that possibly we can learn from your experiences. On the other hand, because of our mission and unique facilities here at the Flight Test Center, I feel you may find some material of value to you in your production."[11]

A little over four weeks after NBC confirmed that *Star Trek* would be renewed for a second season, Roddenberry, James Doohan, who played "Scotty" the ship's chief engineer, and DeForest Kelley, traveled to the Mojave Desert on April 13, 1967, to tour NASA's Dryden Flight Research Center and Edwards Air Force Base. Here some 100 miles northeast of Los Angeles, they were joined by six other production members of the show, including Walter Matthew "Matt" Jefferies Jr., art director and production designer for the series; Herbert Schlosser, who at that time, was head of NBC's programming and oversaw *Star Trek*'s development during the network's production; assistant director, production manager, and associate producer Gregg Peters; director Marc Daniels; assistant director and producer Robert "Bob" Harris Justman; and director Joseph "Joe" Pevney. Daniels and Pevney would go on to direct twenty-eight *Star Trek* episodes (fourteen episodes each), or a little over one-third of the original series. Taking time out of their busy production schedules, the gathering of so many of the show's cast, executives, and production members was unique. Marc Daniels, during an interview, still remembered the gathering many years later recalling that "we went down to NASA and watched the development of the 'X' models. Well, we were very big with NASA and all the sci-fi buffs, who dug it the most."[12]

While at Dryden, "Scotty" met with NASA test pilot Bruce Peterson, and the press took photos of the two standing in front of Northrop's (now Northrop Grumman) M2-F2, one of a group of four experimental wingless research aircraft built to study the feasibility of using an aerodynamic craft to reenter the atmosphere of space. These test vehicles were designed to be carried aloft under the wing of a modified B-52 bomber and released to fly under their own rocket power before gliding down to Earth for landing.[13]

As they continued their tour at Dryden, NASA photos taken at the time showed Kelley speaking with Bill Dana, a NASA test pilot who flew the X-15 rocket plane. Other photos showed Kelley, along with the rest of the visiting cast and crew, looking over one of three X-15s built by North American Aviation for NASA. This was a real spaceship, because at the time of their visit, this particular X-15 had flown to the edge of space just two years earlier. In addition, this vehicle was piloted by Neil Armstrong, who would travel to the moon two years later in another North American–built spacecraft during Apollo 11. It was a moving experience for all the cast and crew of the show to see, up close and in person, hardware that had flown to the final frontier and back.

Because history was being made on a regular basis by astronauts traveling into space during the time that *Star Trek* aired, it is not surprising that there was an organized effort by the show's producers to try to get a real astronaut to appear on the show.[14] Alan Shepard, the first American to travel into space during his historic May 5, 1961, suborbital flight, was asked to play a minor role in Roddenberry's new television series (in 1970, during Apollo 14, Shepard became the fifth man to walk on the moon). Shepard could not accept payment for the part but agreed to have Roddenberry film a one-minute spot for the astronaut's favorite charity in exchange for Shepard's appearance on *Star Trek*. Sadly, the deal never materialized. The studio extended a similar offer to former Mercury astronaut Scott Carpenter, who also declined.[15] Both Shepard and Carpenter eventually did make it to *Star Trek*, though many years later. Shepard appeared in the opening credits to the spin-off series *Star Trek: Enterprise*, which ran from 2001 to 2005. Viewers saw him suiting up for his Apollo 14 lunar mission at the beginning of each episode. Carpenter's voice was also heard in the teaser trailer for J. J. Abrams's 2009 *Star Trek* theatrical release.

Even though *Star Trek* tried but could not get real astronauts to appear in the original series, Roddenberry himself sometimes would speak at functions as a sort of "guest spaceman." In a 1967 article, syndicated newspaper writer Harvey Pack explained how Roddenberry was "often invited to attend scientific meetings sponsored by NASA or companies engaged in aero-space operations." The article described how he "once was used as a guest spaceman invited to present an award to a distinguished colleague. After a straight presentation, he insisted upon showing some films of his own research and it wasn't until Leonard Nimoy and his pointed ears appeared on screen that the scientists realized they were being put on."[16]

NASA Images Seen in *Star Trek*

When *Star Trek* was being made, California had the greatest concentration of defense spending and aerospace industrial capability of any state in the Union. They also had 50 percent of NASA's prime contracts. As a result, the space agency

CHAPTER SIX

established an operations office on the West Coast.[17] In 1966, Roddenberry wrote to Edward A. Orzechowski, a public-information officer at NASA's newly established Western Operations Office in Santa Monica, California, to inquire about support for *Star Trek*. Eddie Perlstein, Desilu's attorney, then had lunch with Orzechowski and reported back, stating that NASA "was very interested in *Star Trek* and most anxious to provide us with all sorts of help including stock footage, valuable models, animation, technical advice, even access to some NASA labs and proving grounds."[18] Roddenberry made use of NASA's generous offer beginning with the show's very first pilot.

In the original *Star Trek* pilot, "The Cage," the crew of the *Enterprise* receives a distress call from the fourth planet in the Talos star system (referred to as "Talos Star Group" in actual dialogue). During the episode, the planet's inhabitants glean records from the *Enterprise*'s computer banks. During this sequence, multiple images flash across one of the bridge's view screens. Many of these images were taken from original NASA documents that were seen and approved by Roddenberry.

In a memo dated November 23, 1964, when "The Cage" was in postproduction, Roddenberry sought ideas for images to show in the aforementioned scene.[19] In a follow-up memo issued to Darrell Anderson, who filmed the special visual effects, Roddenberry provided an annotated list of NASA documents, pamphlets, and papers, complete with instructions on what pages to be filmed for what internal production memos referred to as the "Subliminal, Knowledge Montage" sequence shown in the pilot.[20] Of the nineteen confirmed NASA images that appeared in this "slide show," nearly half of them came from the NASA publication *SPACE: The New Frontier* (1962 edition).[21] *SPACE: The New Frontier* was a heavily illustrated publication that NASA first issued to the general public in 1959 to help explain what the less-than-one-year-old space agency planned to do. The publication proved to be highly popular, and, as a result, NASA ended up reprinting it many times.[22]

NASA photos can be seen in other episodes. In "Court Martial," a first season episode of *Star Trek*, there are scenes from an office on a Federation Starbase where viewers can clearly see NASA photos from the Gemini program hanging on walls. A photo of the launch of Gemini 6 is shown along with docking photos from Gemini 8. It is interesting to note that Gemini 8 flew in March 1966 and filming began for "Court Martial" seven months later so the photos used as set dressing were very new for that time.

Another example of an episode of *Star Trek* using NASA imagery involves a book that Roddenberry received from NASA in 1968 titled *Earth Photographs from Gemini III, IV, and V*. Upon receipt of the book, Roddenberry wrote NASA's Orzechowski, thanking him and saying, "I am amazed by the quality of the pictures, terribly exciting things to see. I will pass it around to all the staff so that they can get all the advantages of it."[23] The timing of the receipt of the book was fortuitous because later that same year, *Star Trek* used one of the book's images in the third-season episode "The Cloud Minders." In this episode, there is a scene in a floating city high above the ground where a man who is arrested tries to escape, jumps from the edge of the city, and falls to his death. The surface of the planet shown below is actually a NASA photo taken by astronauts Ed White and Jim McDivitt during their Gemini IV mission in 1965. As described in the NASA book *Earth Photographs from Gemini III, IV, and V*, "The photo shows a dry river basin in the southern part of Saudi Arabia. This area is composed of gently dipping marine sediments. The dendritic pattern as seen in the foreground is Wadi Hadramawt, a dry river channel and tributaries which eventually join the Gulf of Aden."[24] The overall sepia tint of the original NASA photo allowed *Star Trek*'s artists to easily blend it with the orange clouds of the fictional planet in the episode.

Assignment: Earth

Among the *Star Trek* episodes that make the most-extensive use of NASA imagery is the twenty-sixth and final episode of the second season. "Assignment: Earth" is based on a story originally written by Art Wallace and further developed by Roddenberry. First broadcast on March 29, 1968, the episode involves the crew of the *Enterprise* traveling back in time to conduct historical research of the earth in 1968. In doing so, they encounter Gary Seven, played by Robert Lansing, who is an agent sent from another planet to help prevent twentieth-century Earth from destroying itself. Both Kirk and Spock are uncertain of his motives, especially when he interferes with the launch of a US orbital nuclear weapon. It turns out that Seven's intent was to cause the nuclear weapon to malfunction so as to draw attention to Earth's leaders about the dangers of deploying such weapons, in order to deter their future use.[25]

As part of the protagonist's cover, viewers see close-ups of Seven's fabricated papers, which include convincing-looking IDs not only from the CIA but also from the National Security Agency (NSA). Even though the NSA at that time was a large intelligence agency, it was shrouded in secrecy. The joke was that NSA stood for "No Such Agency." It did not become well known until the 1980s. *Star Trek* was among the first television shows to acknowledge its existence.

NASA's Saturn V rocket is seen multiple times in "Assignment: Earth," but instead of being used to send astronauts to the moon as part of the agency's Apollo lunar-landing program, it was employed to launch a nuclear warhead platform into Earth orbit.

Because the episode was filmed during a time when NASA was using the Saturn V, *Star Trek* producers made extensive use of existing stock footage provided by the space agency. Indeed, in an early document dated December 5, 1967, that pitched the episode, Roddenberry specifically stated that they would be "using authentic NASA space footage," along with "film of our earth shot by United States Astronauts only this year."[26]

Using stock footage and photographs saved time and money in addition to allowing the episode's director Marc Daniels to heighten *Star Trek*'s realism by employing a piece of real NASA space hardware already familiar to most television viewers. Footage of the Apollo 4 Saturn V launch vehicle can be seen throughout the episode. In an early story synopsis, Roddenberry and Wallace stated that "film of a major rocket launch, acquired from NASA and used with their approval and active assistance, including many close-up scenes never before presented on television," will be used. The synopsis also promised that viewers will see "Seven in action as he makes his way into the most carefully guarded acreage in the world—the heart of Cape Kennedy at the peak of launching a nuclear warhead satellite."[27]

In addition to using footage of the Apollo 4 launch, "Assignment: Earth" also showed the vehicle staging as it made its way into space.[28] Onboard film cameras mounted on the Saturn V provided dramatic full-color footage of stage separation. After filming, the footage was ejected from the launch vehicle, from which it then fell back to Earth to be recovered from the ocean and processed. For many television viewers, this was the first time they had been able to see this remarkable footage. This same film was later shown in the opening credits of *Star Trek: Enterprise*.[29]

"Assignment: Earth" also makes use of footage of another Saturn V launch vehicle. This second spacecraft, the Saturn V Facilities Integration Vehicle, also known as the SA-500F or 500F, was the first Saturn V rocket to be built. "Assignment: Earth" used stock NASA footage of both versions of the Saturn V, switching between scenes showing the 500F and Apollo 4 launch vehicles in the episode.[30]

In another challenge to *Star Trek*'s producers, NASA shot all their footage by using the anamorphic format, meaning that all the rocket launch stock footage had to be cropped from the 2.35:1 aspect ratio to television's conventional ratio of 1.33:1. The end result is that observant viewers will notice a somewhat squeezed appearance of the rocket in scenes.[31]

Additional NASA stock footage was spliced into the episode showing ground-tracking dish antennas along with exterior views of the massive Vehicle Assembly Building, where the Saturn V was put together prior to its rollout to the launchpad. The adjacent Launch Control Center is visible, as are cuts to interior shots that show the firing rooms, where large numbers of support personnel can be seen hunched over banks of instrument stations.

Found in Matt Jefferies's personal *Star Trek* files, there is a film clip of an unaired scene that shows Seven as he first beams down to McKinley Rocket Base (NASA's Kennedy Space Center). The clip shows several Paramount back-lot buildings and a small concession truck with people milling about. A towering Saturn V should be shown matted into the top half of the scene, but instead there is a different rocket, an Atlas-Agena.³² Director Daniels may have considered using NASA stock footage to show the much-smaller Atlas-Agena launch vehicle deploying the orbiting nuclear platform rather than the Saturn V. The Atlas-Agena is less than a third the size of the Saturn V, and if it were used instead, it would have resulted in much less dramatic scenes such as the one showing Seven riding the elevator to the top of the Saturn V so he could crawl out on the access arm to sabotage the vehicle.

Perhaps another reason why the Atlas was replaced in the episode may have been out of concern that using an older military weapon would not be as interesting to television viewers as seeing the Saturn V. The Atlas had been around since the mid-1950s. By replacing the Atlas, director Daniels may have sought to capitalize on the public interest in the newer and more powerful Saturn V launch vehicle. Also, in using the Saturn V, Daniels could also make use of the dramatic staging footage. Whatever the reasons, the final decision to show the same rocket that NASA would soon use to send humans to the moon but depicted as a military weapon was risky.

After reading the script, Joan Pearce, one of the script checkers employed by Kellam De Forest's Research Service (see chapter 2), voiced her concerns about this in a letter to Roddenberry: "This would constitute a deliberate violation by the United States of the newly signed—January 1967—Treaty for the Peaceful Exploration of Outer Space[,] which contains a specific prohibition against nuclear weapons in space." She concluded, "To dramatize such an event could bring serious repercussions from the United States government and might lead to discontinuation of NASA cooperation."³³

Despite the concerns voiced by Pearce, Roddenberry received word from the space agency indicating not only that they approved the script but also that they would provide "footage of rocket launchings and general Cape Kennedy stock footage." NASA even allowed the studio to shoot additional footage specifically for this episode through an effort led by NASA's Jerry O'Conner, who helped direct a Technicolor crew sent from Paramount studios in California to Cape Kennedy.³⁴

Days after the episode was finished and a little over a month before it was aired, Edward Milkis, *Star Trek*'s postproduction supervisor, received a letter from NASA's Walter E. Whitaker in which he signed off on the approval for the "use of NASA film scenes" in the episode. Whitaker also reaffirmed provisions that were previously agreed to, including that "the National Aeronautics and Space Administration or NASA will not be identified in the dialogue or picture, or in screen credits or publicity of any type related to the series. If any reference

CHAPTER SIX

must be made, they should be kept fictional such as the 'space agency' or 'federal space agency.'"[35] Even though NASA's Saturn V was depicted as a military weapon, Whitaker seemed very enthusiastic about the episode.[36]

While NASA supported use of their footage, there remained concerns about misrepresenting NASA and the US military, and the letter from Whitaker may have convinced Roddenberry to avoid direct mention of either in "Assignment: Earth." For instance, in an earlier draft of the script, the United States Air Force is referenced in several scenes, along with Cape Kennedy. However, in the final version, neither NASA nor the USAF is mentioned. Also, the Saturn V is never specifically called out by name, only described as a giant rocket "towering forty stories into the air."[37] By 1968, television audiences knew about NASA and were somewhat familiar with the Saturn V, having seen the launch of Apollo 4 on television just four months prior to the episode's airing. Still, Roddenberry had to be careful about how he depicted NASA if he wished to continue counting on the space agency's support. If he kept the script as it was and showed the Saturn V or any other identifiable NASA launch vehicle in a negative or controversial manner, he ran the risk of alienating not only the nation's space agency but also the US government. But that didn't happen. Perhaps motivated by the fact that NBC was interested in making "Assignment: Earth" into a pilot for a new series, Roddenberry ignored the advice given to him by Joan Pearce at De Forest Research and forged ahead with the *Star Trek* episode.

The episode "Assignment: Earth" is unique in the history *of Star Trek* since it was intended to be the pilot of a stand-alone television series. Spin-off series were not an entirely new concept in 1960s television. *The Andy Griffith* show began as an episode of *The Danny Thomas Show*; *Gomer Pyle, USMC* was spun off *The Andy Griffith Show*, and *Green Acres* got its start from *Petticoat Junction*. In November 1967, Roddenberry got the approval from NBC and Paramount to film "Assignment: Earth" as a pilot for a possible new series. However, when NBC viewed the completed *Star Trek* episode that was "Assignment: Earth," they decided not to green-light it as a new series.

It is difficult, from today's perspective, to read the minds of network executives in 1968. Recall that NBC tried to cancel *Star Trek* at the end of its second season, which would have made "Assignment: Earth" the final episode of the series. When the network gave in to fan protests and renewed *Star Trek* for a third season, the fate of "Assignment: Earth" as its own independent series was most likely sealed. It might have worked as a replacement for *Star Trek*, but if *Star Trek* was going to continue, another space series seemed redundant.

Because "Assignment: Earth" was technically a pilot in the eyes of the Writers Guild of America (WGA), writer Art Wallace got paid $14,000, plus residuals, which was considerably more than the going rate of $4,500 for an episode.[38]

Wallace continued dabbling in science fiction television, later writing episodes both for *Space: 1999* and *Planet of the Apes*. In reflecting back on "Assignment: Earth," Walters said, "I must admit that I've been a very busy writer throughout the years, and I didn't know *Star Trek* was going to be rerun 20 times. If I had known that I would have written more."[39a–c]

In a letter to James Webb, NASA's administrator, Roddenberry expressed his appreciation for all of NASA's support: "I have never received friendly and efficient cooperation anywhere near that provided by your Agency."[40] Even though NASA supported the series, it was reluctant to take any credit, since government agencies frown upon being seen participating in commercial ventures. Roddenberry, respecting this view, made it clear to everyone that NASA does "not wish to be mentioned, and we have assured them we will comply with their wishes."[41]

NASA, Alberta Moran, and Leonard Nimoy

Despite NASA's reluctance to take any credit, the space agency continued to work behind the scenes to help the fledgling new series. One of the biggest contributions they made to *Star Trek* began with the assistance of Alberta Moran. In 1967, Moran was the assistant to John F. Clark, then director of NASA's Goddard Space Flight Center in Greenbelt, Maryland. She was also one of the founding members of the National Space Club.

Comprising industry and government representatives along with educational institutions and the press, the National Space Club was originally founded as the National Rocket Club on October 4, 1957, the day Sputnik 1, the world's first artificial satellite, launched into orbit. The National Space Club sought to promote leadership in the burgeoning new field of astronautics, and one way they did that was to sponsor the Goddard Memorial Dinner. Still held annually today in Washington, the dinner is named after American space pioneer Robert H. Goddard and recognizes persons and institutions that have made outstanding contributions to space science and technology during the previous year. Moran chaired the special-events committee that was responsible for assembling the invitation list for the 1967 Goddard Memorial Dinner.

"We were at my house with Bob Hood, the president of the National Space Club," recalled Moran. Hood worked at Douglas Aircraft as a government sales representative and worked his way up to eventually become president in 1989 of McDonnell Douglas Missile Systems.[42] "We were getting the invitation list together for the Goddard dinner, and my oldest daughter, Penny, came in and said, 'Why don't we invite somebody from *Star Trek*?'"[43]

Alberta Moran's daughters—Pamela, Tracy, and Missy—all enjoyed watching *Star Trek*. "My friends and I never really had a *Star Trek* club," recalled Pamela of her interest in the show. She continued, explaining that "while growing up, my friends and I read a lot of science fiction and fantasy. We didn't watch a lot of television, but we did make sure to see *Star Trek*. After watching an episode, my friends and I would get together and figure out who our favorite characters were, talk about what we liked and didn't like."[44]

Moran took her daughter's suggestion and first tried to invite Gene Roddenberry to speak. Unfortunately, he couldn't make it.[45a-b] Neither could William Shatner. But not all was lost! An excited Moran told her daughters, "I couldn't get Captain Kirk, but I got Mr. Spock."[46]

Although Nimoy wasn't Alberta Moran's first choice, in some ways obtaining him for the dinner was actually a bigger coup. Spock was becoming the breakout character on the show, something that for a while rankled Shatner. Spock was particularly popular among women.

On March 14, 1967, just a few days after NBC formally announced that *Star Trek* had been renewed for a second season, Leonard Nimoy landed at Dulles Airport outside Washington, DC. Joining Moran to meet him was daughter Penny and two of her friends, Anne Moroz and Lynn Retter, all of whom were thrilled over the prospect of finally meeting Mr. Spock in person. After picking up Nimoy from the airport, they drove him to the hotel, where he was besieged by fans.

The next morning, Nimoy's wife, Sandy, joined him for a tour of NASA's Goddard Space Flight Center. Upon arriving, both he and his wife discovered that a major part of the population of Goddard, including secretaries and scientists, were there to greet them. During his visit, the tour guide pointed to an actual spacecraft control panel and said, "You will notice, Mr. Nimoy, that this apparatus is the real thing. We know this because it is almost a duplicate of the one you use on your spaceship in the series."[47] Nimoy recalled of his visit, "This was the first real taste that I had of the NASA attitude towards *Star Trek*."[48]

That evening, Nimoy attended the Goddard dinner. All the top space people from NASA and its contractors were at the black-tie affair, including astronaut John Glenn, the first American to orbit the earth. Given the fact that it was barely six weeks after the tragic Apollo 1 fire that took the lives of astronauts Roger Chaffee, Ed White, and Gus Grissom, the mood among those present was tempered by the realization that the moon landing would be delayed. Some even speculated the Apollo program would be canceled altogether.

At the reception, Nimoy was introduced to those who would share the front table with him. In addition to astronaut Glenn, those seated at the dais included NASA administrator James Webb, Robert Goddard's widow Esther C. Kisk Goddard, and Vice President Hubert Humphrey.[49] The memory of the accident was still very much in the minds of everyone that evening. As the vice president launched into his speech, he acknowledged to the 1,500 attending guests that "we cannot help but keep in mind that this is an occasion for remembrance with sorrow, and also with great pride of three gallant men and three of my friends and your friends."[50]

At the end of the vice president's speech, people gathered around Nimoy. "They were all most cordial," wrote Nimoy of the event. "All of them wanted autographs and pictures for themselves and their children."[51] Moran recalled how Webb, Glenn, and Humphrey were very excited to get Nimoy's autograph.[52] "They were not getting Nimoy's autograph," recalled Penny, who worked to help check in all the guests that attended that evening. "They were getting Mr. Spock's. He was so gracious to everyone who approached him."[53]

At the time of the dinner, men had been flying into space for less than seven years. *Star Trek*, on the other hand, showed people exploring the galaxy hundreds of years into the future. Was such a future possible? In the wake of the Apollo 1 tragedy, many questioned if going to the moon was still worth the risk, let alone travel to other planets or beyond.

But that evening, the space community looked beyond the recent disaster. Those shaking Nimoy's hand and receiving his autograph didn't see an actor in their midst, but the one and only Mr. Spock, a representative from the future who knew that NASA's work in the twentieth century may have been difficult, and even painful, but it was necessary for space exploration to move forward. For the half-million workers employed to support the Apollo program during one of the darkest hours in NASA's history, *Star Trek* gave them hope.

Nimoy wrote of that evening, "I do not overstate the fact when I tell you that the interest in the show is so intense, that it would almost seem they feel we are a dramatization of the future of their space program, and they have completely taken us to heart . . . they are, in fact, proud of the show as though in some way it represents them."[54]

The promotional value that Nimoy's attendance at the Goddard dinner gave to *Star Trek* was quickly recognized by the studio. On April 17, 1967, and just a few short weeks before Emmy nominations were formally announced, Desilu Productions took out a full-page ad in *Daily Variety*. The ad appeared next to another full-page ad for Desilu's *Mission: Impossible*. Together, the ads made an eye-catching two-page spread that included the message "Desilu Productions, Inc. respectfully submits these series for consideration by the Members of the National Academy of Television Arts and Sciences." There were two ads for two different shows, but both had the same goal: to win a coveted Emmy.

The ad for *Star Trek* featured an image of the *Enterprise* "warping" out of the page toward the reader. Appearing in the upper left portion of the ad was a reproduction of an award plaque given to Roddenberry and *Star Trek* "For Distinguished Contribution to Science Fiction." This was given to him by Ben

CHAPTER SIX

Jason, the chairman of the 1966 World Science Fiction Convention. Quotes about the show from various individuals were shown against a backdrop of stars. NASA's Alberta Moran was one of these quoted in the ad saying that *Star Trek* is an "Unsurpassed Production."[55]

Like the mention of NASA in the *Variety* ad, the network continued to make use of the stature afforded by the show's recognition by the space agency. In May 1967, *TV Guide* printed an open letter by Paul L. Klein, vice president of audience measurement for NBC, who wrote, "*Star Trek* is the only science-fiction show on television with a scientific basis. . . . I was instrumental in recommending *Star Trek* for the NBC schedule and have been one of the show's staunchest supporters during the agony of renewal time. Messrs. Roddenberry, Coon, and the whole *Star Trek* staff have deserved the public's approval, NBC's faith in them, and, as topping to the cake, were just recently honored by the National Space Club in Washington for their scientific validity."[56]

NBC thought enough about Nimoy's attendance at the Goddard Memorial Dinner to include mention of it in a sales brochure that their marketing team assembled to promote the "2nd Year" of *Star Trek*. This was sent out in June 1967 to network salesmen and managers to help promote the second season. The slick six-page printed color pullout included numerous clips from the series along with market-driven text that included the following: "This unique NBC-TV adventure series was accorded another unusual distinction when Leonard Nimoy was invited by the National Space Club to represent STAR TREK as the guest of honor on a recent extensive tour of the Goddard Space Flight Center in Washington, DC. The tour was followed by a dinner for 1,500 club members, at which Vice President Humphrey was the principal speaker."[57]

Moran continued to stay in touch with both Roddenberry and Nimoy long after the Goddard dinner. "I remember that Mother would keep in touch with them both, especially if they happened to be visiting the DC area," recalled Penny. "One evening Roddenberry came to our house for dinner. It happened to be the same night that *Star Trek* was on. Mom said that he wanted to come over because he didn't want to watch the show alone in his hotel room. It was interesting because he said he wanted to see what our reactions were to the episode. It was the first time that I met him, and it was all very casual and fun. While we were watching, he tried to call the set to have William Shatner speak to us, but they wouldn't let him talk to him because they didn't believe it was Gene Roddenberry on the phone. He ended up calling Majel Barrett [Roddenberry's second wife and the actress who played Nurse Chapel] and said to her, 'Would you please tell them that I actually am Gene Roddenberry!'"[58]

Moran later sent an invitation to Nimoy to have him come as NASA's guest to view a rocket launch at the agency's Western Test Range at Vandenberg Air

Force Base in California.⁵⁹ Nimoy was not able to go, so Roddenberry came instead and he brought his youngest daughter, Dawn, along with him.

"I remember Mom saying, 'Okay, we're going to California to see a launch,'" recalled Penny. "I had not seen a rocket launch up to that point in person, so I thought this was very, very cool. Then Mom added, 'By the way, we've been invited to Dawn's fifteenth birthday party. Oh, and we get to go and visit the *Star Trek* set.'"⁶⁰

Penny fondly recalled "going to Dawn's birthday party, which was my one and only Hollywood party. I spent most of that evening visiting with James Doohan. Gene Roddenberry said that I would be safe with Scotty. I was totally overwhelmed by it all, and the people were all very nice."⁶¹

The rocket launch was early on the morning of August 16, 1968. Roddenberry and Dawn arrived in a limo to tour the range before viewing the launch. Pictures that Penny took show Roddenberry and Alberta talking with NASA officials.

At 3:30 a.m., the Delta rocket lifted off the launchpad, carrying a satellite built by RCA designed to monitor weather and send regular television images of the earth from orbit.⁶² Ironically, RCA owned NBC, which paid Desilu to produce *Star Trek*. RCA wanted to sell lots of new color television sets. It was no coincidence that NBC's colorful peacock mascot was intentionally designed to cater to color television viewers. In addition, the bright and vibrant colors shown in the sets, costumes and props of every episode combined with print ads that featured *Star Trek*, encouraged sales of color televisions, preferably sets made by RCA.

Afterward, they all visited Paramount. "Sometime after the launch we went out to tour the *Star Trek* sets, which was very cool," said Penny. "There were no lights, no cameras, no action. We got to see the chairs and consoles up close. I also got to meet the actor Michael Douglas, who happened to be visiting the set at the same time that day."⁶³

In reflecting back during Moran's time working for NASA, Penny said, "Our mother was a force to be reckoned with. How many other women would have written a letter to Hollywood inviting someone from *Star Trek* to come to the Goddard Memorial Dinner? How many people would have thought that such a person would be well received within that community?"⁶⁴

"I have seldom seen one of our people return from a trip so pleased and so enthusiastic about those he met there," wrote Roddenberry in a letter to Alberta after Nimoy returned from the dinner.⁶⁵ In this letter, Roddenberry acknowledged that "we will be more than happy to work more closely with you folks, with NASA and with others in the space industry. You can be certain we will include in upcoming shows some 'plugs' about the importance of the funds, the planning and the rest of the 'pioneer' effort that made star travel

CHAPTER SIX

eventually possible. We feel sincerely about this and hope our show helps establish throughout the country the certainty of this future and the necessity of adequately planning for it."[66]

Leonard Nimoy's invitation and attendance at the Goddard Memorial Dinner was more than a confirmation of support by the aerospace community and NASA for the new television series. It was an affirmation of Roddenberry's efforts to work with them throughout the series' production. In addition, it appears to have been Nimoy's attendance at this dinner that launched years of support and collaboration between *Star Trek* and the Smithsonian, a partnership that, as we will learn in the next chapter, proved beneficial for both.

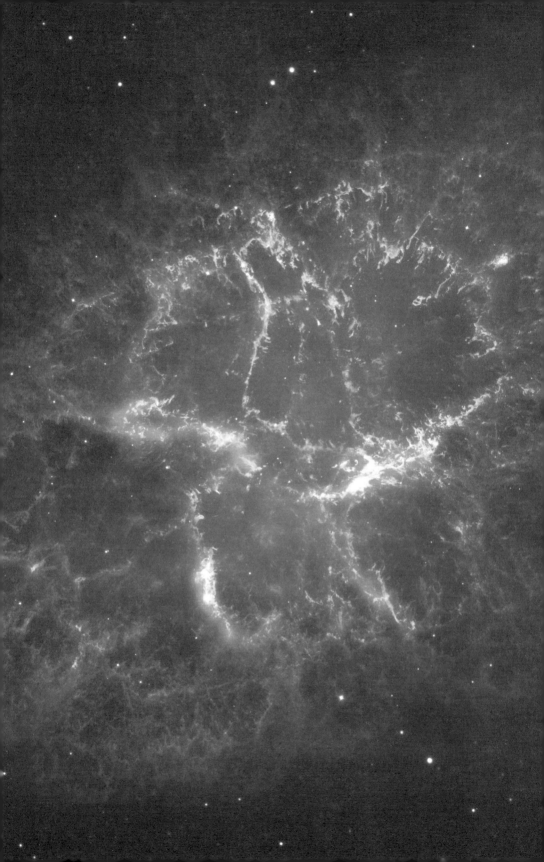

This Franklin Mint twentieth-fifth-anniversary edition of the USS *Enterprise*, which was originally issued in 1991, was aboard the United States Navy's nuclear aircraft carrier USS *Enterprise* (CVN-65). The 15-inch model was transferred to the Naval History and Heritage Command in 2006, where it remains part of their collections. *Photo courtesy of the US Navy and Naval History and Heritage Command*

US Air Force Academy Cadet Squadron 19, now Starship 19, members show the new squadron flag after the redesignation ceremony held at Polaris Hall on August 12, 2023. *Photo by Trever Cokley, courtesy of the United States Air Force Academy*

The original US Air Force Academy Cadet Squadron 19, Starship 19 patch (*bottom*) from 1976 includes the *Star Trek* phrase "Where No Man Has Gone Before." The revised patch (*upper left*) has been modernized with the new motto in Latin, which translates to "Live Long and Prosper." The Operational Camouflage Patch (OCP) version of the patch is shown in the upper right. *Photos courtesy of the United States Air Force Academy*

During the twenty-fifth anniversary of Apollo 11, this giant plaque listing the names of all the aerospace companies that contributed hardware and expertise to the first manned lunar-landing mission was presented by the American Institute of Aeronautics and Astronautics to the Apollo 11 crew on July 20, 1994. Photo was taken on May 11, 2019, by the author at the Richard Nixon Presidential Library and Museum, Yorba Linda, California. At the time that this photo was taken, a special exhibit called *Apollo 11: One Giant Leap for Mankind* featured this plaque. The exhibit ran from April 29, 2019, through January 12, 2020, in honor of the fiftieth anniversary of Apollo 11. *Photo courtesy of the author*

The Thompson, Ramo and Woolridge (TRW) headquarters building at Redondo Beach, California, served as an off-world filming location for *Star Trek*'s twenty-third-century Deneva colony. This aerial photo taken in 1965 shows facilities for data reduction, computing, and research and development. The building in the rear is for manufacturing and testing of spacecraft. *Photo courtesy of the author*

Shown in this scene from the *Star Trek* episode "Operation: Annihilate!," one can see instruments used in radio frequency testing of spacecraft mounted to the rooftop of one of the buildings that occupy TRW's Space Park. A rooftop mast with a polyhedron-shaped object attached to its tip can be seen erected behind Mr. Spock. This object has the profile of a satellite from the Vela program, on which TRW served as the main contractor. *Photo courtesy of CBS/Paramount and NASA*

Scenes of TRW's Space Park campus during the time of *Star Trek* and as it looks today. Several photos were taken on February 15, 1967, when the *Star Trek* cast were on location filming the episode "Operation: Annihilate!" The upper left photo shows *Star Trek* fan Rob McFarlane on the Space Park campus with an unexpected visitor hanging nearby. The lower left photo is Schoenberg Hall, a building located on the campus of the University of California, Los Angeles (UCLA), which was used as Sam Kirk's laboratory in the episode. *Photos courtesy of David Arland, Mickey Snelson, Rob MacFarlane, and the author.*

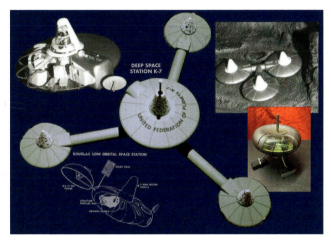

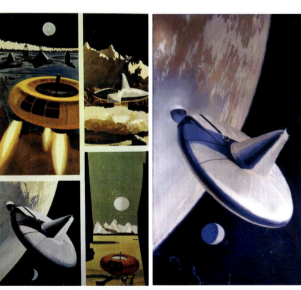

A 1959 Douglas Aircraft report features artwork by George Akimoto. His work also appears on the cover of the proceedings of a Manned Space Stations symposium that was cosponsored in 1960 by NASA and RAND. Roddenberry eventually got ahold of the models of the spacecraft design, which were then used to build the K-7 space station filming miniature that first appeared in "The Trouble with Tribbles." The eight tribbles are all screen-used. The large, long-haired, brown-and-gray tribble shown in the upper right matches the one used by Kirk to discover the Klingon spy Arne Darvin. The rust-colored tribble directly beneath the brown-and-gray tribble has a windup beetle toy inside and matches the tribble in the scene that crawls over to eat Chekov's grain. Documents courtesy of the author, with tribbles from the personal collection of Bill Kobylak. The K-7 model is the author's AMT model kit, which he filmed in 1978 at age fifteen for a planetarium show. *Photos courtesy of Bill Kobylak, Mickey Snelson, and the author*

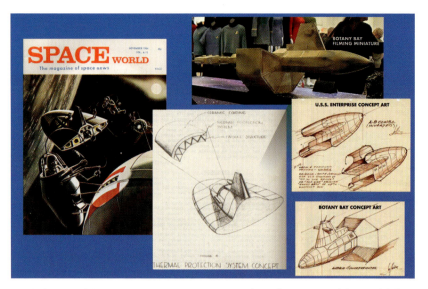

The same 1960 Manned Space Stations symposium proceedings that contained the model that would become the K-7 also included a paper that may have inspired the design of the *Botany Bay* as seen in the episode "Space Seed." The November 1964 issue of *Space World* magazine also features artwork using a similar design. *Photos courtesy of the author and Karl Tate*

Above right: The first page of a two-page promotional letter sent out by Desilu through McFadden, Strauss, Eddy, and Irwin (MSEI), the studio's public-relations agency. The photo inset to the left of the letter shows an actual tektite. Pieces of this "genuine antique from outer space" were sent out along with the two-page letter to help promote *Star Trek*. Photos courtesy of the author's collection.

Above left: One of the cardboard stand-ups that MSEI sent out in 1966 to select NBC affiliates as part of an advance media campaign to promote the new series. The fact that this promotion item featured a physical cutout of the USS *Enterprise* and not something depicting cast members is an interesting choice. Matt Jefferies's design was unique and looked contrary to what television viewers were accustomed to seeing at the time. This stand-up belongs to the University of Wyoming's American Heritage Center archive collections and was photographed by the author during a research visit in June 2024.

Launched in June 1967, Mariner V flew within 4,000 kilometers of Venus. Here Mariner V spacecraft managers Ted Parker (*left*) and Dan Schneiderman (*right*) are shown wearing what appear to be "Spock" ears in JPL's mission control during the spacecraft's closest approach to the planet on October 19, 1967. Colleague Dave Shaw looks over their shoulders. *Photo courtesy of NASA/JPL, photo P-8218A*

The cast and production crew from *Star Trek* toured NASA's Dryden Flight Research Center at Edwards Air Force Base, California, on April 13, 1967. Shown are, *left to right*, unknown suited person; actor James "Jimmy" Montgomery Doohan, who played chief engineer Montgomery Scott or "Scotty"; Walter Matthew "Matt" Jefferies Jr., art director and production designer for the series; Herbert Schlosser, who at that time was head of NBC's programming and oversaw *Star Trek*'s development during the network's production of the series; *Star Trek* creator Gene Roddenberry; assistant director, production manager, and associate producer Gregg Peters; series director Marc Daniels; assistant director and producer Robert "Bob" Harris Justman; actor Jackson DeForest Kelley, who played chief medical officer Leonard "Bones" McCoy; director Joseph "Joe" Pevney; and unknown. Shown behind the cast is NASA's HL-10 experimental lifting body. *NASA photo E67-16643-4*

An unknown suited official (*to the left*) speaks to Roddenberry, Doohan, and Kelley in front of Northrop's HL-10 experimental lifting body. *NASA photo E67-16643-3, https://images.nasa.gov/details/E67-16643-3*

A slightly different shot of the same photo as shown above. This photo appeared in Stephen Whitfield's *The Making of Star Trek*. *NASA photo from author's collection*

De Forest Kelley is shown talking to NASA test pilot Bill Dana in front of X-15-1 (vehicle no. 56-6670). *NASA Photo E67-16643-1, https://images.nasa.gov/details/E67-16643-1*

James Doohan is shown talking with NASA test pilot Bruce Peterson. Behind them is the Northrop M2-F2, one of several experimental wingless research aircraft developed for NASA. *NASA Photo E67-16643-2, https://images.nasa.gov/details/E67-16643-2*

De Forest Kelley is shown standing next to Herbert Schlosser with X-15-1 (vehicle no. 56-6670) in the background. Both are listening to an unknown suited official giving a briefing at Dryden. *NASA photo E67-16643-6, https://images.nasa.gov/details/E67-16643-6*

Unknown person standing to the far left is joined by Herbert Schlosser, Joe Pevney, DeForest Kelley (*standing in foreground*), Gene Roddenberry, Matt Jefferies, and Robert Justman. All are listening to a briefing given by an unknown suited individual at NASA's Dryden Flight Research Center on April 13, 1967. *NASA photo E67-16643-7*

DeForest Kelley is shown inspecting X-15-1 (vehicle no. 56-6670) with Gene Roddenberry, Joe Pevney, Bob Justman, and Matt Jefferies (*back to camera*). *NASA photo E67-16643-5, https://images.nasa.gov/details/E67-16643-5*

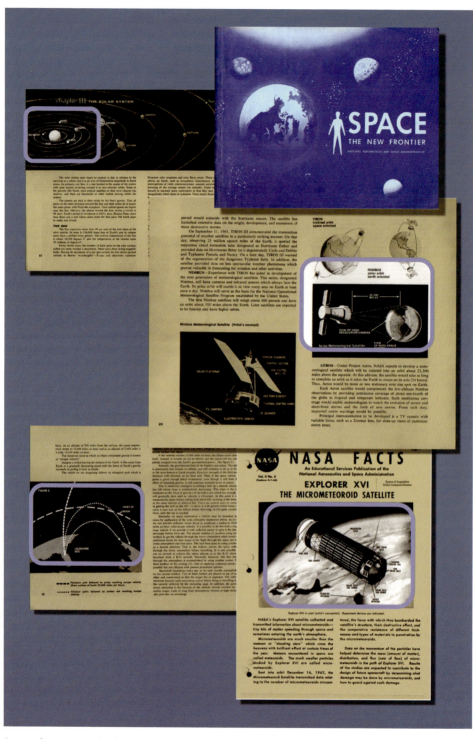

Scenes from *Star Trek*'s first pilot, "The Cage," featured numerous images from various NASA publications. *Photos courtesy of NASA*

Leonard Nimoy is shown on March 14, 1967, after arriving at Dulles International Airport outside of Washington, DC, to tour NASA and attend the tenth annual Goddard Memorial Dinner. Shown left to right are: Anne Moroz, Penny Moran, Lynn Retter, Alberta Moran, and Leonard Nimoy. *Photo courtesy of NASA, photo G-67-6896*

Bottom left photo shows Leonard Nimoy and his wife Sandy touring NASA Goddard while listening to Robert C. Baumann (*right*), head of the Spacecraft Integration and Rocket Division and Paul Henley. Upper left photo is astronaut Col. John Glenn and his wife Annie talking with NASA administrator James Webb during the tenth annual Goddard Memorial Dinner held at the Sheraton Park Hotel in Washington, DC, on March 15, 1967. Large photo shows Alberta Moran briefing Leonard Nimoy before speaking at NASA Goddard on March 15, 1967. Photo below right is Vice President Hubert H. Humphrey, along with the Chairman of the National Aeronautics and Space Council, talking with Mrs. Robert H. Goddard, and Congressman George P. Miller, US House of Representatives at the Goddard Memorial Dinner. *All photos courtesy of NASA and Alberta Moran.*

The episode "The Cloud Minders" makes use of a photograph taken by Gemini 4 astronauts, which was published in the NASA book *Earth Photographs from Gemini III, IV, and V*. The photo is of a dry river basin in the southern part of Saudi Arabia. *Photos courtesy of NASA photo S65-34658, and the author*

This unaired clip shows a different scene from "Assignment: Earth." Using footage shot at Cape Kennedy, the still shows an Atlas-Agena ready for launch instead of a Saturn V. The composite image is shown blended with a scene filmed on Paramount's lot. You can see a soundstage just off to the left. *Unaired film trim from the Matt Jefferies collection, courtesy of Gerald Gurian*

NASA's Alberta Moran and Gene Roddenberry are shown being briefed by NASA officials at Vandenberg Air Force Base prior to viewing the launch of the ESSA-7 meteorological satellite during the early-morning hours of August 16, 1968. *Photo courtesy of Alberta Moran*

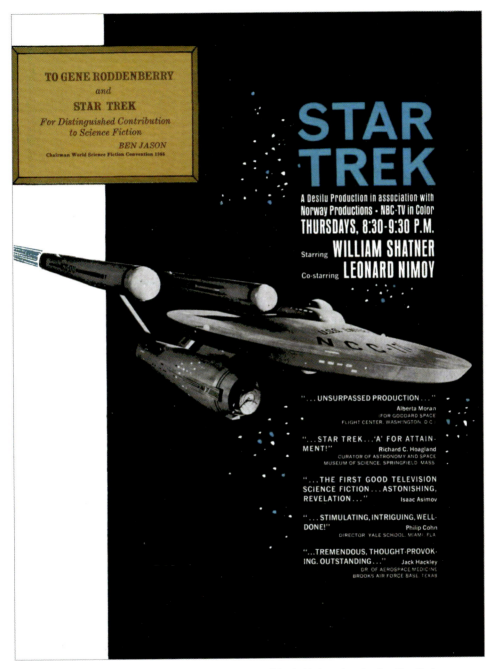

A full-page ad taken out in the March 14, 1967, issue of *Daily Variety* shortly after *Star Trek* was renewed by NBC for a second season. The same ad also served to promote the series for Emmy consideration. *Photo courtesy of the* Daily Variety

a thinking man's adventure series

Returning for its second year as an NBC-TV color attraction, STAR TREK will take off for its weekly adventures in interstellar space from a new 8:30-9:30 PM/NYT Friday launching pad. In this location, it will benefit from the lead-in strength of the Tarzan series, which paced NBC's new entries in the 1966-67 ratings sweepstakes.

A viewing favorite among young adults (18-34) and teenagers, STAR TREK has averaged a 36 Q-score this past season to rank in the TV-Q Top 20 among these groups (and also in the total sample). The NTI Full Analysis Report also shows that it slants strongly toward better-educated, middle and upper income families in urban areas.

one of TV's most loyal followings

Second only to The Monkees among NBC programs in the volume of fan mail it attracts, STAR TREK has been the recipient of more than 28,000 letters of support and encouragement from one of the most loyal and articulate viewer followings currently attracted by any television series. This mail comes from a wide diversity of sources—space technicians, college professors, college students, businessmen, housewives, and high school and grade school students. For the most part, the letters of these STAR TREK enthusiasts are distinguished by their thoughtful and literate content and by their stress on the believability of the series and the high quality of the acting and writing.

five emmy nominations

Singled out for Emmy nominations in five different categories, including "outstanding dramatic series," STAR TREK was one of the most honored new series of the past season. Other show and talent categories in which it received 1966-67 award consideration were "outstanding supporting performance by a dramatic actor" (Leonard Nimoy as Mr. Spock), Photographic Special Effects, Mechanical Special Effects, and Film and Sound Editing.

This unique NBC-TV adventure series was accorded another unusual distinction when Leonard Nimoy was invited by the National Space Club to represent STAR TREK as the guest of honor on a recent extensive tour of the Goddard Space Flight Center in Washington, D.C. The tour was followed by a dinner for 1500 club members, at which Vice President Humphrey was the principal speaker.

NBC thought enough about Leonard Nimoy's 1967 tour of NASA's Goddard Space Flight Center and his attendance at the tenth Annual Goddard Memorial Dinner to include mention of them both in a sales brochure that their marketing team assembled to promote the second season of *Star Trek*. *Photo courtesy of NBC/Desilu and Bill Kobylak*

Irwin Allen standing in front of the Apollo 8 command module in Downey, California. The spacecraft, which carried astronauts Frank Borman, Jim Lovell, and Bill Anders around the moon in 1968, had just arrived in California in the fall of 1970 after having been on display at the US pavilion during Expo '70, the Japan World Exposition in Osaka. Allen was a film and television producer who created and directed the CBS television series *Lost in Space* which ran from 1965 to 1968.

NASA astronaut Gordon Cooper visits the cast and set of Irwin Allen's television series *Land of the Giants*, which ran on CBS from 1968 to 1970. Shown to the left is Don Marshall. Standing to the right of astronaut Cooper are Irwin Allen and actress Deana Lund. *Photos courtesy of Synthesis Entertainment*

Richard K. Preston, the Smithsonian messenger who in 1967 helped secure a copy of *Star Trek*'s second pilot for the National Air and Space Museum and later helped start the campaign to change the name of NASA's first space shuttle from *Constitution* to *Enterprise*

Preston in his later years. *Photos courtesy of Richard Preston's children, Larkin, Richard, and Bannon*

Gene Roddenberry (*third from left*) is shown standing with Fred Durant to his right during an April 24, 1973, black-tie reception held at Baird Auditorium in the Smithsonian's National Museum of Natural History as part of the opening ceremonies for "The Nature of Scientific Discovery" symposium. This was held at the Shoreham Hotel in Washington, DC, in honor of the five hundredth anniversary of the birth of Nicholas Copernicus. The event was just one of many held during the weeklong observance (April 22-26, 1973) of the history-making Polish-born scientist that was sponsored by the National Academy of Sciences and the Smithsonian Institution in cooperation with the Copernicus Society of America and the US National Commission for UNESCO. *Photo courtesy of Smithsonian Institution Archives, acc. 11-009, image 73-3981-26A*

Smithsonian Advisory Board members to the National Air Museum (NAM) are shown on May 2, 1966, viewing a planning model of the proposed National Air and Space Museum building. The building opened in 1976. *Left to right*: museum director S. Paul Johnston, astronaut Col. John H. Glenn Jr., and NAM assistant director of astronautics Frederick C. Durant III. *Photo courtesy of Smithsonian Institution Archives, acc. 11-008, box 04, image OPA-8762R2-C*

On the evening of March 29, 1979, Gene Roddenberry spoke at the University of Maryland. Earlier that day he stopped by the Smithsonian. Shown here is Roddenberry (*left*) standing inside the backup Skylab orbital workshop on display at the National Air and Space Museum. Also pictured is Smithsonian staff member Greg Kennedy (*center*) and Fred Durant (*right*). *Photo courtesy of Greg Kennedy*

On August 28, 1967, during a formal ceremony held at the Smithsonian's Arts and Industries building, S. Paul Johnston, director of the Smithsonian's National Air and Space Museum, is presented with a copy of *Star Trek*'s second pilot, "Where No Man Has Gone Before." Gene Roddenberry, the series creator and executive producer, is shown along with his first wife, Eileen-Anita Rexroat, handing over a 16 mm copy of the film. *Photo courtesy of Smithsonian National Air and Space Museum, NASM photo 00163636*

Newspaper articles announcing the *Star Trek* pilot going to the Smithsonian

CHAPTER 7

The Smithsonian and *Star Trek*

As described in the last chapter, on the evening of March 15, 1967, the National Space Club's annual Goddard Memorial Dinner was held in Washington with Leonard Nimoy at the head table. After Vice President Hubert Humphrey concluded his speech at the gala event, NASA officials, congressmen, and the vice president approached "Mr. Spock" for pictures and autographs. Among those in attendance at the black-tie affair was Richard Preston. An administrative assistant with the Smithsonian's National Air and Space Museum, Preston patiently waited his turn to approach Nimoy and deliver an important message: the Smithsonian wanted a copy of *Star Trek*'s second pilot.

The story of *Star Trek*'s struggle to get on the air is well known. Roddenberry first tried to sell his new series to network executives with the production of a pilot called "The Cage." It failed to impress NBC executives, who reportedly called it too cerebral, too intellectual, and too slow. Rather than reject the series outright, the network took the rare but not unprecedented step of ordering a second pilot.[1]

The show's second pilot was called "Where No Man Has Gone Before" and, for the first time, introduced William Shatner as James T. Kirk, the ship's captain. This new pilot was accepted by NBC, thereby allowing Desilu Productions, the studio responsible for *Star Trek*, to move forward with making the series. Since the first pilot never aired intact and Kirk appeared only in the second one, this may have been among the reasons why the Smithsonian chose to add this pilot to its collections rather than the first. What is clear is the fact that prior to this time, no other show in the history of television had ever had a copy of its pilot requested by the Smithsonian. Having *Star Trek* added to the "nation's attic" was, by all accounts, a unique and distinguished honor.[2]

Richard K. Preston

Richard Knowlton Preston Sr. was born in Winthrop, Massachusetts. Growing up, he joined the ROTC and upon graduation from high school went to the University of Maryland, where he earned an undergraduate degree in education.

After three years of service in the Marines, Preston became a civil servant, working first for NASA and then for the Department of the Interior. In January 1967 he came to work for the Smithsonian, serving as an administrative assistant under S. Paul Johnston, the director of the Smithsonian's National Air and Space Museum.[3]

"Dad loved to learn and was a good student in spite of having an extreme learning disability," said daughter Larkin during an interview about her father. "He had a form of dyslexia but was extremely smart. All through college, my mom tutored him orally and wrote all of his papers that he would dictate to her. I should point out that my mom also went to college at the same time as my dad to get her law school degree."[4]

Preston's son, Richard K. Preston II, recalled growing up in a very cultured and creative environment. "He very much wanted to fly planes, but his height combined with his extreme learning disabilities prevented that" said Richard II during an interview. "He was an artist and very creative. He did mixed media, mostly sketches. He did some painting and sculpting and took his sketchbook everywhere. He also enjoyed photography. He was always interested in mechanical things, avionics, rocketry, and space travel and delving into something that he found fascinating."[5]

"I think he was a frustrated teacher," said Richard II. "He wanted us to know about the planets and would read us space stories. I remember that he took us into a closet with a screw-in lightbulb. When he was talking about Mars, he put in a red lightbulb. When he was talking about Pluto, he put in a maroon lightbulb. I think he was passionate about space and encouraging us to use our imaginations."[6]

Preston was also an avid collector of science fiction. He had a huge collection of vintage science fiction toys and thousands of books. It then seemed natural that he should take an interest in *Star Trek*. "We all watched *Star Trek* together growing up," said Richard II. "All three of us kids have probably seen every single episode at least a dozen times."[7] Bannon, one of the elder Preston's daughters said, "I have some of the shows memorized. My ex-husband was also a big *Star Trek* fan."[8]

Because of Preston's interest in *Star Trek*, it seemed that he was the obvious choice to attend the Goddard Memorial Dinner and approach Nimoy on behalf of the Smithsonian with the request for the show's second pilot. But Preston was only the messenger. The person who initiated that message was Frederick C. Durant III, the National Air and Space Museum's assistant director for astronautics.

Frederick Clark Durant III

Before his death in 2015 at age ninety-eight, Frederick Clark Durant III was fortunate enough not only to witness the birth of the space age but also to play an active role in its development.

CHAPTER SEVEN

After receiving a degree in chemical engineering from Lehigh University in 1939, Durant enlisted in the Navy. A peptic ulcer prevented him from being deployed overseas during World War II, but that did not keep him from serving at home. Trained as a naval aviator, he worked as a flight instructor, teaching new cadets how to make carrier landings on training ships in the Great Lakes. Upon retiring from the Navy, Durant went on to serve as a consultant and engineer in various research labs, including stints with the Department of Defense. His new arena of combat during the Cold War was outer space, where he worked for the CIA analyzing Soviet missiles.

Durant was an ardent promoter of spaceflight and served in a variety of positions in newly formed organizations that supported the nascent field of astronautics. He was president of the American Rocket Society, now known as the American Institute of Aeronautics and Astronautics, and spearheaded the growth of the International Astronautical Federation, an international organization devoted to space advocacy. He became a fellow of innumerable other astronautical societies, many of which he helped form. He was good friends with noted scientist and author Arthur C. Clarke and helped Wernher von Braun launch Explorer 1, America's first artificial satellite, in 1958.

In 1965, Durant joined the staff of the Smithsonian Institution's National Air and Space Museum after serving as a consultant.[9] Durant's extensive knowledge and experience impressed many of his fellow museum workers. Over the next fifteen years until his retirement in 1980, Durant became a valuable resource as the museum grew and expanded.

Durant was a close personal friend of space artist Chesley Bonestell, having hosted him many times as a guest in his Chevy Chase, Maryland, home. There the two would talk about the future of space travel. Durant, like the notable painter Bonestell, also enjoyed collecting art.

"He had a side business called Space Art International that promoted, displayed, and arranged shows and exhibitions for artists in the space genre," recalled Durant's son Steve during an interview. "He was always interested in art. Back when I was fourteen, during our annual summer trek through Massachusetts, we would go up into Vermont. One night we had dinner with some old man named Norman Rockwell. It didn't sink in until I was a little bit older how important that dinner was. To actually sit down with Norman Rockwell in his studio and talk to him was amazing."[10]

Steve described how Fred had paintings everywhere in their home. "We'd go to bed at night, and there, in the hallway, would be a blank wall devoid of any sort of particular painting that had been there for months prior. You'd come down and there's something else in its place because that one had been shipped off with others to an art show in Japan or on display in New York."[11]

Durant had excellent collecting skills, and he had been around long enough to know lots of people . . . important people. Aerospace historian Randy Lieberman, who knew Durant during his later years, recalled that "with a telephone call, he could bring together top brass or captains of industry."[12]

"My dad was a very old friend of Arthur C. Clarke, and it may have been through him that he first learned about Gene Roddenberry," said Steve when asked how his father first came to meet the creator of *Star Trek*. "Dad was very well connected. He always had notable individuals visiting him at his office as well as coming over to our home."[13] During the fifteen years that he worked at the Smithsonian, Durant kept several guest registry books. After Durant died, Steve kept one, and it is an incredible artifact. Its ninety-six pages contain the signatures and comments of more than two thousand people who passed through Fred's office and home. "This book captures thirty years of space history. Everyone who was anyone in this field knew my dad at one time and visited with him,"[14] said Steve. Those who signed it include such notables as Charles Lindberg, Norman Rockwell, and Carl Sagan, along with many actors, celebrities, and astronauts. Arthur C. Clarke signed it many times, as did Gene Roddenberry.

If Durant and Roddenberry did not know each other previously, they most certainly came to know each other as a result of Richard Preston's meeting with Nimoy during that evening of the Goddard Memorial Dinner in 1967. A few days after the dinner, NASA's Alberta Moran, who was discussed at length in the previous chapter, wrote to Nimoy, reminding him that "during the lecture you talked with someone from the Smithsonian Institution relative to an interest in obtaining perhaps a pilot film from *Star Trek* to be placed in the Air and Space Museum." She also included Preston's mailing address.[15]

As a result of the Goddard dinner and Preston meeting Nimoy, a series of letters and phone calls were exchanged between Durant and Roddenberry to secure a copy of the show's second pilot for the Smithsonian.[16] "We believe the *Star Trek* pilot would be a valuable addition to the archives of the National Air and Space Museum. Science Fiction forms a definite segment of astronautics chronology,"[17] wrote Preston in a June 1, 1967, letter to Roddenberry.

During further exchanges of letters, final details of the film's official transfer to the Smithsonian began to emerge. The media also picked up on the news about the Smithsonian request.[18] In an August 16, 1967, letter to Preston, Roddenberry indicated that he had sent Durant a copy of the AMT *Enterprise* model kit and was looking forward to the pilot's presentation ceremony later that month.[19]

On August 28, 1967, Roddenberry presented a 16 mm color print of *Star Trek*'s second pilot to S. Paul Johnston,[20a–b] then director of the National Air and Space Museum. The formal press event was held in the Smithsonian's Arts

◼ CHAPTER SEVEN ▶

and Industries Building against a backdrop of a full-sized command module mockup from NASA's Apollo Program.[21a–g]

The news that *Star Trek* was enshrined in the Smithsonian, as well as the story leading up to that, was good publicity for the series. Roddenberry knew that the Smithsonian's request for the second pilot that Durant helped orchestrate could bolster the credibility of the show, and it did. The series was renewed for a second season, and, after additional help through an organized letter-writing campaign in which fans inundated NBC with tens of thousands of letters in support of the series, the show made it to a third season.

On March 1, 1968, at 9:28 p.m., following the first-run airing of the *Star Trek* episode "The Omega Glory," an NBC announcer came on the air telling viewers that the series would continue to be seen on television that fall. Seven days after NBC announced *Star Trek*'s third-season renewal, Roddenberry sent Fred Durant the following telegram: "THE HONOR THE SMITHSONIAN BESTOWED ON US HELPED GREATLY IN THE RENEWAL OF STAR TREK FOR ANOTHER SEASON. THANKS FOR ALL YOUR HELP. WE'LL DO OUR BEST TO MAINTAIN QUALITY SHOWS. OUR MOST SINCERE THANKS. WARMEST PERSONAL REGARDS GENE RODDENBERRY STAR TREK."[22] The honor that Roddenberry referred to was not only the Smithsonian's request that the second *Star Trek* pilot be enshrined in their institution but also the bragging rights that *Star Trek* earned from that request.[23]

The *Life in the Universe?* Exhibit

In 1970, less than six months after the crew of Apollo 11 returned to Earth, astronaut Michael Collins retired from the Air Force as a major general, left NASA, and took a job with the Department of State as assistant secretary of state for public affairs. In 1971, he left that position to become director of the Smithsonian's National Air and Space Museum, which would soon be building a brand-new building. As director, Collins took up the formidable task not only of convincing Congress to provide funds for a new, much-larger building to house the museum, but also, once it was successfully funded, of overseeing its construction and staffing.

Collins never really carried the "steely-eyed missile man" persona. As an astronaut, he was much more subdued and reflective both in person and when talking about his achievements. This trait even extended to the type of car that he drove. While most astronauts liked to drive muscle cars, Collins favored a much-humbler set of wheels. When Collins first reported to his new job at the Smithsonian on April 12, 1971, a memo sent out a week prior to his arrival requested that "an appropriate parking space be assigned" to him. The three-line typed memo ended by mentioning that Collins "drives a Volkswagen."[24]

Among Collins's many duties were to oversee what would be put into the new museum. In a memo issued on April 27, 1971, only weeks after Collins came on board, Fred Durant presented guidelines to aid in exhibit planning of the new building. In this memo, Durant pointed out how S. Paul Johnston, led an effort that began in late 1964 to produce an operational plan. Known as the "1964 Plan," one of the significant recommendations of this report was that the museum "would not be concerned—as many museums are—simply with the past. Rather, because of the rapid rate of technological progress of flight, a significant amount of professional effort would be spent on contemporary programs and accomplishments and trends for the future." Durant went on to note that with the new building being constructed on the National Mall, "NASM [National Air and Space Museum] will have just one opportunity to spend the initial capital investment in the new building exhibits."[25]

Because exhibits take time to create, ideas for how to populate the new museum were well underway when construction formally began on the new building in September 1972. New exhibits were still being built and displayed in the Arts and Industries Building, where the original National Air and Space Museum had previously been housed. Keeping the old museum open while the new one was being built permitted less of an interruption for visitors who came still expecting to see something during construction. This also allowed various exhibits to be tested in the old building prior to their installation in the new museum.

With no new *Star Trek* episodes being filmed after 1969, except for a short-lived Saturday morning animated series by Filmation that aired from 1973 to 1975, Durant was eager to secure items from the canceled television series for the Smithsonian's collections. Durant wrote Roddenberry in 1970 asking, "When *Star Trek* officially closed, were there posters, still photos, models, uniforms, ZAP pistols, small artifacts, etc. which we might be able to acquire for the collection?" He even asked if they might obtain 16 mm copies of other episodes as well as another copy of the show's pilot, since "we don't like to screen our archival copy."[26] At one point, Durant considered securing some of the original set pieces from the show, including the massive bridge set and the full-sized studio prop of the Galileo shuttlecraft. Due to space limitations, he had to turn these down.[27]

In 1972, Durant wrote to Roddenberry thanking him for helping the Smithsonian document *Star Trek*. "I am hopeful of giving proper recognition to the *Star Trek* story in our new building," wrote Durant. "In one sense the story belongs to the rich history of books and motion pictures on the theme of space flight. However, I am interested in your view on the importance of symbolic legends and heroes relating to national effort and interest."[28] Durant was thinking about not only how to acquire various artifacts from the now-defunct television series, but also ways to interpret and incorporate them into the museum's exhibits.

CHAPTER SEVEN

With the new building under construction, now was the time to try to populate it with items that would reflect not only Durant's interests but those desired by the Smithsonian in general as well as of the public, who he felt not only enjoyed watching *Star Trek* but were also influenced by its positive image of the future.

Once word got out that the Smithsonian was interested in gathering *Star Trek* material, Durant's office was inundated with letters from those seeking to donate. To help with these queries, Roddenberry and Durant drafted a more formal letter explaining why the Smithsonian was requesting items from the show:

> The National Air and Space Museum of the Smithsonian Institution is interested in acquiring documentation and artifacts relating to *Star Trek*. Their interest is two-fold. First, it's the story of *Star Trek* as a highly successful science fiction production. Secondly, they are interested in the role of science fiction as a stimulator for imagination and creative thinking. F. C. Durant III, the Museum's Assistant Director for Astronautics, points out that the three great rocket pioneers, Tsiolkovsky, Goddard and Oberth[,] all gave credit to Jules Verne as a stimulus to their contemplating space exploration. Many prominent space scientists and rocket engineers today will tell you that works of science fiction influenced them toward their life's work. Mr. Durant writes that he is developing reference files on science fiction, importantly including *Star Trek* which will be useful as source material for consideration of exhibits as well as for study of the sociological and cultural aspects of science fiction and its influence. Mr. Durant would be interested in receiving for the Astronautics collection documents about *Star Trek* which may be significant, and which would be useful for reference now, as well as in the future.[29a-b]

Durant's letter was sent out not only to the general public but also to serious devotees of the series, including Bjo Trimble, David Gerrold, Jacqueline Lichtenberg, who later authored the bestselling book *Star Trek Lives!*, and Michael McMaster, who went on to create a series of popular blueprints that depicted the Klingon D7 Battlecruiser, the *Enterprise* Bridge, and Romulan "Bird of Prey" Cruiser.[30]

Even Stephen Whitfield in his bestselling 1968 book *The Making of Star Trek* seemed to predict that *Star Trek* would someday be enshrined in the nation's museum. In the opening pages, Whitfield writes, "This history text is an authentic contemporary record of early STAR TREK, written with the cooperation of its original producer, with reproductions of much of the actual material used—memos, letters, etc. It is not available on audio or videotape, although students making field trips to Earth may wish to visit the Smithsonian Institute where one of STAR TREK's visualizations (flatscreens) may be viewed."[31]

As previously mentioned, Whitfield's book was first published in September 1968, a little over a year after the Smithsonian requested and received a 16 mm copy of the show's second pilot. The "history text" that Whitfield refers to is a reference to his book *The Making of Star Trek*. The "STAR TREK's visualizations (flatscreens)" that he mentions, which "may be viewed," is most likely a reference to the show's second pilot, which the Smithsonian received for its collections. Whitfield assumed that it would be shown to visitors at the museum even though the original 16 mm print copy was far too fragile to survive frequent screenings.

One visitor seemed to take what Whitfield wrote as factual. After touring the museum in 1973 and not seeing what Whitfield described, *Star Trek* fan Russell A. Kirby wrote a letter voicing his concerns:

> Recently, I visited your institute [the Smithsonian] and marveled at all the wonderous displays which have been collected. I would like to let you know how much I appreciate your museum. I will visit them again, for a third time, next fall. I have but one question, so I shall be right to the point, please excuse me if I am rude. Sir, I was very much disappointed when I learned that the display of *Star Trek* was not in your museum, as was stated in the book *The Making of Star Trek* by Steve E. Whitfield and Gene Roddenberry, or at least none of your employees knew where it was. Please inform me as to what has happened to it. I know of many people who have gone there and have been disappointed in not seeing it. . . . Please accept my apologies if I have scolded you for something which has never existed.

Most *Star Trek* correspondence received at the National Air and Space Museum during this time, was handled by Fred Durant, who wrote back to Kirby in a letter dated March 26, 1973, saying, "We regret that on your recent visit that you were disappointed in not finding an exhibit of *Star Trek*. There never has been an exhibit, actually, and Mr. Whitfield was incorrect in so stating in the book *The Making of Star Trek*. However, we are interested in the *Star Trek* 'story' and its part in the role of science fiction and its influence. Some years back Mr. Gene Roddenberry donated a print of the pilot film for *Star Trek*. This film is archived, however, and is not available for screening."[32]

Kirby wrote his letter five years after Whitfield's book was first published and over a year prior to the opening of the Smithsonian's first exhibit that would include *Star Trek*. He did not realize that Whitfield's comments were simply the author projecting into the future where he saw a world where *Star Trek* was real and not just a fictional television show. Soon, however, *Star Trek* would be featured in a display at the Smithsonian that would be seen by millions of visitors.

With construction of the new National Air and Space Museum on its way to a 1976 opening, the Smithsonian was in a position of having not only a brand-new building to showcase more artifacts, but also new exhibits.

CHAPTER SEVEN

In the history of the Smithsonian, there were two *Life in the Universe?* exhibits built. The first, completed in 1974 and opened with an invitation-only, black-tie gala on September 19, was displayed in the Arts and Industries Building.[33a-d] In an interview that year, the curator of the exhibit, Melvin Zisfein, noted that "ten years ago we could not have developed such an exhibit, but the last decade has produced a vast amount of knowledge about the universe."[34] Two years later this exhibit would be updated, refurbished, and moved to the new National Air and Space Museum just down the street. After Roddenberry visited the new museum during its construction, he offered some suggestions. In a letter to Roddenberry, Collins wrote, "Mel Zisfein and Fred Durant have been keeping me informed of your visit here and subsequent discussions which appear to offer the potential of substantial input to our forthcoming *Life in the Universe?* exhibit."[35]

When the first *Life in the Universe?* exhibit opened in the Arts and Industries Building in 1974, visitors began their tour through the exhibit by entering a reproduction of a cave, where they first learned how life began. There they were confronted with floor-to-ceiling photographs and black-light-lit graphics that showed the vastness of space. Additional special effects helped visitors trace the theories of the origins and development of the universe. One of the more popular features of the exhibit was called "Powers of Ten." Here, visitors watched a short film that depicted the relative scale of the universe according to an order of magnitude that increased by a factor of ten. The film began by showing a man resting in a park on a blanket. The film quickly moved out from there at a rate of 10^{10} meters per second until the entire universe was surveyed. The film then reversed itself, taking viewers back to the resting man, and then continued shrinking into the microscopic universe until a single atom was observed in his hand. The film was written and directed by the famous husband-and-wife industrial-and-graphic-design team of Charles and Ray Eames.[36]

The Life on Earth section of the exhibit dealt with the origins and nature of life on our own planet. A display explaining the research of Cyril Ponnamperuma of the University of Maryland showed how his work built on the pioneering experiments of Harold Urey and Stanley L. Miller, who demonstrated in the 1950s how amino acids, the building blocks of life, could be synthesized by firing electric sparks into a solution made up of their components, thereby mimicking the assumed conditions found on the early earth.

Following this, a short eight-minute film called *Primordial Soup* was shown in which Julia Child, famous for her popular television show *The French Chef*, carefully measured and mixed "the chemicals of life" to show how life might have formed on Earth.

The third section of the exhibit gave information about the solar system as learned by various robotic space probes. It also highlighted then-future projects such as the two Viking missions to Mars, both of which looked for signs of life after landing on the Red Planet in 1976.[37]

The last part of the exhibit covered efforts then being made by cosmologists, astronomers, astrophysicists, chemists, and biologists to seek out and communicate with what may be other intelligent life in the universe. This section also included push-button shows called "Pick a Planet" and "Pick a Star," where visitors could manipulate various factors to create hypothetical planetary systems and extraterrestrial life-forms. The resulting creations were then shown in original artwork done by artist Bonnie Dalzell. It took Dalzell six weeks to create twenty-five alien conceptions for the exhibit.[38] Among the life-forms that Dalzell created was a six-legged Hexalope; a two-headed, three-legged Puppeteer; a huge, armored, multilegged herbivore that fed on desert plants; and a Cthulu "larva," which Dalzell said was "a bottom dwelling, largely motionless vertebrate whose main characteristic is that it is disgusting." In describing her creations, she insisted that "these animals are created on a very reasonable basis. We didn't just stick legs on a piece of romaine lettuce. . . . They just seem weird because people don't understand terrestrial animals, much less extra-terrestrial ones."[39]

Unidentified flying objects (UFOs) of alien origin were never directly mentioned in the exhibit. "It's such a singular subject," said Zisfein, that "we didn't want to dilute what we had to say." Exhibit curator Alexis "Dusty" Doster III said in a 1974 press interview, "We did indeed consider it. But there isn't enough scientific evidence to say yes or no about it."[40]

Hanging above the exhibit near the exit were two large model spacecraft. One represented a vehicle that actually flew into space, while the other was a model that flew there only on television. The model of the real spacecraft was an early depiction of what eventually became the Hubble Space Telescope, which was launched by NASA in 1990 and provided astronomers with decades' worth of discoveries. Facing the model of the early Hubble was the original 11-foot studio filming model of the fictitious starship *Enterprise* as seen in the original *Star Trek* television series.

The *Life in the Universe?* exhibit was one of twenty-five that the Smithsonian moved from the old Arts and Industries Building to the newly built National Air and Space Museum when it opened to the public on July 1, 1976.[41] The *Life in the Universe?* exhibit that opened in the new museum proved to be just as popular as the original. "I thought that was one of the best galleries in the building when we opened, if not the best," said Smithsonian curator emeritus Tom Crouch in an interview recalling that time in 1976 when the new building opened. "I just loved that gallery, because in talking about the possibility of life in the universe, you had to talk about the universe, and we did an exceptional job in telling both."[42]

The new *Life in the Universe?* exhibit was similar to the old one, with some changes made by using recommendations based on a pilot study that was conducted on the original exhibit.[43] According to Crouch, such studies were common. "We used them a lot, especially after we opened the new building."[44] One of the findings reported in the study was that visitors who heard about the *Life in the Universe?* exhibit came because "they wanted to see *Star Trek*."[45]

Like the first exhibit, the new *Life in the Universe?* exhibit also featured the original 11-foot filming model of the *Enterprise* from *Star Trek*. In the original 1974–75 exhibit, the model was displayed deep within the exhibit. That changed when the exhibit opened in the new building in 1976. Now the *Enterprise* hung above the exhibit's exit, making it more visible, which helped draw visitors into the exhibit.[46]

In both the old and new *Life in the Universe?* exhibits, blueprints of the famous starship were hung directly beneath the displayed production model. Visitors were shown large, printed sheets of scale drawings that revealed the complete exterior and interior of the filming model that hung above them. For most visitors, seeing detailed drawings of a fictitious television spaceship was an unusual experience. For fans of *Star Trek*, it was awe inspiring, for the drawings proved to be just as popular as the model itself. For the first time, details of the *Enterprise* were presented in a highly organized and professional manner. As a result, Smithsonian officials were inundated with requests about how museumgoers could obtain copies of the blueprints for themselves from the man who drew them—Franz Joseph.

Franz Joseph Schnaubelt

Born on June 29, 1914, in the suburbs of Chicago and christened Francis Joseph Schnaubelt, he later preferred to be called Franz Joseph. According to his daughter, Karen, in a 1999 interview, "He adopted the name Franz sometime in his twenties, probably because he thought it sounded more sophisticated."[47]

Interestingly, the man who would go on to create the detailed blueprints coveted at the Smithsonian exhibit failed the Palmer method, which is a form of penmanship, because even though his letters were written perfectly, in the eyes of his teachers he did it with his left hand, and that was frowned upon by his teachers at the time.

He could not only write but also draw. With an avid interest in art, Joseph became a technical illustrator and design engineer. During World War II, he worked at Consolidated Vultee Aircraft in San Diego (now General Dynamics). According to his daughter, Karen, he was involuntarily laid off by the company at age fifty-nine in 1969, shortly after receiving his twenty-three-year service pin.

Joseph was married for nearly fifty years to Hazel Van Kampen, whom he met after moving to San Diego. "They were literally married until death parted them," said Karen during an interview. "He was a dashing young man on a motorcycle, and he offered her a ride. They rode around the block and were married three months later, which was just amazing to me because my parents didn't seem that impulsive to me as older adults. And finding this out it's like, 'Really?'"[48]

"My dad started watching *Star Trek* first. It was on Thursday nights opposite *Bewitched*, so I was watching *Bewitched* at the start of that first season in 1966. We only had one black-and-white television in the house. Finally, six or seven weeks in, my dad put his foot down and said, 'We will watch *Star Trek*.' So, I started watching it and went, 'Oh, this is okay.'"[49]

In 1973, at age eighteen, while attending San Diego State University, where Karen studied anthropology, she went to her first *Star Trek* convention. Soon she and her friends became involved in a local *Star Trek* club, where they talked about the characters, shared 35 mm film clips or trims from the episodes, and displayed simple replicas of props that they had made. It was during one of these gatherings that Karen brought along her dad. After looking over some of the drawings and props that her friends had made, Paul Newitt, in an interview that he conducted in 1982, wrote that Joseph said, "I thought they could do better. And they said, 'Show us.'"[50]

Using some *Star Trek* slides provided by his daughter along with Stephen Whitfield's *The Making of Star Trek* as references, Joseph began making three-view drawings of the props, starting with the communicator and phaser. Soon Karen had a bunch of her friends over to show them what her dad had done. "We sat in the dining room, and I showed them the drawings I'd made," said Joseph. "They went wild over them and then began to write lists of all the things they wanted to see and wanted me to convert to plans," said Joseph.[51]

Joseph knew he was making something of interest to fans, and so he continued working on his drawings, creating about a dozen views of other props from the show. He also began working on deck plans of the ship itself.

As an experienced artist and technical illustrator who worked in the aerospace industry, Joseph knew about copyright and licensing. He saw other drawings, albeit of inferior quality, that were being sold through fan groups as well as Roddenberry's own commercial interest, Lincoln Enterprises. He knew that if he wanted to reproduce his work on a commercial scale, he would have to get permission.

Joseph started by writing to Gene Roddenberry. In a May 14, 1973, letter, Joseph explained what he had been doing. He also included samples of his drawings. "I have no intention to plagiarize your creation," wrote Joseph to Roddenberry. "I do not want to steal anything from you, or any others who have a legitimate right to *Star Trek* property . . . but I wish to retain propriety to what I believe is my original work."[52] This initial letter began a protracted

CHAPTER SEVEN

exchange between the two in which Roddenberry sought to try to acquire Joseph's drawings for his own Lincoln Enterprises business.

Joseph then approached Paramount Television to try to secure licensing for his work. They ended up giving him a one-shot-sale licensing agreement to produce five hundred sets of his *Enterprise* blueprints, now referred to by Joseph as the *Booklet of General Plans*. These quickly sold out at a local *Star Trek* convention called Equicon, which was held in April 1974.

Around this same time, Roddenberry informed the Smithsonian about the drawings. "Franz Josef [*sic*] Schnaubelt, retired industrial designer and engineer, has concluded a remarkable set of drawings which shows in extraordinary detail every deck of the USS *Enterprise*, every stateroom, elevator, storage facility, corridor, laboratory, head, weapons battery," wrote Roddenberry in a letter to Fred Durant dated March 29, 1974. "Really, Fred, you will be almost unable to believe your eyes when you see this page after page of beautiful diagram-drawings."[53]

A month earlier, the Smithsonian received the 11-foot filming model of the *Enterprise* from Paramount. Durant knew that Joseph's blueprints would be a great accessory to display with the model. Upon receiving Roddenberry's letter, Durant contacted Joseph about securing a set of his plans for the museum. In an April 17, 1974, letter, Durant wrote Joseph, "We are indeed impressed by the fine quality and creativity of your set of *Enterprise* drawings. They will be of great assistance in the refurbishment and understanding of the 14-foot[54] model obtained from Paramount."[55]

Joseph shipped copies of his blueprints to the Smithsonian, which were soon put on display beneath the hanging model of the *Enterprise* in the first *Life in the Universe?* exhibit. This set of drawings came from the original print run of five hundred uncut copies that Joseph made for Equicon '74. Indeed, the drawings still had printed on them "SPECIAL EQUICON '74 FIRST EDITION."

In delivering the plans to the Smithsonian, Joseph gave specific details about the history of the drawings, along with suggestions on how they should be displayed.[56] Joseph also asked that the museum use "Franz Joseph Schnaubelt," instead of "Franz Joseph Designs," when crediting the drawings, explaining that "first, my family and relatives would be pleased by the family association . . . and while my professional associates are familiar with my free-lance identity, thousands of people in the Navy Department and the military establishment in Washington know me by my name or initials 'FJS.'"[57]

As work progressed on getting the first *Life in the Universe?* exhibit ready for opening by fall 1974, Joseph sent Durant a slide depicting artwork that he created for a proposed postage stamp that would commemorate the new exhibit.[58a-b] Joseph gave the original artwork to the postmaster general, but the US Postal Service chose not to produce it. The original artwork was completed on June 28, 1974, and sent to the postmaster general shortly thereafter. Since that time, nobody knows what happened to it. All that remains is a 35 mm slide housed in the Smithsonian archives that was taken of the original artwork.[59]

Roddenberry was invited to the 1974 black-tie reception held for the opening of the first *Life in the Universe?* exhibit but could not attend.[60] Those attending included NASA's deputy administrator, George Low. Photos taken of the event show Low and his wife, Mary, exploring the exhibit, including studying the 11-foot model of the starship *Enterprise*.[61] Several years later, as NASA's deputy administrator under Jim Fletcher, Low played a role in the renaming of America's first space shuttle after *Star Trek* fans launched a successful letter-writing campaign to have the White House change its name from *Constitution* to *Enterprise*.[62]

The same week that the first *Life in the Universe?* exhibit opened to the public, Franz Joseph received a phone call from Lou Mindling, vice president of Paramount Television, informing him that Judy-Lynn Del Ray from Ballantine Books was in his office. They all were very excited about Joseph's blueprints and were serious about getting a contract in place to publish them. By the end of the year, Joseph formally signed a contract and sent the master art to Ballantine.

After previous protracted efforts by Joseph to get a formal publishing contract for his drawings in place, the whole process moved very quickly once his plans went on display at the Smithsonian in 1974. It seemed that public interest in his work after being shown at the National Air and Space Museum finally made a formal contract happen. Only one week after the exhibit opened, Joseph wrote Durant, "I'm now beginning to receive queries from people who have been to see your exhibit. Since there has really been no advertising, I think the response I've received is really remarkable." Indeed, the number of visitors asking the museum about where they could purchase copies of Joseph's blueprints was so great that Durant asked Joseph if he could send postcards with his address on them so that museum officials could give them out to visitors. Since Ballantine would not have commercial copies of the blueprints available until nearly six months after the exhibit had opened, this seemed like a good solution, since Joseph could then start keeping a list that would allow him to inform those wishing to purchase the drawings when they finally would be available in stores. Joseph sent the museum a box of five hundred cards and noted that "although these cards were prepared for Equicon '74 when the plans were first introduced, the fans have accepted them for post-Equicon usage."[63a-d]

After Joseph's postcards were given out, visitor requests for copies of his plans grew even greater. In a November 18, 1974, letter to Durant, Joseph wrote, "Your exhibit 'Life in the Universe?' must be quite a show because I've been receiving a steady stream of inquiries from people who have seen the plans on display in it. Even from Congress, the Federal Communications Commission, The American Bureau of Shipping, and Fleet Aviation Specialized Operations

CHAPTER SEVEN

Training Group Atlantic, USN. It is beginning to look like this exhibit will be a stronger 'drawing card' than *Star Trek* fandom."[64]

Interest in Joseph's plans did not stop when the first *Life in the Universe?* exhibit closed in early January 1975. On February 25, Ballantine reported presale orders of 20,000 copies for his soon-to-be-available blueprints.[65] When his blueprints finally did hit store shelves that spring, booksellers reported swift sales. Nedra Reposh, manager of a Waldenbooks at Oakdale Mall in Binghamton, New York, said people would steal Joseph's blueprints, especially around Christmas. The blueprints were sold as twelve sheets folded and packed in a clear vinyl pouch that snapped shut. Reposh described how fans would come into her store, open the pouch, grab a few of the preferred sheets, and walk out with them. As a result, management ended up keeping them behind the cash register.[66]

Because of large sales volume, the blueprints made headlines. During a 1982 interview, Joseph explained that "they should have been on the bestseller lists but they couldn't, because those lists are for hardcovers or paperbacks, and the plans are classified as a production in the publishing business. So the editors of the newspapers were writing editorials about why the plans should be on the bestseller list but couldn't be put there."[67] One such account appeared in the July 13, 1975, issue of the *New York Times Book Review*. "There's one publication that's been selling so furiously in bookstores during recent weeks that it would be included on the list above except for one fact. It's not a book," explained the reviewer in the column "Paper Back Talk." "'Star Trek Blueprints' is a set of 12 reproductions of precise designs by Franz Joseph Schnaubelt showing 'every foot of every level of the fabulous starship *Enterprise*.' Since mid-May, Ballantine Books has sold 150,000 sets, enclosed in a plastic and leatherette portfolio, at $5. This week it goes back to press for 100,000 more."[68] B. Dalton's, one of the nation's leading booksellers, also reported that "the Blueprints continue to sell at an incredible rate. . . . With this kind of movement and publicity, we have gone back to press for the third time with a 60,000-copy reprint. This brings our total printing to 210,000!"[69] By the end of the year, the *New York Times Book Review* reported that Ballantine had sold over 300,000 sets since the blueprints first appeared for sale in May.[70]

Riding on the wave of success from his blueprint sales, Joseph signed another contract with Ballantine in August 1975, this time for the *Star Fleet Technical Manual*. This new book contained multiple starship drawings along with other illustrations and documents that expanded upon Joseph's universe. The book proved to be just as popular as his *Star Trek* blueprints and quickly sold out after first appearing in bookstores over the 1975 Thanksgiving holiday. In Franz Joseph's annual Christmas letter, he wrote that his *Technical Manual* "became the No. 1 bestseller just four weeks after it appeared in the bookstores and held that spot for something like two months."[71] By December 17, only three weeks after publication, his book became number 1 on B. Dalton's bestseller list, an unheard-of achievement for a technical publication. By the end of that year, Ballantine Books sent Joseph a bottle of champagne as the book climbed to number 1 on the *New York Times* bestseller list for trade paperbacks during the week ending January 4, 1976.[72] It held that spot for seven straight weeks before finally dropping to number 2 on February 22.[73]

Even though it was no longer number 1, the book remained among the top five trade paperback bestsellers for four more weeks as it competed for the number one spot with such literary greats as Alex Comfort's popular *The Joy of Sex* and *More Joy of Sex*, along with Mable Hoffman's *Crockery Cookery* (the world was a very different place for a kid growing up in the midseventies). The editors of both the *New York Times* and *Publisher's Weekly* told Joseph that this was the first time that purely "technical" material had ever become a popular bestseller.[74] As the *New York Times* reported, "Few sci-fi books sell more than 400,000 copies. 'Star Trek' books account for close to 500,000 copies each." In response to Joseph's phenomenal sales of his work plus other *Star Trek* books that began appearing at the same time, Roddenberry wrote, "We suspected there was an intelligent life form on the other side of the tube. . . . Never in our wildest imaginings did we expect the volume and intensity."[75]

According to Joseph, the blueprints required 252 hours of research and 248 hours to draw. The *Technical Manual* required four hundred hours of research and one thousand hours to illustrate. Joseph had strict control over the content of his work, something he remained firm about when he negotiated contracts. "In fact, both of my contracts with Ballantine state that nothing may be altered from the manner in which it was submitted without my consent," said Joseph. "I even supplied a chip of the precise shade of red that was to be used for the covers." As to why both the blueprints and *Technical Manual* were packaged the way they were, Joseph said in a 1982 interview, "Since my contracts also state that the name 'Star Trek' may not appear on my work, Ballantine had to devise a means to apply it and still adhere to the contract terms. The 'throwaway' insert was the answer for the plans, and the 'over-jacket' was the answer for the *Manual*."[76] Indeed, if you look closely at the dozen printed sheets that make up each blueprint set as well as examine the *Technical Manual*, nowhere does it say "Star Trek" on either product. The words "Star Trek" appear only on the sheet inside the clear-view pouch containing the blueprints; the same on the insert that was slipped into the front of the oversize vinyl jacket that protects the *Technical Manual*.

In an interview with Joseph's daughter, Karen, she was asked about the inserts as well as some of the inconsistencies that both the blueprints and the *Technical Manual* have when compared with the actual television series. These included Joseph's decision to print "Starfleet" as two words instead of one. As fans of the series know, the word "Starfleet" appears in the second part of the two-part first-season episode "The Menagerie." In that episode, Commodore Mendez presented Kirk with a classified document about Talos IV that has clearly imprinted on the cover "TOP SECRET FOR EYES OF STARFLEET COMMAND ONLY." When asked about this, Karen replied, "He wanted it to be the *Star Fleet Technical Manual* as if it were a real thing. So, if it were a real thing, it would not say 'Star Trek' on it because *Star Trek* is the whole fictional show. Ballantine's compromise was 'Well, hey, we'll ship this card in that says 'Star Trek,' and if people want to remove that, they can then join the fictional world so it just says 'Star Fleet.'"

Karen went on to explain that "the whole conceit of the *Technical Manual* was that it had been accidentally broadcast to the computers on Earth during the 'Tomorrow Is Yesterday' episode. That is why we had the manual in this form, because it postulated that the whole *Star Trek* universe was real, and they time-traveled, and this is why we have this document. He made up this whole fictional persona for himself who was some minor bureaucrat at Star Fleet. He had this stationery made up, and when people wrote him fan letters, he would write them back on that stationery. People wanted to sign up with Star Fleet. They thought that because there's this book, Star Fleet must be real." This was his "Star Fleet" (two words) and not the show's "Starfleet" (one word).

While Joseph was trying to distance his work from *Star Trek*, Ballantine was trying to link the work to the show. He could have used the one-word spelling of Starfleet but chose not to, since he wanted his work to seem real and be uniquely his even though it was about a fictitious television show. "That's what made *Star Trek* special," said Karen. "Once you saw it, you wanted to know everything about that world, even to live in it. There was nothing else like it on television at the time, and the reason that most people didn't 'get it' was because it was just another television program to them."[77]

As a result of the financial success of the blueprints and the *Technical Manual*, Judy-Lynn Del Rey, who led the efforts that eventually resulted in Joseph securing a formal contract for his work, was promoted to editor at Ballantine Books. Both she and her husband, Lester, then went on to create Del Rey Books, an imprint of Ballantine that specializes in science fiction, fantasy, and horror. In addition, since the original *Life in the Universe?* exhibit that the Smithsonian ran from 1974 to 1975, as well as the one in their new building that ran from 1976 to 1979, included displays of Franz Joseph's blueprints, millions of visitors had an opportunity to see them. Once the blueprints and the subsequent *Technical Manual* became commercially available, others were inspired to create similar products.

For example, in 1975, Stephen V. Cole created Task Force Games, which came out with a board game called *Star Fleet Battles* that featured ships taken from Joseph's original *Technical Manual*.[78] Fans created, marketed, and sold their own blueprints and *Technical Manual* spin-offs. These varied in content and quality. Among the standouts were a series of expertly drawn plans by *Star Trek* enthusiast Michael McMaster that included blueprints of the Klingon D7 (1975), the Bridge (1976), the Romulan Bird of Prey (1977), and a Size Comparison Chart (1978), which showed various ship profiles from the series all drawn to scale. Around the same time, Geoffrey Mandel produced several sets of blueprints including those for the USS Independence-class freighter that was featured in the *Star Trek* animated series *The Pirates of Orion* (1976). Mandel also worked with another fan, Kenneth Altman, to create the Space Station K-7 blueprints (1976). Except for Cole's board game, all of these were unlicensed products.

On the basis of the success of Franz Joseph's work, other publishers jumped on the bandwagon to produce blueprint sets, especially following the premiere of *Star Wars* in 1977. That same year, Ballantine published the *Star Fleet Medical Reference Manual*. This time, Franz Joseph was not involved in its production. Instead, it contained contributions by Doug Drexler, who, as mentioned in chapter 3, had been running a successful *Star Trek* store in New York City called the Federation Trading Post.

There were commercial blueprints sold for *Star Wars* (1977) and *Battlestar Galactica* (1978). There was even a set for *Superman: The Movie* (1979). But unlike Franz Joseph's original blueprints, these were not the in-universe kind but rather the practical sort used in the construction of movie sets. Even so, fans still ate them up because they provided valuable details that could not be found anywhere else.

In 1979, *Star Trek: The Motion Picture*, a big-screen adaptation of the original NBC television series, was released. Among the products marketed to promote the film was a set of blueprints done by David Kimble and Andy Probert. These included views of the starship *Enterprise* refit, the Klingon warship, and a variety of auxiliary vehicles. The drawings were in a style like Franz Joseph's, with spacecraft parts labeled according to their science-fictional functions. The careful viewer will notice that the *Star Trek* films and the various television spin-offs feature many of Franz Joseph's original designs, as production artists snuck in images from his work.

The Smithsonian and the Starship *Enterprise*

As mentioned previously, it was Fred Durant who led the Smithsonian's effort to obtain artifacts from *Star Trek*. This effort intensified once the television series ended production in 1969. Among the items that he sought to acquire was the original 11-foot production filming model of the USS *Enterprise*.[79]

CHAPTER SEVEN

By the time construction began on the new National Air and Space Museum in 1972, Durant and Roddenberry knew each other well. After Durant had secured a copy from Roddenberry of *Star Trek*'s second pilot in 1967, the two kept in touch. Shortly after the last *Star Trek* episode aired on NBC, Durant sent off a letter to Roddenberry asking if any models from the show could be acquired for the museum. In early May 1973, Roddenberry wrote Durant, "Let me know when you are requesting the 14-foot model of the USS *Enterprise* so that I can arrange proper press coverage out here with an aim to forcing Paramount to be cooperative. As mentioned to you last week, I think they will find it hard to turn such a request down if handled properly."[80]

At Durant's urging, Michael Collins sent a formal letter on May 7, 1973, to Frank Yablans, the president of Paramount Pictures, expressing interest "in acquiring the 14-foot model of the *Star Trek* spaceship, *Enterprise*." The model would be put on display first in the *Life in the Universe?* exhibit in the Arts and Industries Building and then transferred to the new building once it was completed.[81]

At first, Paramount ignored Collins's letter. In an August 22, 1973, letter to Durant, Roddenberry wrote, "Sorry you haven't heard from Paramount about the model of the *Enterprise* but not greatly surprised as a lack of simple courtesy has characterized most of their dealings with me and others." Roddenberry went on in his letter to indicate that he would try to help, stating, "In case the Smithsonian request never reached the desk of the president of the company, I will see that an indication of your interest in the model reaches him thru other channels."[82] Part of the reason for the delay in responding to Collins's request was because Paramount's publicity office and all of its files were being moved from the West Coast to New York.[83]

It took several months, but Paramount finally agreed to donate the model. On February 7, 1974, Dick Lawrence, the executive vice president in charge of domestic television for Paramount, confirmed in writing that Paramount would donate the model to the museum.[84a]

The model was broken down into five sections and shipped from Paramount to the Smithsonian at an estimated cost of between $350 and $500.[84b] Three shipping crates left Los Angeles International Airport (LAX) via Emery Freight on February 26, 1974 (airbill #LAX 12001), and were received in Washington two days later.[85] After unpacking, Smithsonian officials went about assessing the model's condition and making plans for its restoration.[86]

The Smithsonian was not the first place where the 11-foot filming model had been put on public display. Craig O. Thompson, a teacher at Golden West College in Huntington Beach, California, was able to show the model during two space-themed exhibitions that he organized beginning in 1972. His first space expo was held April 9–15, 1972, and featured a lecture by noted scientist and author Arthur C. Clarke.[87a–c] Other speakers also gave talks including noted rocket engineer and aerospace writer Krafft Ehricke. Film director George Pal also spoke and gave a special showing of his 1953 film *The War of the Worlds*.[88] A traveling NASA exhibit called "America in Space" was present and featured a real moon rock.[89a–c] The highlight of the expo was the first public appearance of the 11-foot filming model *Enterprise* from *Star Trek*. People lined up to see it. "The *Enterprise* is definitely our number one attraction," said Thompson in an April 11, 1972 news article.[90]

Based upon the success of that first effort, Thompson held a second space expo the following year. This time he wasn't able to get Clarke but he was able to secure the original spacecraft that carried astronauts John Young, Charlie Duke and Ken Mattingly to the moon and back during NASA's Apollo 16 mission.[91] The first space expo concluded on April 15, 1972, the day before the launch of Apollo 16, so it was fitting to have the capsule brought to the campus the following year after having completed its mission. In addition to displaying the Apollo spacecraft, Thompson was able to once again bring out the 11-foot production model *Enterprise* for public viewing. From March 26 through April 7, 1973, both the Apollo 16 command module and the original starship *Enterprise* could be seen for free at Golden West College, along with a real moon rock and other models and props including the original twenty-two-inch shuttlecraft filming model from Star Trek and miniatures from *Earth II*, a pilot for a television series about a colony that orbited the earth.[92a–c]

Prior to his employment at the college, Thompson had worked as a post-production office manager for the original *Star Trek* television series.[93] Officials at Paramount allowed him to take the *Enterprise* model out of storage and bring it to Golden West College, where Thompson reassembled it for display at his expos. In a 1997 interview in *Star Trek Communicator* magazine, Thompson recalled that after the two space expos were done, "props would have paid me to keep it," said Thompson. "I kick myself every time I tell this story, because I probably could have had it for fifty bucks, if not for free. But my garage was too small to store it and even I didn't have the foresight to recognize what an icon it would be in the future. So, it slipped from my hands."[94]

During the time it was displayed at the two space expos, the model was relatively intact and had only one piece, the deflector dish, missing. The impulse deck was scuffed up so Scott Steidinger, one of Thompson's students at the time, hand-brushed the deck with paint that was darker than the original color on the model. Some of the Christmas tree lightbulbs that were originally installed to light up the domes of the warp engines, were burned out so Steidinger replaced them.[95] The metal stand that originally supported the model for filming was used again to mount the model for display at the space expos but the stand

CHAPTER SEVEN

arrived from Paramount bent, which is why, in photos taken during the two space expos, the model is shown displayed at a very steep angle.

The appearance of the 11-foot model at West Coast College proved to be a popular attraction. During an interview with Steidinger, he stated that the 1973 space expo was better attended than the first. Both expos were free to the public except for the Arthur C. Clarke talk in 1972. "They didn't charge any admission during either space expos," said Steidinger. "The college had a large community services budget that had to be spent each year. I suspect that the underwriting for the event came from that money. The second year we had many school field trips from the local schools. It could be rather crowded at times."[96]

After the Smithsonian acquired the model, museum officials knew that it would be a unique artifact that would generate a lot of public interest. Even though the 11-foot *Enterprise* had been well cared for after the series ended, it still needed some attention. By the end of April 1974, Smithsonian conservators drew up a two-phase plan to restore the model. Phase I involved getting the model ready for the new *Life in the Universe?* exhibit, which would open in September 1974. Contractor Rogay, Inc., was put in charge of doing the initial cleanup, which involved, among other things, replacing the two missing hemispheres at the forward ends of the warp engines. The main sensor dish and center spike on the front of the secondary hull also were missing and had to be replaced. Missing hull windows were glued into place, lettering was retouched, and the model was thoroughly cleaned, "using cloth, mild detergent and warm water."[97a-b] Retouching of the black lettering on top of the main hull was done, along with painting any "chafed and other minor injuries to reasonable point." All the exposed wiring on the port (left) side of the model was covered with "three-inch, silver colored, pressure-sensitive (refrigeration) cloth tape."[98]

When installed in the first *Life in the Universe?* exhibit, the model of the *Enterprise* was hung in the northeast corner of the open portion of the East Hall of the Arts and Industries Building. Because the left side of the model was where all the wiring connections were present, the model was normally filmed during production of the series from the right side. When first hung for display by the Smithsonian, the model had the left side pointed away from the main hall. Combined with tipping the ship at a 20-degree angle, this helped make the wires less visible. In an effort to help visitors identify what they were seeing, a simple cardboard card with "USS Enterprise" printed on it hung beneath the model.[99]

When the *Enterprise* model went on display at the Smithsonian's first *Life in the Universe?* exhibit in 1974, Paramount Television published a news release about its appearance. "It is not just a coincidence that the 'Star Trek' spaceship was chosen for the exhibit," Paramount said. "'Star Trek' is a series that deals with man's adventure in space hundreds of years in the future. Based on projections of today's best scientific minds the series deals with the higher dramatic stories of occurrences which take place when man is able to travel in space far beyond his present boundaries."[100a-b]

In late 1975, in preparation for its display at the new *Life in the Universe?* exhibit that would open in the new museum, the model underwent additional restoration that included some minor repainting and installation of lights in the engine pods and body of the model.

When the new National Air and Space Museum building opened during America's bicentennial in 1976, the world's most famous starship and one of television's most iconic cultural icons was in public view once again. Many fans of the series expressed delight upon seeing the model on display with other notable pioneers in the history of air and space travel. Authors of the book *Star Trek Lives*, which focused on fandom generated by the original series, shared how personally thrilled they were at the model's appearance, "In the Smithsonian Institution in Washington, DC, where tribute is paid to man's efforts to go where no man has gone before—there in the halls that celebrate man's reaching for the sky, the Moon, the stars, there among the memories of the Wright Brothers at Kittyhawk [*sic*] and the Apollo astronauts leaving footprints on the face of the moon, there beside Lindbergh's *Spirit of St. Louis*, is *Star Trek*'s starship *Enterprise*. No higher tribute could be paid to the spirit of *Star Trek*."[101]

While most fans agreed that having the *Enterprise* in the Smithsonian was a significant achievement, others objected to how it was displayed. Some were turned off after seeing wires duct-taped to the exterior left side of the ship, blinking red dome lights on the front ends of the warp engines, and a crudely fashioned replacement "salad bowl" deflector dish on the front of the secondary hull.

The model's revised look, as unpopular as it might have been with some fans, was done at the suggestion of Roddenberry himself. Durant explained in a letter to a concerned fan that "the original technique had been to install a spinning disc in each pod, covered with broken glass. As this was not feasible for long-term use, the flashing lights were suggested as an alternative." In this same letter, Durant went into considerable detail about the condition of the filming model, stating, "I can assure you the 'Enterprise' has not been damaged in any way. When we received the model, we had two choices. We could have repainted and refurbished the craft completely, so that it would have resembled the craft as seen on TV. We chose instead to exhibit the 'Enterprise' for what it is, not a starship, but a slightly marred studio shooting model."[102a-g]

The second *Life in the Universe?* exhibit closed in February 1979. Over the next forty-five years, the *Enterprise*, traveled to different locations within the museum. Its first stop was in the History of Rocketry and Space Travel Gallery, where the model hung above that exhibit entrance. In 1984 it then moved to the second floor of the museum, where it hung, angled underneath a stairway, as part of a yearlong exhibit of space art by artist Robert McCall.[103]

Prior to moving the model to a special twenty-fifth-anniversary *Star Trek* exhibit that opened at the Smithsonian on February 28, 1992, the museum contracted to do another renovation on the *Enterprise* model. Ed Miarecki operated a model shop called Science Fiction Modelmaking Associates (SFMA) and performed services for the *Star Trek* franchise. This included being asked by the Smithsonian to refurbish their *Enterprise* along with models of the D7 Klingon and Tholian ships so they could be displayed in this new exhibit. Miarecki's interpretation of the *Enterprise*'s new paint scheme heavily accentuated the ship's grid lines.

After being seen by nearly one million visitors, the Smithsonian's special anniversary *Star Trek* exhibit ended on January 31, 1993. Museum officials then moved the exhibit to the Hayden Planetarium in New York. After the model's return to the Smithsonian in 1993 , the *Enterprise* was placed in storage, where it remained until the museum completed construction of the model's new home. This time, rather than being hung from the ceiling as the model had been displayed before by the Smithsonian, the *Enterprise* was now exhibited on a stand inside a clear enclosure in a basement gift shop that the museum opened in March 2000. Throughout all of these relocations, the model underwent multiple restorations, of varying levels of success.

Beginning in 1974, when the *Enterprise* went on display at the National Air and Space Museum, people saw the model and learned about Franz Joseph's blueprints, which helped nurture public interest even more in the show. The fact that a filming model from a television series set in the future shared floor space with real space hardware from the past was not overlooked by the millions of people who visited the museum every year. To help educate those who saw the model and may have felt it to be out of place in the Smithsonian, museum officials sought to impress how *Star Trek* fit in with the institution's message. One way they did this was through the use of interpretive signage. In the case of the *Life in the Universe?* exhibit, signs near the *Enterprise* helped explain not only the model but how *Star Trek* "dealt specifically with the problems and results of human interaction with extraterrestrial lifeforms and civilizations."[104] Items in the museum's gift shop also helped carry home the message on why the model was important. Souvenirs that were sold included *Star Trek* floaty pens. These special ballpoint pens encased an image of the starship in a liquid-filled barrel that, when tipped, allowed the *Enterprise* to warp through space by "floating" back-and-forth across a printed backdrop. The message "'USS *Enterprise*' from '*Star Trek*' television series spurred interest in space travel" was printed on each pen and served as a reminder of the show's importance and impact.

In studying the role that the Smithsonian played in *Star Trek* since the museum requested and received a copy of the show's second pilot in 1967, the institution has justified its interest over the years of not only collecting artifacts pertaining to the series but in displaying and interpreting those items to the public. In addition to the early floaty pen, during the period of the 1992 *Star Trek* exhibit, the Smithsonian licensed a copy of toy manufacturer Ertl's 4.5-inch die-cast metal miniature of the refit *Enterprise* from *Star Trek: The Motion Picture* to sell at its museum gift shops. This toy was first issued in 1984 in conjunction with the film *Star Trek III: The Search for Spock* as part of a three-ship series that included the Excelsior and the Klingon Bird of Prey. With the special Smithsonian-licensed edition, the blister pack was marked "Smithsonian Museum Shops" for a line of collectibles. On one copy of the toy, a card containing revised text was taped over the original printed packaging. The revised text included the start and end dates of the 1992 exhibit along with a statement under the title "'STAR TREK' AT THE SMITHSONIAN" that read: "The National Air and Space Museum of the Smithsonian Institution encourages space research and exploration. It acknowledges the role of *Star Trek*, the popular television show and movie series, in promoting public interest in space exploration." The card then states that "all income from the Museum Shops sales supports the chartered educational activities of the Smithsonian Institution." It is obvious that the Smithsonian wanted to be clear not only about the role that *Star Trek* played in the museum but also where the funds went from licensed *Star Trek* items that they sold.[105]

Fred Durant knew that science fiction influenced those visionaries who made air and space travel a reality, and by including the *Enterprise* model alongside those milestones of flight, it might inspire future generations. Somewhat unexpectedly, Gene Roddenberry himself played down his role in the Smithsonian's exhibition of the *Enterprise*. "When Roddenberry would come to the museum, he was always much more interested in the airplanes than the rockets and spacecraft. He wanted to see and talk about airplanes much more than space," said Smithsonian's curator emeritus Tom Crouch during an interview conducted about the significance of the Smithsonian acquiring the *Enterprise* model. Crouch continued, "Fred is really the only reason there was a connection to *Star Trek* at the Smithsonian. Fred was always interested in the popular response to the notion of spaceflight and grabbed on to *Star Trek* early by securing the second pilot and then later, when the opportunity to get the *Enterprise* came up, he seized upon that."[106]

By 2014, the *Enterprise* model was in dire need of work. The model was originally filmed hung by wires early in production. In addition, after the Smithsonian received the model from Paramount, it was displayed hung at the Museum from 1974 until 2000. The model was never designed for such long-term suspension and, as a result, it began showing its age. As a result, the *Enterprise* moved to the National Air and Space Museum's Steven F. Udvar-Hazy Center, the National Air

CHAPTER SEVEN

and Space Museum's annex that first opened at Washington's Dulles International Airport in 2003. There curators began a multiyear restoration effort that was completed in time for the fiftieth anniversary of *Star Trek*'s network television premiere in 2016.

Most people when they hear of the Smithsonian know it as group of museums clustered around a large patch of grass in Washington, DC. Here is where eleven of the twenty museums and galleries that make up the Smithsonian are located, anchored on one end of the National Mall by the Capitol Building and on the other end by the Washington Monument. Few recall that the Smithsonian was founded as a research institution whose charter is "to increase and diffuse knowledge." It was here where the Wright brothers first turned to for information about how to build an airplane, and where Dr. Robert H. Goddard received funding to support his early rocket experiments.

Having a fictitious spaceship such as the *Enterprise* from *Star Trek* take up residence in the Smithsonian alongside real flight hardware such as the Wright brothers' original "Wright Flyer," Charles Lindbergh's "Spirit of St. Louis," and the Apollo 11 astronauts' command module *Columbia* is a big thing.[107]

The 11-foot *Enterprise* model is not just a relic from a canceled television show, but an important cultural artifact. *Star Trek* has inspired others by helping promote public interest in science and space exploration, while the show's presence in the Smithsonian helped elevate the series to a greater level of respectability.

When the *Enterprise* first landed at the Smithsonian in 1974, it was during uncertain times both for the nation and for *Star Trek*. That year, the Watergate scandal was coming to a head, and President Nixon would become the first president to resign from office. But there were bright spots during that decade. We recall that 1976 is when America celebrated its bicentennial and the Smithsonian opened its new National Air and Space Museum, which featured the *Enterprise* in its new *Life in the Universe?* exhibit.

Around this same time, *Star Trek*'s Nichelle Nichols traveled across the country speaking to minorities and women, encouraging them to join NASA to fly their new space shuttle.

The decade of the 1970s also saw the launching of spacecraft with bold-sounding names such as *Pioneer*, *Viking*, and *Voyager*, designed to explore the solar system. Some of these carried specialized instruments to help us answer the question about other life and are we alone in the universe.

Though support for *Star Trek* remained strong during this time, as proven by the huge number of fans who showed up at conventions, studio executives remained skeptical. By 1979, *Star Trek: The Motion Picture* (*STTMP*) premiered, and, though not as big a draw as *Star Wars* that premiered two years prior, the film did respectably well at the box office. *STTMP* was welcomed by fans who, for the first time, saw all their favorite cast members from the original series perform on the big screen.

Today, the *Enterprise* can be seen on the main floor of the National Air and Space Museum near the Independence Avenue entrance, in front of the 146-foot-long uniquely L-shaped mural called *The Space Mural, a Cosmic View*. The mural, done by noted space artist Robert McCall, who also worked on *STTMP*, depicts the first humans who landed on the moon, while at the same time showing planets beyond the earth and the birth of the universe. Like the *Enterprise* model, McCall's work depicts an exciting past but also suggests a promising future.

The *Enterprise* model remains a popular attraction even after it was first put on display by the Smithsonian over fifty years ago. When Fred Durant's son Steve was asked about the enduring popularity of the model and the show, he replied that his father would often mingle with visitors at the museum. When they gathered around the *Enterprise*, he would tell them, "The three most asked questions by visitors are 'Where is the bathroom?,' 'Where is the cafeteria?,' and 'Where is the *Enterprise*?'" Such a story was very much in character, since Durant thought a great deal about *Star Trek* and its influence on popular culture.[108]

When visitors gather around the *Enterprise* model, there is an almost universal understanding of what they are looking at. Nobody has ever protested in front of the model or thrown paint or tomato soup at it, which says a lot not only about the acceptance of the *Enterprise* as an important cultural icon but also that *Star Trek* is something that everyone remembers in a good way.

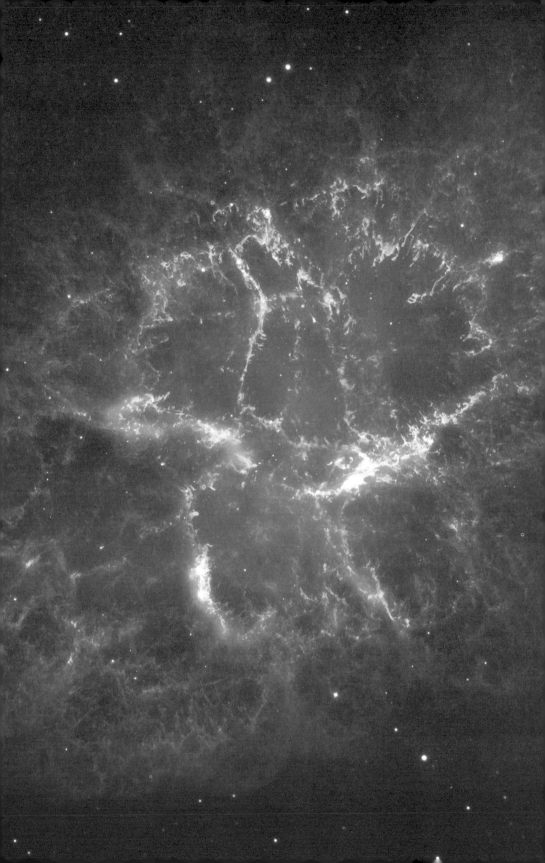

CHAPTER 8

Vindication through Syndication

Like most fans of the series, I was introduced to *Star Trek* after its original network television production run had ended in 1969. Too young to fully understand what I was watching when the show originally aired on NBC, I discovered the series through reruns while growing up during the 1970s. To better understand how *Star Trek* became so widely shown long after its original network run had ended, we need to take a brief look at the technology and network practices that helped define television viewing of that time.

When *Star Trek* first aired in the 1960s, watching your favorite television shows required the use of either an outdoor antenna or an indoor set of "rabbit ears." Either way, these had to be connected to the TV set. Viewers had a very limited number of channels to watch. Local stations typically carried three major networks (ABC, CBS, NBC), which broadcast nationwide mostly on channels 2 through 13 using VHF, or "very high frequency." Independent stations used UHF, or "ultra high frequency," and broadcast on channels 14 through 59. Most UHF stations were generally not affiliated with a specific network and had a limited range of signal that required a separate antenna to be attached to your television.

In the world of 1960s television, when a network agreed to produce a show, they paid a licensing fee to the studio. However, this fee did not cover the entire cost of production. If the show was successful, the network would pay more, usually half of the original fee, which then gave the network permission to air reruns of certain episodes. For those episodes that aired again, the studio would break even. The longer a series remained in production, the more episodes the network could choose to rerun, resulting in more profits for the studio.

The normal rule of thumb was that a series needed to stay on the air for a minimum of two or more seasons to turn a profit. If not, the studio almost always lost money, because anything less would not be enough for something called stripping. Stripping was used in television to ensure consistency and coherency in program scheduling. In stripping, a television series was given a regular weekly time slot, thereby appearing as a "strip" straight across the weekly schedule. This allowed broadcasters to deliver consistent content to targeted audiences because they would know when a certain demographic would be watching their programs, and would then make sure to air them at that time. Strip scheduling applied to programs that were shown on multiple consecutive days of the week, most commonly Monday through Friday. It was normally used for those programs that aired on a weekly basis during their original network run.

For much of the 1960s and into the early 1990s, stripping was the main model used in syndication by production studios to make money. A studio had a greater chance of turning a profit if it succeeded in producing at least three full seasons (about seventy-five episodes). The more episodes a show had, the more weeks could be stripped before repeating, thereby allowing a greater chance for the show to hold a viewer's attention and generate a following. With *Star Trek*, it took sixteen weeks to get through all three seasons before needing to show them again from the beginning.

Writer John D. Black was an associate producer during *Star Trek*'s first season. In commenting on the show's opening narration, which he helped write, he said, "There was a reason for it being five years. Sure, the Navy—and this, in a sense was the US Navy in space—will send you on a tour of duty, but not for five years. Truth is Gene was hoping the show would last five years! If you could get five seasons done, you were assured a long run in syndication."[1] In a newspaper article published five days before *Star Trek*'s network premiere, Charles Witbeck, a columnist for the *Buffalo Evening News*, wrote, "An expensive series to begin with, *Star Trek* must run two or more years in order to recoup the original investment for co-owners Desilu Studios and NBC."[2] Again, these comments help reinforce the television industry rule at the time that the more seasons a series filmed, the better its chances of being sold into syndication, resulting in more profits to the studio that created the show.

By the fall of 1967, the chances of *Star Trek* lasting long enough to survive through its second season were not looking good. The series moved from its original time slot of Thursday nights from 8:30 to 9:30 p.m. to the same time slot on Fridays. This new time proved to be a struggle for *Star Trek*'s demographic. As Roddenberry noted in an August 13, 1967, interview about the time change and its effects, he said, "We were making out fine where we were, but now we may lose many of the young people who've been watching, because Friday is the night they like to go out."[3]

Even with a poor time slot, NBC picked up *Star Trek*'s option and guaranteed its viewers a full second season. With a full twenty-six episodes secured for a second year, Herb Solow, *Star Trek*'s executive producer in charge of production,

CHAPTER EIGHT

was optimistic. "The likelihood of post network syndication became more of a reality," said Solow in the 1996 book *Inside Star Trek: The Real Story*, which he wrote alongside the show's coproducer, Robert Justman. Solow continued, "There would now be fifty-five one-hour shows to sell to local stations. And if the series was lucky enough to go into a third or fourth year, the possibility of financially breaking even became a small light, yet a light nonetheless, at the end of the tunnel."[4]

Unfortunately for *Star Trek*, the attainment of a second season didn't increase its viewership enough to make a difference. By late 1967, when it looked like the series would be unable to secure a third season, *Star Trek* fans attempted to save the show by engaging in a massive letter-writing campaign. The novelty and passion of this fan uprising generated a lot of attention.

Depending on whom you talk to, the number of letters that NBC received during this 1967–68 letter-writing campaign to save the third season seems to vary greatly, with no definitive count available. In a 1967 *Mailcall* pamphlet issued by NBC that included figures on internal fan mail, the network reported that "115,893 letters had been received as a result of the fan campaign, 52,358 during the month of February."[5] On the other hand, in David Gerrold's 1973 book *The World of Star Trek*, he quotes *Star Trek* superfan Bjo Trimble (Bjo is a nickname for Betty JoAnne), who helped spearhead the massive letter-writing effort: "'NBC told TV Guide that they had gotten two hundred thousand pieces of mail,'" said Trimble.[6] The *Hartford Courant* newspaper reported on March 17, 1968, that the network "received 114,667 pieces of mail in support of *Star Trek* between December and March with 52,151 arriving 'in the month of February alone.'"[7] Years later, in her 1982 book *On the Good Ship Enterprise*, Trimble said, "NBC admitted, unofficially, that over one million letters had crossed their desks. Within a year NBC was to announce during an interview on the mail campaign that only 500,000 had come in, and ten years later, NBC claimed that only 50,000 letters had been received!" Trimble also points out that each piece of correspondence that the network received counted only as "one letter" even if it included a petition signed with many names.[8] Adding to the confusion are the numbers tossed about by Roddenberry himself, who was often fast and loose with facts and figures.

However many there actually were, the letters sent during the campaign helped change the network's attitude about *Star Trek*. The grassroots activism combined with the honor given by the Smithsonian through its acquisition of the show's second pilot, and the support shown for the series by NASA and the larger aerospace community, all served to strengthen the credibility of *Star Trek* by conveying a level of respect, approval, and a certain cachet. Roddenberry was right when he said during a 1967 interview that "*Star Trek* is the 'in' show with the people who work at NASA, Caltech and the space plants." Boasting that NBC received an average of over 4,500 letters per week, Roddenberry went on to say that many of these came from "curators of museums and college presidents, all of whom appreciate our serious attempt to portray what space travel will be like."[9]

As mentioned in the previous chapter, on March 1, 1968, at 9:28 p.m., following the first-run airing of the *Star Trek* episode "The Omega Glory," an NBC announcer came on the air telling viewers that the series would continue to be seen on television that fall.[10]

Even though Roddenberry won the battle for a third season, *Star Trek* lost the war.

For its third season, *Star Trek* would end up still airing on Friday nights, but at an even later time period. As Roddenberry noted above, Friday evenings were date nights for the fourteen-to-twenty-one-year-old age group that made up most of *Star Trek*'s viewers. The show had already begun to lose its audience when the series switched to 8:30 p.m. Friday evenings for the second season. Now it would be on even later, with the show starting at 10:00 p.m. for the third season.

On June 3, 1969, "Turnabout Intruder," the last episode of *Star Trek*, aired on NBC. A little over a month later, NASA launched astronauts Neil Armstrong, Edwin "Buzz" Aldrin, and Michael Collins on the first mission to land men on the surface of another world. The success of Apollo 11 convinced the globe that travel to other worlds was now possible. Things that seemed limited to one's imagination or science fiction were now becoming everyday facts.

Although *Star Trek* was not lucky enough to make it into a fourth season, it did survive for three, and that was important. As Bjo Trimble said, "We knew back in those days, that anything that didn't have at least three seasons was never going to be rerun."[11]

"Television's Most Successful Unsuccessful Series"[12]

In 1967, Gulf+Western purchased Desilu Studios for $17 million. The studio that Desi Arnaz and Lucille Ball originally founded now became part of Paramount Pictures. Shortly after the buyout, a meeting took place at the historic Warwick Hotel in midtown Manhattan. Here is where the Beatles first stayed when they came to the US. It was here where Richard Block, president of Kaiser Broadcasting, and Bob Newgard, head of Paramount's Domestic Syndication, met over Saturday-morning breakfast in 1967 to talk about *Star Trek*.

Founded in 1958 by Henry K. Kaiser, the Kaiser Broadcasting Corporation owned and operated television and radio stations throughout the US. In the 1960s, Kaiser Broadcasting expanded its holdings by buying up or building new UHF stations in major cities throughout the country, such as Philadelphia,

CHAPTER EIGHT

Boston, Cleveland, Detroit, and San Francisco. UHF was viewed negatively because VHF had greater range.[13] Block owned six UHF stations and needed programming to fill the air.

He sought to buy *The Twilight Zone* but was talked out of it, a decision he later regretted when that series became a highly successful syndicated television show. Block, who was a big science fiction fan, wanted *Star Trek*. The only problem is that this was 1967, and Roddenberry's show had not yet completed its original network run on NBC.

In an unprecedented move, Block proposed to Newgard to buy the syndication rights to *Star Trek* before the series had completed its second season. Newgard was surprised at this offer because even Paramount hadn't yet decided whether *Star Trek* would survive. Maybe they'd even cancel it during the first season. Block was shrewd. He knew that by making an offer this early, he would pay less than the going rate because he was guaranteeing Paramount the sale and the money.

"I kept pushing Bob, saying, 'I want to buy *Star Trek*,' said Block, and he would say, "We don't even know if we're going to syndicate it." I still pushed, and eventually we scribbled the deal out on a napkin or menu for us to play the show."[14]

Feeling that both men had done well, the two shook hands on the deal, since no contract could be prepared or signed until after *Star Trek*'s network cancellation by NBC.

That Saturday-morning breakfast handshake proved critical.

After *Star Trek* was officially canceled in 1969, it had seventy-nine episodes in the can. The Kaiser New York breakfast handshake deal of two years earlier now resulted in a contract that helped ensure that *Star Trek* would live on through syndication.

The UHF stations that Kaiser had in the early 1970s were of limited range, and as a result, fewer audience members could be reached. To survive and compete against the more powerful VHF stations, Kaiser had to take a different approach.

KBHK, a Kaiser station in Oakland, California, decided to schedule *Star Trek* reruns at 6:00 p.m., directly opposite the time that their competitors were showing the evening news.[15] They found that the typical *Star Trek* viewers were young males who were not heavy news watchers. In addition, they also ran each episode in its original uncut network form, sacrificing valuable commercial time for content. Normally, when a series is run in syndication, each episode is cut by the individual stations to allow the insertion of commercials. The syndicated shows were on 16 mm, and these episode cuts produced film pieces or trims. These were gathered up and sold by fans at various *Star Trek* conventions. Roddenberry also found that these trims could be marketed and sold through his Lincoln Enterprises catalog.

Finally, the station ran them in the order that they were originally shown on NBC, even going so far as to include announcements for each episode informing viewers of the exact time and date of each initial NBC telecast. They ran the series daily for sixteen weeks and then, when that was done, started over. Audience response was phenomenal. Other Kaiser stations got word of the success of this new format and began doing the same thing.

Star Trek was syndicated not just on UHF stations. VHF channels also began to carry it. But it was the Kaiser programming approach that became the model that ensured the success of *Star Trek* in syndication. Soon, word spread to other stations, and the original seventy-nine *Star Trek* episodes were run, rerun, and rerun again all over the country and, eventually, the world. Within a few short years, *Star Trek* became the first television series to become more popular in syndication than it had been in its initial run.

The success of *Star Trek*'s syndication in the 1970s resulted in an explosion of fan activity. The first major convention devoted to the series occurred in 1972, bringing thousands of *Star Trek* fans to the Statler Hilton Hotel in New York City. Other conventions that followed drew equally impressive attendance. Spurred on by fan support for a canceled television show, Filmation Studios produced *Star Trek: The Animated Series*, which aired on Saturday mornings from 1973 to 1975 and resulted in the franchise's first Emmy award when it won the 1975 Outstanding Entertainment, Children's Series.

Star Trek took off through syndication like no other show in the history of television. Fans, increasing exponentially because of daily access to the series, refused to let the show die. The syndication of the original television series, coupled with the new animated show, served to promote the production of merchandise, including books, blueprints, and models. Frank Tupti, vice president for marketing and planning at New York's WPIX-TV, said in a 1975 interview that syndicated *Star Trek* "gives us double-digit ratings every Saturday and Sunday evening, and that's mostly adults."[16]

Paramount Television's syndication offices used publications such as *Variety* and *Broadcasting* to promote the phenomenal success of *Star Trek* by printing ads boasting the large number of viewer markets. These ads featured headlines such as "Take Off with Star Trek," "Star Trek Is Out of This World," "Star Trek Ratings Orbit on Any Heading," "Star Trek Keeps on Growing," and "No Other TV Series in Syndication History Has Taken Off This Way." These are just a few examples of the marketing tools used to catch the attention of television stations to encourage them to purchase *Star Trek*.[17] These ads clearly worked, and viewers could now watch the series on multiple outlets at multiple times over multiple days and even in multiple countries. This market saturation created a new generation of fans who missed watching the series during its original three-season network television run. Herb Solow once said that *Star Trek* was the "most successful unsuccessful series."¹⁸ The fact that interest in *Star Trek* lasted so long after the series formally ended in 1969 can be attributed not only to the enduring quality of the show but also to the wide viewership it achieved through syndication.

By the 1970s, *Star Trek*'s popularity continued to be felt even in the White House. NASA asked President Ford for help after the public began to press them on the name for their new "space shuttle." As it turns out, Richard Preston, whom we first met in chapter 6 when he asked Leonard Nimoy for a copy of the show's second pilot for the National Air and Space Museum archives, was now working alongside space activist Richard Hoagland on a letter-writing campaign to change the name of the space shuttle. Originally, NASA was going to name the world's first reusable spaceship "*Constitution*," but *Star Trek* fans thought otherwise. They sought to name it after their favorite starship and petitioned the White House through a massive letter campaign to change the name to "*Enterprise*." After a somewhat disorganized start, longtime *Star Trek* fan Bjo Trimble and her husband, John, took over the letter-writing effort.

"NASA has received hundreds of thousands of letters from the space-oriented 'Star Trek" group asking that the name 'Enterprise' be given to the craft," wrote William Gorog, a senior economic advisor in a September 3, 1976, White House memo to President Ford.¹⁹ Gorog continued, "This group comprises millions of individuals who are deeply interested in our space program. The name 'Enterprise' is tied in with the system on which the Nation's economic structure is built. Use of the name would provide a substantial human-interest appeal to the rollout ceremonies scheduled for this month in California, where the aeronautical industry is of vital importance." Gorog concluded, "In short, this situation could provide the same public interest as the CB radio provided for Mrs. Ford."²⁰ The CB (short for citizen band) radio that Gorog mentions turned out to be a successful public-relations tool for the White House. First Lady Betty Ford was caught up in the CB craze that was popular during the, and often used CB to talk to supporters during motorcades with her husband.

On Wednesday, September 8, 1976, at a press conference that included NASA administrator James Fletcher, President Gerald R. Ford made the following statement:

> A great many people have written to me in recent months, suggesting one name in particular for this spaceship, which will carry us not only into space but into the future. It is a distinguished name in American naval history, with a long tradition of courage and endurance. It is also a name familiar to millions of faithful followers of the science fiction television program "Star Trek." To explore the frontiers of space, there is no better ship than the space shuttle and no better name for that ship than the *Enterprise*.²¹

Four days later, a follow-up memorandum that included opinions of White House officials was sent from James Connor, the secretary of the cabinet, to President Ford. In that memo, James L. Cannon, a domestic-policy advisor to Ford, approved of the name, stating for the record that "it seems to me 'Enterprise' is an excellent name for the space shuttle. It would be personally gratifying to several million followers of the television show 'Star Trek,' one of the most dedicated constituencies in the country."[22]

Jack Marsh, who served as the assistant secretary of defense and was a cabinet member under President Ford, voiced support for the name but disapproved of the reasons behind it. "I have no objection to this selection of the name; however, I am not enthusiastic about the rationale for the selection," noted Marsh in his recorded comments about Ford's decision. "'Enterprise' is a famous name for vessels since the early days of the Republic. I think that is a far better reason than appealing to a TV fad."[23]

Others were more open to the idea behind the name. Myron Malkin, head of the space shuttle program at NASA, accepted the name change in good humor. "This way," he said in a telephone interview by the *Washington Post* the day after Ford announced the name, "we get a ready-made public." He pointed out that "there may be some closet 'Star Trek' fans (in his department) but nobody goes around with an arm band saying, 'I'm a 'Star Trek' fan.'"[24]

On Constitution Day, September 17, 1976,[25] with *Star Trek*'s theme song playing in the background, the nation's first space shuttle, OV-101 *Enterprise*, was unveiled at Rockwell's assembly plant in Palmdale, California.[26] Among the many invited guests attending the rollout ceremony was Gene Roddenberry, along with cast members of the original show. Also present was NASA's associate administrator George Low, who, as we read in the previous chapter, had attended the 1974 opening of the Smithsonian's first *Life in the Universe?* exhibit, where he saw *Star Trek*'s 11-foot production model *Enterprise* in the Arts and Industries Building.[27] Did seeing that television show model of a fictitious spacecraft have any influence on Low approving the name change for NASA's real spaceship? We can only speculate.

Star Trek's influence continued to be felt in the 1970s, not only in the naming of the space shuttle but in the recruitment for engineers and astronauts to help build and fly it. Nichelle Nichols, the actress who played *Star Trek*'s Lieutenant Uhura, was hired by NASA to recruit minority and female personnel to pursue careers in engineering and work as astronauts in the space agency's new 1978 astronaut class. Among those recruited through her efforts were Dr. Sally Ride, the first American female astronaut, and Guion Bluford, a US Air Force colonel who became the first African American astronaut. Also recruited were Dr. Judith Resnik and Dr. Ronald McNair, both of whom flew successful missions aboard the space shuttle prior to their deaths during the 1986 *Challenger* accident.

CHAPTER EIGHT

Additional recruits included future NASA leaders such as Charles Bolden, who would go on to be a veteran of three shuttle missions and become a NASA administrator, and Frederick D. Gregory, who went on three shuttle missions and worked as a NASA deputy administrator.

Gene Roddenberry died on October 24, 1991. On January 30, 1993, his widow, Majel Barrett, accepted NASA's Distinguished Public Service Medal on behalf of her husband from the space agency's administrator Dan Goldin during a ceremony held at the Smithsonian's National Air and Space Museum. The citation accompanying NASA's highest award read "For distinguished service to the Nation and the human race in presenting the exploration of space as an exciting frontier and a hope for the future."

In 1992, NASA flew some of Roddenberry's ashes aboard the space shuttle *Columbia*. Five years later, the private aerospace company Celestis flew more of his cremated remains aboard their Founders Flight, during which an Orbital Sciences Pegasus XL booster launched its payload into low earth orbit and then returned. Subsequent Celestis flights carried aloft into space the ashes of other *Star Trek* cast members, including Jimmy Doohan, the actor who played the ship's engineer, "Scotty." On January 8, 2024, a Celestis mission called the *Enterprise* flight, named in honor of the famous starship, carried into deep space the cremated remains from Roddenberry and his wife, Majel Barrett, and Doohan, along with the ashes of Nichelle Nichols and DeForest Kelley. The "passengers" on this flight will eventually orbit the sun forever, truly going where no one has gone before.[28]

Today, Gene Roddenberry's original vision of *Star Trek* still remains deeply relevant. Now owned by CBS/Paramount, the franchise continues to thrive as the series and its many offshoots have taken on new life with online streaming. With multiple new television series in production as well as talk of more cinematic versions, *Star Trek* continues to transcend generations, mediums, and even science itself thanks to the initial help and support given by NASA, the Smithsonian, and those within the aerospace community who believed in it.

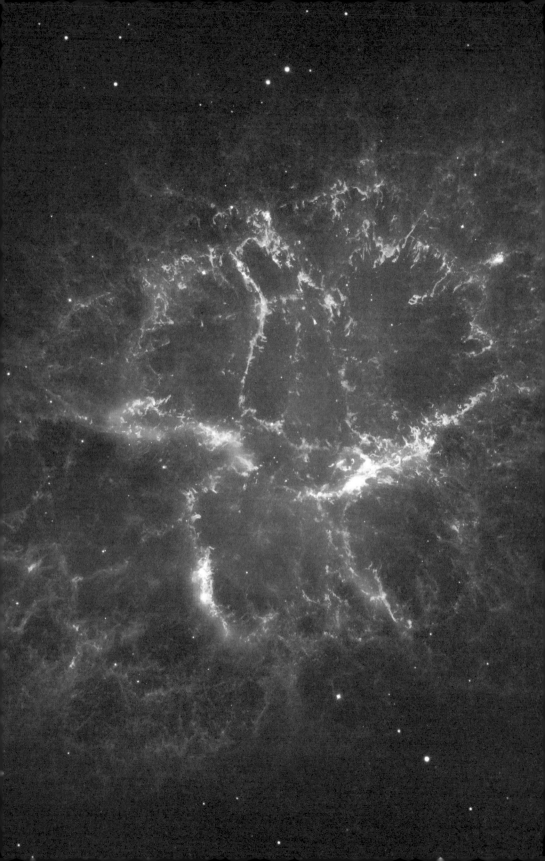

AFTERWORD

Space, the final frontier . . .

Roddenberry was not necessarily a lover of science fiction. Notable writers in the field were often surprised over how little he knew about the genre. But what he lacked in that subject, he more than made up by his knowledge about the human condition. It also helped that he surrounded himself with others who knew even more. Roddenberry recognized that in *Star Trek* he could tell stories that could not be easily told in any other way. At the time that *Star Trek* first aired, spaceflight was often in the news, and Roddenberry realized that here was a field where he could disguise his messages while still telling a story. The space race was an anomaly in history, one that formed out of a unique set of conditions involving post–World War II geopolitics and technologies. Rockets that emerged during this period had the ability to both explore and destroy. This dichotomy created challenges within the aerospace community. One could not explore space without the funds and the technology driven by the military industrial complex needed to get you there. As was often said in the business, "No bucks, no Buck Rogers." Some enlightened folks both in government and industry saw *Star Trek* as a means to promote space exploration. Though not entirely symbiotic, Roddenberry knew that the two entities could help each other. As we have seen, *Star Trek* was influenced by the aerospace community. At the same time, *Star Trek* influenced the aerospace community by nurturing existing public interest and support.

These are the voyages of the starship *Enterprise* . . .

Star Trek was not just about people, but also about the ship. Every episode begins with the opening lines that include "These are the voyages of the starship *Enterprise*." When you first approach the 11-foot production model on display at the Smithsonian, you see a shape like no other. Upon closer inspection, you notice the many portholes that remind you of just how big it really is. The starship *Enterprise* allowed viewers to travel every week through space via the medium of television. It really was the wagon train that took us to the stars.

Its five-year mission . . .

Star Trek did not last for five years, but the show did give us three full seasons, which helped secure its future in syndication. Through reruns, *Star Trek* endured long enough to make it to 1977, when the success of such blockbuster films as *Star Wars* and *Close Encounters of the Third Kind* caused Hollywood to finally take notice of science fiction. As a result, *Star Trek* was brought back to appear in a series of major motion pictures and television spinoffs that extended well beyond its original five-year mission. Now multiple generations of kids who grew up with *Star Trek* became the engineers, scientists, and explorers that gave us the Space Shuttle, the Hubble Space Telescope, the International Space Station, and rovers on Mars.

To explore strange new worlds . . .

During the time of *Star Trek*'s original airing, nations were sending spacecraft to other planets, some for the very first time. NASA was getting ready to land men on the moon, which they did only a little over a month after the show ended. Roddenberry saw space as a place where he could tell stories about other worlds. Were these other worlds like our own and, if so, how did their inhabitants overcome their differences? *Star Trek* was shown on one channel while real space missions made the news on another. Television could not help but see the future unfold before them as astronauts went into space and walked on the moon. NASA showed us what we could do today, while *Star Trek* showed us what we would do tomorrow.

To seek out new life and new civilizations . . .

At the time of *Star Trek*, we had the technology to help us answer the question of if there is other life in the universe. How would life be different on other worlds and how would we react if we encountered it? Roddenberry showed us through *Star Trek* that different was not necessarily bad. Acceptance of differences in appearances, cultures, ethnicities, and genders remained a central theme throughout *Star Trek*, which is one characteristic that has made the show so enduring today. Many of those mentioned in this book who helped create *Star Trek* faced challenges themselves. Stephen Whitfield grew up poor. After his father was killed, his mother was forced to raise him and his siblings by herself during the height of the Depression. Harvey Lynn also came from a family of limited means who saw a way out of his surroundings through the military.

AFTERWORD

Kellam De Forest was born premature, which resulted in coordination issues that also stunted his growth. Richard Preston was dyslexic, and Fred Durant had ulcers. Perhaps these people saw *Star Trek* as a way to rise above their circumstances and, in so doing, served as motivation for them to contribute to the show. To them, *Star Trek* presented a future that they saw as being better than the present.

To boldly go where no man has gone before . . .

On Halloween 2023, a bombshell hit the *Star Trek* world with fallout that circled the globe. Someone had posted on eBay what appeared to be the first filming model of the starship *Enterprise*.

It turns out that the model was indeed original and made for the television series. The 3-foot (33-inch model) was built in the 1960s by Richard Datin, who was also involved in the making of 11-foot production model. Both *Enterprise* models were used in the filming of the show's first two pilots. Shots of each can be seen throughout series, including the main title and opening credits, where the smaller model was used for the "whooshing" flybys.

Fans went nuts over the news, and the internet was abuzz. Roddenberry owned the model for years, but later it went missing and was assumed lost. Now it had been found.

The found 3-foot model represents one of the earliest physical representations of a cultural icon. It remains a unique part of the franchise's history, and, as such, it is hoped that one day it will be preserved and, like the Smithsonian's 11-foot *Enterprise* model, displayed as another reminder of the show's enduring influence in popular culture.

During a visit to the Smithsonian's National Air and Space Museum in 2023, I studied the people looking at the 11-foot production model *Enterprise*. It seemed strange to watch people gather around a prop that shares floor space with other notable artifacts such as the original Wright Flyer, the *Spirit of St. Louis*, and the Apollo 11 *Columbia* spacecraft.

The *Enterprise* model remains important not only for what it is, but for what it represents . . . inspiration. Konstantin Tsiolkovsky of Russia / Soviet Union, Hermann Oberth from the German-speaking province of Transylvania (now part of Romania), Eugen Sanger of Austria, Wernher von Braun from Germany, and Robert Goddard from the US all gave credit to writers such as Jules Verne as being the stimulus for contemplating space exploration. Similarly, many prominent space scientists and rocket engineers of today will tell you that it was science fiction that influenced them in deciding their life's work.

For some, *Star Trek* was that inspiration, since the show clearly sends a message that our transformation into a spacefaring civilization is not only possible, but inevitable. *Star Trek*, for many kids like me who grew up in the 1970s, was a vision of tomorrow validated by the success of the Apollo lunar landings. It seemed like we were boldly going for sure into the twenty-first century, where we would see regular flights into space, permanent bases on the moon, and routine visits to other planets. This seemed like a viable future, one that we would experience in our lifetime.

Andrew Chaikin, author of the bestselling book *A Man on the Moon: The Voyages of the Apollo Astronauts*, said of the Apollo lunar landings, "How could the most futuristic thing humans had ever done be so far in the past. In the narrative of the Space Age, Apollo is a chapter that is jarringly out of sequence." Gene Cernan, the last American to walk on the moon, said, "It's almost as if Kennedy grabbed a decade out of the 21st century and spliced it into the 1960s."[1]

We have not gone far from Earth's cradle. As of this writing, the International Space Station is scheduled to burn up in the atmosphere within the next five to six years. Even after nearly three decades of being in orbit, it has not gotten us closer to the future in space that I had imagined.

Rest assured, there will be more humans who will fly into space. Thanks to billionaires who grew up on *Star Trek*, such as Jeff Bezos, an acknowledged fan of the original television series who now builds rockets, these people are fulfilling the dream of routine access to the final frontier, a dream that was once promised by the space shuttle but never really happened. But these people are not going where I want them to go.

Space tourism has become real. But access to space remains stubbornly restricted to low earth orbit. Except for brief forays during Apollo, humans have not ventured further. The last lunar landing ended in 1972.

Even before the twenty-first century arrived and I began working for NASA in 1998, I had not given up hope that our future might still resemble the one envisioned by Stanley Kubrick and Arthur C. Clarke in the film *2001: A Space Odyssey*.

During the years since leaving NASA and in researching and writing this book, it was a privilege to talk to individuals who either worked directly on the original *Star Trek* television series or just grew up watching it. Listening to them tell me how the show impacted their lives or how they impacted *Star Trek* was both inspirational and motivational.

Is *Star Trek* an anomaly like the Apollo program? Both were products of the same decade. Apollo was the opening act to a period of time when our species could and did leave this world to venture out and explore another. Apollo was our first giant leap into space, and *Star Trek* was there to remind us not to get

AFTERWORD

discouraged. We need to continue moving forward. In the end, when we boldly go where no man has gone before, the fact that it took us so long to do so will hardly matter at all. *Star Trek* reminds us to still dream. When I sit down and watch an episode from the original series, I feel like a kid again, full of inspiration and hope that someday, there will be another great age of human space exploration and, this time, it will last.

◀ INSPIRED **ENTERPRISE** ▶

The new National Air and Space Museum building opened to much fanfare on July 1, 1976. The Smithsonian commissioned this artwork done by artists Thomas Hall and Melinda Mason, which was sold as a poster that helped illustrate the location of all the galleries and exhibits. The *Life in the Universe?* exhibit is located on the ground floor near the middle of the building. If you look closely, you can see the 11-foot production model *Enterprise* hanging above the exhibit's exit. I bought this poster at age fifteen, when I first visited the museum in 1978. *Photo courtesy of the author*

ENDNOTES

Foreword

1. Daniel Leonard Bernardi, *Star Trek and History: Race-ing Toward a White Future* (Rutgers University Press, 1998), 11.

2. See, for example, Bernardi, *Star Trek and History*; Andre M. Carrington, *Speculative Blackness: The Future of Race in Science Fiction* (University of Minnesota Press, 2016); Robin Roberts, *Sexual Generations: "Star Trek: The Next Generation" and Gender* (University of Illinois Press, 1999); Jennifer E. Porter, ed., *Star Trek and Sacred Ground: Explorations of Star Trek, Religion and American Culture* (State University of New York Press, 1999); and Nancy R. Reagin, ed., *Star Trek and History* (John Wiley and Sons, 2013).

3. Just a few examples include: Henry Jenkins, III, "*Star Trek* Rerun, Reread, Rewritten: Fan Writing as Textual Poaching," *Critical Studies in Mass Communication* 5, no. 2 (June 1988): 85–107; Lincoln Geraghty, "A Network of Support: Coping with Trauma through *Star Trek* Fan Letters," *Journal of Popular Culture* 39, no. 6 (2006): 1,002–24; and Bruce E. Drushel, ed., *Fan Phenomena: Star Trek* (Intellect Books, 2013).

4. See, for example, Allan Asherman, *The Star Trek Compendium* (Simon and Schuster, 1981); J. M. Dillard, *Star Trek, "Where No One Has Gone Before": A History in Pictures* (Pocket Books, 1994); Herbert F. Solow and Robert H. Justman, *Inside Star Trek: The Real Story* (Pocket Books, 1996); Marc Cushman with Susan Osborn, *These Are the Voyages: TOS; Season One* (Jacobs/Brown, 2013); Marc Cushman with Susan Osborn, *These Are the Voyages: TOS; Season Two* (Jacobs/Brown, 2014); and Marc Cushman with Susan Osborn, *These Are the Voyages: TOS; Season Three* (Jacobs/Brown, 2014).

5. See, for instance, Roberta Pearson and Máire Messenger Davies, *Star Trek and American Television* (University of California Press, 2014).

Chapter 1

1. Stephen E. Whitfield and Gene Roddenberry, *The Making of Star Trek* (Ballantine Books, 1972 [originally 1968]), 22.

2. Gene Roddenberry, *Star Trek is…* (first draft, March 11, 1964), 2.

3. Roddenberry, *Star Trek is…*, 3.

4. Ellen Torgeson, "*Star Trek* Is Science Fact, Says Its Creator," *TV Times*, November 15, 1967, 7. This was originally published in Australia as their equivalent to *TV Guide* here in the United States. The article misspells Ellen Torgerson Shaw's name.

5. Giuseppe Cocconia and Philip Morrison, "Searching for Interstellar Communications," *Nature* 184, no. 4690 (September 19, 1959): 844–846.

6. Frank Drake, "The Drake Equation Revisited: Part I," *Astrobiology Magazine*, September 29, 2003, https://www.spacedaily.com/reports/The_Drake_Equation_Revisited_Part_I.html.

7. Whitfield and Roddenberry, *The Making of Star Trek* (1972), 17.

8. "A Brief History of RAND," https://www.rand.org/about/history/a-brief-history-of-rand.html.

9. Stephen H. Dole, *Habitable Planets for Man* (New York: Blaisdell, 1964), vii.

10. Dole, *Habitable Planets for Man.*

11. Roddenberry, *The Star Trek Writers/Directors Guide*, Third Revision, April 17, 1967, 2.

12. Whitfield and Roddenberry, *The Making of Star Trek* (1972), 44.

13. Gene Roddenberry, *Star Trek Studio Pitch,* first draft, March 11, 1964. Not every alien world had to look entirely different from Earth. Some could look like ancient Egypt and others like the Roman Empire. Other times, similarities could result from Earth contamination like the Nazi-ruled planet seen in the *Star Trek* episode "Patterns of Force." Still others could look like Earth, due to natural evolution, such as the planet seen in the episode "Bread and Circuses." This "natural evolution" idea reached its peak in the episode "Miri," where coastlines of this supposedly distant world (appearing in the pre-title sequence) were somehow the same as those we have on Earth. No explanation for this is given for this coincidence, and it doesn't play a role in the plot. Indeed, this may simply be a case where the scriptwriter was taking the metaphor too literally. It makes for a great opening "sting" but ultimately leads nowhere. Fortunately, the idea that the geographic features of Earth would be replicated on another planet became less popular in the series over time. However, the idea did make sense when the planet was seen as an alternative universe version of Earth, a theme that was briefly explored in the episode "Mirror, Mirror."

14. Joan Schmitt "Male Call" *Los Angeles Citizen News*, January 30, 1965. It should be noted that in casual correspondence, Roddenberry and others, including the press, tend to use the spelling "Rand" instead of the correct spelling "RAND" which stands for **R**esearch **AN**d **D**evelopment. Jeffrey Hunter never got to play Captain Pike again, even though he was offered to do so. He turned it down when Roddenberry asked him to reprise his role for a second pilot that NBC had approved after turning down the first one as being "too cerebral." William Shatner stepped in to play the lead as Captain James T. Kirk in "Where No Man Has Gone Before." After executives viewed it, the second time proved to be the charm and NBC gave the green light to begin filming *Star Trek*.

15. Whitfield and Roddenberry, *The Making of Star Trek* (1972), 202. In addition, Stephen Dole, in the first published edition of *Habitable Planets for Man* (1964), notes on p. 103 that "since the Galaxy has a volume of about $1.6 \times 10^{(12)}$ cubic parsecs, this means that the total number of habitable planets in the Galaxy is about 600 million."

16. Torgeson, "*Star Trek* Is Science Fact, Says Its Creator," 7.

17. After the war, Prickett went on to study nuclear physics at Ohio State University, then worked at the Pentagon doing research and development in nuclear power. Prickett was the USAF liaison officer for the Project Orion nuclear powered spacecraft. See https://en.wikipedia.org/wiki/Don_Prickett.

18. Donald Prickett to Gene Roddenberry, May 25, 1964, Gene Roddenberry Collection, Correspondence/General Files, Correspondence, Miscellaneous, box 27, folder 17, Gene Roddenberry Papers, (Collection PASC 62), UCLA Library Special Collections, Charles E. Young Research Library, University of California, Los Angeles.

19. Donald Prickett to Jack Whitener, May 25, 1964, *Antiques Roadshow*, PBS, season 23, episode 23, air date July 8, 2019. This episode was originally filmed in 2012 in Seattle. Also comments made by Theresa Whitener during the PBS show.

20. Theresa Whitener during PBS *Antiques Roadshow*, Season 23 Episode 23, Air Date July 8, 2019. This episode was originally filmed in 2012 in Seattle, Washington. Also comments made by Theresa Whitener during the PBS episode.

21. Whitener, *Antiques Roadshow.*

22. Gene Roddenberry to David Jones, June 18, 1964, Gene Roddenberry Collection, Correspondence/General Files, Lynn, Harvey, box 29, folder 1, Gene Roddenberry Papers, (Collection PASC 62), UCLA Library Special Collections, Charles E. Young Research Library, University of California, Los Angeles.

◀ ENDNOTES ▶

23. David Jones to Gene Roddenberry, hand-written note, June 26, 1964, Gene Roddenberry Collection, Correspondence/General Files, Lynn, Harvey, box 29, folder 1, Gene Roddenberry Papers, (Collection PASC 62), UCLA Library Special Collections, Charles E. Young Research Library, University of California, Los Angeles.

24. Jack Whitener to Gene Roddenberry, June 30, 1964, Gene Roddenberry Collection, Correspondence/General Files, Lynn, Harvey, box 29, folder 1 Gene Roddenberry Papers, (Collection PASC 62), UCLA Library Special Collections, Charles E. Young Research Library, University of California, Los Angeles.

25. San Angelo Central High School 1937 class yearbook.

26. Dennis Lynn, interview by Glen E. Swanson, October 29, 2022.

27. Lynn, interview by Swanson.

28. Lynn, interview by Swanson.

29a. Lynn, interview by Swanson.

29b. Glen E. Swanson, email message to Dennis Lynn, March 7, 2023.

30. Lynn, interview by Swanson.

31. Lynn, interview by Swanson.

32. "One Angeloan Is Promoted, Another on Duty in Korea," *San Angelo (TX) Evening Standard*, January 17, 1956, 7.

33. Gene Roddenberry to Oscar Katz, memo, July 10, 1964, Gene Roddenberry Collection, Correspondence/General Files, Lynn, Harvey, box 29, folder 1, Gene Roddenberry Papers, (Collection PASC 62), UCLA Library Special Collections, Charles E. Young Research Library, University of California, Los Angeles.

34. Gene Roddenberry to Harvey Lynn, July 10, 1964, Gene Roddenberry Collection, Correspondence/General Files, Lynn, Harvey, box 29, folder 1, Gene Roddenberry Papers, (Collection PASC 62), UCLA Library Special Collections, Charles E. Young Research Library, University of California, Los Angeles.

35. Harvey Lynn to Gene Roddenberry, July 16, 1964, Gene Roddenberry Collection, Correspondence/General Files, Lynn, Harvey, box 29, folder 1, Gene Roddenberry Papers, (Collection PASC 62), UCLA Library Special Collections, Charles E. Young Research Library, University of California, Los Angeles.

36. Lynn, interview by Swanson.

37. Harvey Lynn to Gene Roddenberry, September 14, 1964, Gene Roddenberry Collection, Produced Episodes, "The Menagerie" & Materials From "The Cage," box 7, folder 6, Gene Roddenberry Papers, (Collection PASC 62), UCLA Library Special Collections, Charles E. Young Research Library, University of California, Los Angeles.

38a. Lynn to Roddenberry letter (September 14, 1964); and Whitfield and Roddenberry, *The Making of Star Trek* (1972), 90–99. Note this is one of the earliest appearances of the term "space shuttle."

38b. Helen T. Wells, Susan Whiteley and Carrie E. Karegeannes, *Origins of NASA Names*, NASA History Series NASA SP-4402, (NASA Scientific and Technical Information Office, 1976), 111–115. Here we have an early description of what became the "Tractor Beam." It should be noted that Lynn's mention of the term "space shuttle" reflects a familiarity with some of the newer terminology being used by the technical and trade publications of the time. By 1964, the term "space shuttle" was beginning to appear more frequently. During the previous year, the newsletter *Defense Space Business Daily* had begun to refer to Air Force and NASA lifting-body tests as "Space Shuttle." The newsletter's editor-in-chief, Norman L. Baker, said at the time that "the newsletter had tried to reduce the name 'Aerospaceplace' to 'Spaceplane' for that project and had moved from that to 'Space Shuttle' for reusable, back-and-forth space transport concepts."

38c. Walter J. Dornberger, "Space Shuttle for the Future: The Aerospaceplane," *Rendezvous* 4, no. 1 (1965): 2–5; and Dornberger, "The Recoverable, Reusable Space Shuttle," *Astronautics and Aeronautics*, November 1965, 88–94. Dr. Walter Dornberger was one of the original members of Wernher von Braun's German rocket team in Peenemunde who, after WWII, came to the United States and became Vice President for Research of Textron Corporation's Bell Aerosystems Company. He authored a study entitled "Space Shuttle for the Future: The Aerospaceplane" which was released in 1965 by Bell Aerosystems. That same year, Dornberger gave a talk at the University of Tennessee entitled "The Recoverable, Reusable Space Shuttle."

39. The spelling of "Starfleet" evolved over time. See Chapter 4 for a more detailed explanation of the term's history.

40. Harvey Lynn to Gene Roddenberry, September 14, 1964, Gene Roddenberry Collection, Correspondence/General Files, Lynn, Harvey, box 29, folder 1; Harvey Lynn to Gene Roddenberry, letter, September 14, 1964, Gene Roddenberry Collection, Produced Episodes, "The Menagerie" & Materials From "The Cage," box 7, folder 6, Gene Roddenberry Papers, (Collection PASC 62), UCLA Library Special Collections, Charles E. Young Research Library, University of California, Los Angeles; and Whitfield and Gene Roddenberry, *The Making of Star Trek* (1972), 94.

41. Edward Gross and Mark A. Altman, *The Fifty-Year Mission: The Complete, Uncensored, Unauthorized Oral History of Star Trek: The First 25 Years* (St. Martin's, 2016), 83.

42. Harvey Lynn to Gene Roddenberry, September 14, 1964, Gene Roddenberry Collection, Correspondence/General Files, Lynn, Harvey, box 29, folder 1, Gene Roddenberry Papers, (Collection PASC 62), UCLA Library Special Collections, Charles E. Young Research Library, University of California, Los Angeles.

43. Harvey Lynn Remittance Advice, October 15, 1964, Gene Roddenberry Collection, Correspondence/General Files, Lynn, Harvey, box 29, folder 1, Gene Roddenberry Papers, (Collection PASC 62), UCLA Library Special Collections, Charles E. Young Research Library, University of California, Los Angeles.

44. Harvey Lynn to Roddenberry, March 15, 1966, Gene Roddenberry Collection, Correspondence/General Files, Lynn, Harvey, box 29, folder 1; Harvey Lynn to Roddenberry, letter, June 10, 1966, Gene Roddenberry Collection, Correspondence/General Files, Lynn, Harvey, box 29, folder 1, Gene Roddenberry Papers, (Collection PASC 62), UCLA Library Special Collections, Charles E. Young Research Library, University of California, Los Angeles.

45. Gene Roddenberry to Harvey Lynn, June 24, 1966, Gene Roddenberry Collection, Correspondence/General Files, Lynn, Harvey, box 29, folder 1, Gene Roddenberry Papers, (Collection PASC 62), UCLA Library Special Collections, Charles E. Young Research Library, University of California, Los Angeles.

46. Gene Roddenberry to Coon, memo, August 15, 1966, Gene Roddenberry Collection, Correspondence/General Files, Lynn, Harvey, box 29, folder 1, Gene Roddenberry Papers, (Collection PASC 62), UCLA Library Special Collections, Charles E. Young Research Library, University of California, Los Angeles; Ed Perlstein to Dave Atwood, memo, April 21, 1967, Gene Roddenberry Collection, Correspondence/General Files, Lynn, Harvey, box 29, folder 1, Gene Roddenberry Papers, (Collection PASC 62), UCLA Library Special Collections, Charles E. Young Research Library, University of California, Los Angeles;

47. Dennis Lynn, interview by Glen E. Swanson, August 28, 2024.

48. Harvey Lynn to Michael Ellington, Jr., memo, December 21, 1966, Gene Roddenberry Collection, Correspondence/General, Fan Letters, 1966, box 28, folder 1, Gene Roddenberry Papers, (Collection PASC 62), UCLA Library Special Collections, Charles E. Young Research Library, University of California, Los Angeles.

49. Ed Perlstein to Dave Atwood, memo, April 21, 1967, Gene Roddenberry Collection, Correspondence/General Files, Lynn, Harvey, box 29, folder 1, Gene Roddenberry Papers, (Collection PASC 62), UCLA Library Special Collections, Charles E. Young Research Library, University of California, Los Angeles.

50. Ellen Torgeson, "*Star Trek* Is Science Fact, Says Its Creator," 7.

51. Torgeson, "*Star Trek* Is Science Fact."

52. *Star Trek* promotional brochure issued by Desilu Sales, ca 1966.

53a. "New Space Adventure Series Starts in July BBC-1 to Screen Top American Series 'Star Trek,'" BBC Press Service.

53b. David Kaufman, "'Trek' Into Future; Reagan Win Would Cue 'Death' Exit," *Daily Variety* 132, no. 2, June 7, 1966, 9.

54. "Frequently Asked Questions," RAND, https://www.rand.org/about/faq.html#ive-heard-a-lot-of-rumors-about-.

55. Whitfield and Roddenberry, *The Making of Star Trek* (1972), 398–399.

56. Michael Okuda and Denise Okuda, *The Star Trek Encyclopedia: A Reference Guide to the Future,* rev. and exp. ed., vol. 1, *A–L* (Harper, 2016), 218.

57. Michael Okuda, email message to Glen E. Swanson, September 2, 2022.

58. Gene Roddenberry to Kellam de Forest, memo, April 11, 1966, Gene Roddenberry Collection, Correspondence/General Files, Whitney, Grace Lee, box 31, folder 13, Gene Roddenberry Papers, (Collection PASC 62), UCLA Library Special Collections, Charles E. Young Research Library, University of California, Los Angeles.

59. The choice of using Rand as a character's last name as a nod to the RAND Corporation was not without precedent. Commander Christopher "Kit" Draper, the lead character in the 1964 film *Robinson Crusoe on Mars* played by Paul Mantee, seems to have been a subtle way to acknowledge Draper Labs, another research-and-development organization similar to RAND. Founded in 1932 by Charles Stark Draper at the Massachusetts Institute of Technology (MIT) to develop aerospace instrumentation for guidance, navigation and control, it was originally called the MIT Instrumentation Laboratory. Draper was a well-known name in the aerospace industry during the 1950s and 1960s, responsible for building many guidance systems for missiles and rockets. One of their best-known achievements in this field was developing the Apollo Guidance Computer, the first silicon integrated circuit-based computer. By 1970, the MIT Instrumentation Laboratory was renamed Draper Labs after its founder and separated from MIT in 1973 to become an independent, non-profit organization.

60. Email exchange between Greg Tyler and Harvey P. Lynn III, August 2, 2002, http://www.trekplace.com/harveyplynnjr.html.

61. Harvey Lynn's reports on episodes of the *Star Trek: The Animated Series* have been posted by Roddenberry.com. Specifically, reports on "The Lorelei Signal" (May 13, 1973) and "Mudd's Passion" (May 14, 1973).

62. Dennis Lynn, interview by Glen E. Swanson, October 29, 2022. Glen E. Swanson email exchange with Dennis Lynn, March 13, 2023.

Chapter 2

Portions of this chapter were published in Glen E. Swanson, "In Memoriam: Kellam de Forest, Who Gave Us Stardates and the Gorn." *Space Review*. January 25, 2021. https://www.thespacereview.com/article/4110/1.

1. Ann de Forest, interview by Glen E. Swanson, March 24, 2023.

2. Huston directed *The African Queen,* and James Agee wrote the screenplay adapted from a novel by C. S. Forester.

3. De Forest, interview by Swanson, 2023.

4. Gene Handsaker, "Kellam de Forest Runs TV's Reference Service," *Janesville (WI) Daily Gazette*, June 7, 1967, 4A.

5. Kellam de Forest, interview by Glen E. Swanson, December 22, 2019.

6. Handsaker, "TV's Reference Service," 4A.

7. Michael Kmet, "Script Clearance and Research: Unacknowledged Creative Labor in the Film and Television Industry," Mediascape: UCLA's Journal of Cinema and Media Studies, August 28, 2012.

8. R. A. Lee, "Value of Studio Libraries: Closings Called False Economy," *Weekly Variety*, vol. 260, no. 8, (January 3, 1973): 8, 46.

9. Stephen Farber, "Before the Cry of 'Action!' Come the Painstaking Effort of Research," *New York Times*, March 11, 1984, section 2, 17.

10a. Ernest Sloman, "LOOKING FOR A PEARL PEELER?" *TV Guide*, 13, no. 32, (August 7, 1965), Issue 646, 12–14.

10b. Handsaker, "TV's Reference Service," 4A.

11. "De Forest at Desilu, *Daily Variety*, 100, no. 33, (July 23, 1958): 11.

12. Kmet, "Script Clearance and Research."

13. De Forest, interview by Swanson, 2019.

14. Ann de Forest interview by Glen E. Swanson, March 24, 2023.

15. De Forest, interview by Swanson, 2023.

16. Gene Roddenberry to Kellam de Forest, memo, August 25, 1964, Gene Roddenberry Collection, Correspondence/General Files, Art Direction, box 27, folder 4, Gene Roddenberry Papers, (Collection PASC 62), UCLA Library Special Collections, Charles E. Young Research Library, University of California, Los Angeles.

17. De Forest, interview by Swanson, 2019.

18. De Forest, interview by Swanson, 2019.

19. De Forest, interview by Swanson, 2019; and Roddenberry to Kellam de Forest, memo, February 22, 1967, Gene Roddenberry Collection, Correspondence/General Files, De Forest Research, box 27, folder 21, Gene Roddenberry Papers, (Collection PASC 62), UCLA Library Special Collections, Charles E. Young Research Library, University of California, Los Angeles.

20. "LOOKING FOR A PEARL PEELER? The first thing you do is call that fellow Kellam de Forest," *TV Guide*, 13, no. 32, (August 7–13, 1965), 12–14.

21. Peter Sloman, interview by Glen E. Swanson, January 22, 2021.

22. Sloman, interview by Swanson, 2021.

23. Sloman, interview by Swanson, 2021.

24. De Forest, interview by Swanson, 2019.

25. Sloman, interview by Swanson, 2021.

26. Gene Roddenberry to Herb Solow, memo, July 6, 1966, Gene Roddenberry Collection, Correspondence/General Files, box 29, folder 1, Gene Roddenberry Papers, (Collection PASC 62), UCLA Library Special Collections, Charles E. Young Research Library, University of California, Los Angeles.

27. Gene Roddenberry to Kellam de Forest, memo, November 3, 1966, Gene Roddenberry Collection, Correspondence/General Files, De Forest Research, box 27, folder 21, Gene Roddenberry Papers, (Collection PASC 62), UCLA Library Special Collections, Charles E. Young Research Library, University of California, Los Angeles.

28. Dorothy Fontana to Kellam de Forest, memo, December 19, 1966, Gene Roddenberry Collection, Correspondence/General Files, De Forest Research, box 27, folder 21, Gene Roddenberry Papers, (Collection PASC 62), UCLA Library Special Collections, Charles E. Young Research Library, University of California, Los Angeles.

29. Handsaker, "TV's Reference Service," 4A.

ENDNOTES

30. Kellam de Forest to Roddenberry, memo, February 23, 1967, Gene Roddenberry Collection, Correspondence/General Files, Dc Forest Research, box 27, folder 21, Gene Roddenberry Papers, (Collection PASC 62), UCLA Library Special Collections, Charles E. Young Research Library, University of California, Los Angeles.

31. Kellam de Forest research report on "The Trouble With Tribbles," August 11, 1967; and David Gerrold (UCLA), "The Trouble with Tribbles," Gene Roddenberry *Star Trek* Television Series Collection (1966–1969). Also "Kellam de Forest," Fact Trek, January 26, 2021, https://www.facttrek.com/blog/kellam.

32. Solow and Justman, *Inside Star Trek*, 334.

33. Kellam de Forest research report on "The Trouble With Tribbles," August 11, 1967; and Gerrold, "The Trouble with Tribbles," Gene Roddenberry Star Trek Television Series Collection (1966–1969). Also, "Kellam de Forest," Fact Trek, January 26, 2021, https://www.facttrek.com/blog/kellam.

34. Joseph T. Sullivan, "Science Drama to Debut Sept. 15," *Boston Herald*.

35. Gene Roddenberry to Kellam de Forest, memo, November 3, 1966, Gene Roddenberry Collection, Correspondence/General Files, De Forest Research, box 27, folder 21, Gene Roddenberry Papers, (Collection PASC 62), UCLA Library Special Collections, Charles E. Young Research Library, University of California, Los Angeles.

36. Herb Solow to Roddenberry, memo, December 14, 1966, Gene Roddenberry Collection, Correspondence/General Files, De Forest Research, box 27, folder 21, Gene Roddenberry Papers, (Collection PASC 62), UCLA Library Special Collections, Charles E. Young Research Library, University of California, Los Angeles.

37. Gene Roddenberry to Kellam de Forest, memo, December 15, 1966, Gene Roddenberry Collection, Correspondence/General Files, De Forest Research, box 27, folder 21, Gene Roddenberry Papers, (Collection PASC 62), UCLA Library Special Collections, Charles E. Young Research Library, University of California, Los Angeles.

38. Gene Roddenberry to Herb Solow, memo, January 30, 1967, Gene Roddenberry Collection, Correspondence/General Files, De Forest Research, box 27, folder 21, Gene Roddenberry Papers, (Collection PASC 62), UCLA Library Special Collections, Charles E. Young Research Library, University of California, Los Angeles.

39. Kellam de Forest, interview by Glen E. Swanson, December 22, 2019.

40. Farber, "Before the Cry of 'Action!'" section 2, 17.

41. Gene Roddenberry to Kellam de Forest, memo, May 14, 1968, Gene Roddenberry Collection, Correspondence/General Files, De Forest Research, box 27, folder 21, Gene Roddenberry Papers, (Collection PASC 62), UCLA Library Special Collections, Charles E. Young Research Library, University of California, Los Angeles.

42. Lee, "Value of Studio Libraries," 8, 46.

43. "De Forest Research Available to 20th," *Daily Variety*, Vol. 154 no. 52, February 15, 1972, 2.

Chapter 3

Portions of this chapter were published in Glen E. Swanson, "To Boldy Go Where No Model Has Gone Before… the USS *Enterprise*, Star Trek and the AMT Corporation," *Michigan History Magazine*, 105, no. 1, (January/February 2021): 18–23.

1. "Star Trek Merchandise," Gene Roddenberry to Gene Coon, memo, May 8, 1967, Juliens' Auctions Item #3359, April 22–23, 2023.

2. Susan Skorpus, "'Vision' Quest: Reno Writer Publishes Second Behind the Scenes 'Star Trek' Book," *Reno (NV) Gazette*, March 8, 1998, 16. Because few *Star Trek* fans know him by his birth name of Stephen Edward Poe, I chose to use Whitfield. Even though he later began using his birth name, fans better know him today as Stephen Whitfield.

3. "Jack Poe Dies Here of Injuries from Fall," *Tulare (CA) Daily Advance Register*, 58, no. 13, (October 1, 1937): 1.

4. Bobbi Moore, interview by Glen E. Swanson, November 11, 2020.

5. Moore, interview by Swanson, 2020. Palpate meaning to examine by touch, especially medically.

6. Moore, interview by Swanson, 2020.

7. "Three Rivers Girl Weds Lieutenant in Community Presbyterian Church," *Tulare (CA) Advance Register*, June 14, 1956.

8. "Ann Jaren's Troth to S. E. Whitfield Is Announced*,*" *The Fresno (CA) Bee the Republican*, June 3, 1956, 63.

9. Susan Whitfield, interview by Glen E. Swanson, November 11, 2020.

10. Whitfield and Roddenberry, *The Making of Star Trek*, (1972), 11.

11a. Mat Irvine, *Creating Space: A History of the Space Age Told Through Models* (Apogee, 2002), 9. Beginning with the 1950s, the plastic modeling industry helped meet the demand for public interest in the latest rockets, missiles and spacecraft that emerged from Cold War technologies by producing accurate scale model kits you could build. Model kits were often made available before the actual vehicles took flight, and companies recruited spokespersons noted for being experts in their field to endorse and promote their models. Monogram hired Willy Ley, a German-born science writer and early rocket pioneer who ended up in the United States as a result of World War II. He became a well-known expert in the field of rocketry and space travel and extensively lectured and published on the subject. Monogram hired him to convert some of his spacecraft designs into models. Ley was delighted with the attention and would use Monogram's models as displays during his public presentations. Strombecker hired Krafft Ehricke, a visionary engineer and protégé of Walter Thiel at Peenemunde, to promote a series of model designs that they produced based upon Ehricke's drawings. Born in Berlin, Ehricke came to the United States with other German rocket scientists at the end of WWII as part of Operation Paperclip, the secret United States intelligence program that brought German scientists, engineers, and technicians from Nazi Germany to the United States for employment in America's missile and rocket program. While working for Convair Aerospace in the 1950s, Ehricke developed early concepts for nuclear and other advance spacecraft propulsion. The model department at Convair made large replicas of Ehricke's designs for promotion and, as a result, both he and his employer received a lot of public acclaim. His work was published in *Life* magazine, in which he and his factual futuristic designs competed with those of Willy Ley and Wernher von Braun. Noted spaceflight historian and space model collector Mat Irvine uses the phrase "factual futuristic" to describe "serious designs that were invariably the product of well-known and eminent scientists and engineers as part of astronautical design studies. These types of designs were especially popular at the beginnings of the modern 'space age,' which, perchance, also coincided with the beginnings of the modern model kit industry—the 1950s." Irvine also notes that the distinction between "fact" and "fiction" can be extremely blurred, since these futuristic designs often found their way into the popular media.

11b. Thomas Graham, *Monogram Models* (Schiffer, 2013), 37–38. In 1958, a year after the launch of Sputnik, Monogram produced a model called the "U.S. Missile Arsenal." The 1:128-scale kit comprised thirty-one scale model missiles mounted on a clear base, with a cardboard insert identifying each missile. Monogram took out a four-page advertisement in *Craft, Model, Hobby Industry Magazine* to promote the model. The ad said, in part, "The most important hobby kit produced in the Space Age! Ready now at the height of the world's consciousness and anxiety over the use of missiles and men in space… They are authentic in every way but do not violate any restricted information." Many of the missiles depicted in the model were brand new then. Obtaining accurate drawings was a challenge, as most of the missiles were classified. As a result, the model suffered from some scaling problems.

11c. The Genie was one of the models included in the US Missile Arsenal kit. Technically, the Genie is not a missile, but a rocket. A missile has a guidance system, and the Genie had none. It was an unguided air-to-air rocket that carried a nuclear warhead. To ensure simplicity and reliability, the weapon was designed to be unguided, since the large blast radius made precise accuracy unnecessary. The United States Air Force used it from 1957 to 1985. Before the release of the US Missile Arsenal kit, Monogram thought to increase publicity by sending a public-relations representative to Washington to show the model to a high-ranking Air Force general. Things did not go as planned. As model historian Thomas Graham recalled of the incident, "the general spotted the tiny model of the Genie air-to-air atomic missile on the display… [and] declared that Monogram could not release the model that included this top-secret missile."

11d. Graham, *Monogram Models*, 37–38. "Monogram already had built up a stockpile of 100,000 kits to send to distributors and was producing hundreds more kits daily." Monogram was in a pickle. They flew out the company attorney to try and reach a compromise but got bounced from one department to the next without results. Monogram then turned to Ley for his advice about the Missile Arsenal kit. Ley told them to go ahead and ship the kits. The kits made it to store shelves and stories on the Genie missile appeared in magazines and newspapers. Monogram never heard anything more from the Air Force.

11e. In 1969, Monogram released an updated version of the kit. Now called "US Space Missiles," the 1:128-scale kit contained thirty-six missiles and added two scale figures. They removed the cruise missiles (Nark, Matador, and Regulus II) along with both satellite launchers, the Jupiter C, and Vanguard. In their place, they added ten newer missiles: the Hound dog, ASROC, Poseidon, Minuteman, Pershing, Phoenix, Spartan, Titan II, Lance, and SUBROC. Monogram changed the mold for the Tartar, but other than that, all the flaws of the original 1958 kit were carried over to the 1969 version. In 1983, Monogram reissued the model as the "Heritage Edition" kit. The molds and decals remained the same, but the instructions were changed to a language-neutral format, with code letters added for the colors. They also removed the "Know the Missiles" poster. Atlantis models reissued the kit in 2022, using the same molds and decals. They include a reprint of the original "Know the Missiles" poster but made from a poorly scanned copy that has severe Moiré patterns. For an excellent overview of this kit and its various versions along with a detailed guide on how to build it, I highly recommend *Space in Miniature* Tech Report 6 "US Space Missiles" edited by Michael Mackowski. This can be obtained from www.spaceinminature.com or email Mike at mike@spaceinminiature.com for more information.

12a. It should be noted that the military in general, especially the Air Force, remained an early and enthusiastic supporter of model kits. "We feel that this hobby is a real investment in the future of American aviation, the conquest of outer space and the defense of our nation," said Major General Elvin S. Ligon Jr., in a 1959 *New York Daily News* article about the injection molded model industry. "It's an industry and educational project in which the USAF takes an extreme interest and to which it will continue to lend its fullest support." Ligon acknowledged that many early scientists and engineers in the field of astronautics worked with models. "Leonardo da Vinci and Otto Lilienthal, through the Wright brothers, Langley, Sikorsky, and Kraft Ehricke, first experimented—then expressed his conceptual ideas—with working models." Ligon concluded that "Men and kids who enjoy putting together these little planes and rockets have proven repeatedly, through psychological tests, to have IQs and aptitudes above the norm. They learn and comprehend most subjects more readily than the average person—and, in human terms—they represent a most important segment of our nation's intellectual resources." SOURCE: Eckert Goodman, "The Little Model Builders They Grow into Space-Age Giants," *New York Daily News*, February 23, 1959. Smithsonian National Air and Space Museum Archives, NASM OS-503400-03, folder: "Societal Impact, Memorabilia & Toys, General," IMG_3344-3345.

12b. In an article that appeared in the June 1969 issue of *Toys and Novelties*, the Hobby Industry Association of America (HIAA), the leading trade organization that includes the model and toy industry, said the models and toys dealing with space themes that first emerged during the decade following the Second World War were short-lived. According to the article, "The timing just wasn't right the first time around," said Dick Schwartzchild of Aurora models. Schwartzchild said the first round of space merchandise failed because they emphasized missiles and rockets. "People weren't involved," he said. "A rocket is like a bullet—it's a cold piece of hardware. But in such projects as Apollo or Gemini—as well as in science fiction properties—there are people with whom the child can identify." Saul Robbins of Remco Toys, makers of early *Star Trek* toys, agreed. "A few years ago, the space toys on the market were considered far out—the Buck Rogers type of science fiction—and the youngster just couldn't identify with it too well. Today, however, we have children identifying with actual living Americans; our astronauts, our men in space." Fred Pierce of Topper Toys, makers of Johnny Astro, said that through the "massive national TV coverage of our space efforts, the astronauts became the idols of small boys, who are, of course, going to become eager consumers of space toys." Bill Silverstein a former executive with Aurora added that, "as our space program accelerated, and television coverage increased, kids probably thought, 'Hey, it's real!'—and practically every successful toy is based on some real counterpart in the world around us." According to the article, a sense of identity is not all that is needed to make a great toy. Nick Underhill of Eldon Toys, makers of Billy Blastoff, said, "Space can be a big category, but if you are going to produce a space item, it had better not just have 'space' hung on it like a tag. You'd better be sure it has a lot of play value attached to it." Source: "Space Toys: second Attempt at a Full Orbit," *Toys and Novelties*, 68 (June 1968): 64–66. Smithsonian National Air and Space Museum Archives, NASM OS-503400-01, Folder: "Societal Impact, Memorabilia & Toys, General," IMG_3288-3290.

13. Ed Perlstein to Allan Stone, November 16, 1966, Gene Roddenberry Collection, Correspondence/General Files, Merchandising, box 29, folder 4, Gene Roddenberry Papers, (Collection PASC 62), UCLA Library Special Collections, Charles E. Young Research Library, University of California, Los Angeles.

14. Ed Perlstein to Don Beebe, September 14, 1966, Gene Roddenberry Collection, Produced Episodes, The Galileo Seven, box 5, folder 4. Gene Roddenberry Papers, (Collection PASC 62), UCLA Library Special Collections, Charles E. Young Research Library, University of California, Los Angeles.

15. Ed Perlstein to Don Beebe, October 28, 1966, Gene Roddenberry Collection, Correspondence/General Files, Merchandising, box 29, folder 4, Gene Roddenberry Papers, (Collection PASC 62), UCLA Library Special Collections, Charles E. Young Research Library, University of California, Los Angeles.

16. AMT ALL MAGNIFICIENT TROPHIES, dealer trifold issued by AMT Corporation on October 10, 1966.

17. Perlstein to Stone, (1966).

18. Ed Perlstein to Don Beebe, October 28, 1966, Gene Roddenberry Collection, Correspondence/General Files, Merchandising, box 29, folder 4, Gene Roddenberry Papers, (Collection PASC 62), UCLA Library Special Collections, Charles E. Young Research Library, University of California, Los Angeles.

19. AMT ad appearing in the April 1967 issue of *Hot Rod Magazine,* 51, promoting the sale of the AMT *Enterprise* model kit.

20. "Toys and Novelties Marketing Trends," August 1, 1967, Gene Roddenberry Collection, Correspondence/General Files, Merchandising, box 29, folder 4, Gene Roddenberry Papers, (Collection PASC 62), UCLA Library Special Collections, Charles E. Young Research Library, University of California, Los Angeles.

ENDNOTES

21. Bryce Russell to Allan Stone, August 11, 1967, Gene Roddenberry Collection, Correspondence/General Files, Merchandising, box 29, folder 4, Gene Roddenberry Papers, (Collection PASC 62), UCLA Library Special Collections, Charles E. Young Research Library, University of California, Los Angeles.

22. Stephen Whitfield to Gene Roddenberry, August 23, 1967, Gene Roddenberry Collection, Correspondence/General Files, Merchandising, box 29, folder 4, Gene Roddenberry Papers, (Collection PASC 62), UCLA Library Special Collections, Charles E. Young Research Library, University of California, Los Angeles.

23. Bob Justman to Gene Roddenberry, memo, October 19, 1967, Gene Roddenberry Collection, Correspondence/General Files, Merchandising, box 29, folder 4, Gene Roddenberry Papers, (Collection PASC 62), UCLA Library Special Collections, Charles E. Young Research Library, University of California, Los Angeles.

24. Justman, to Roddenberry, memo (October 19, 1967). The claim that AMT sold a million kits of the *Enterprise* during the first year of their production is not without merit. John Mueller was an employee of AMT from 1963 to 1982. During an interview I conducted with Mueller in 2023 from his home in Iowa, I asked him if such a claim could be true. "Yes, it is true especially if we had several sets of tooling running at the same time over multiple shifts," said Mueller. "AMT had a 1957 Thunderbird model car that became so popular, they had to make another tool, a duplicate tool, to keep up with demand." Mueller was there when they made the first *Star Trek* kits. "At the time of the AMT model of the *Enterprise*, we were running three shifts a day to keep up with demand. We could easily produce over a million kits of that model in one year and we did." Source: John Mueller, interview by Glen E. Swanson, October 22, 2023. On p. 11 of the pilot synopsis dated December 5, 1967 for a proposed *Star Trek* spin-off pilot, a blurb about Gene Roddenberry mentions that "His 'imagineering' knack is equally well known, underscored by the fact that his spaceship conception, *Star Trek*'s USS *Enterprise*, is now the largest selling model kit ever to appear on the toy market." Source: MissionLogPodcast.com, Roddenberry Archives.

25. Justman, to Roddenberry, memo (October 19, 1967).

26. Ed Perlstein to Lou Mindling, August 2, 1966, Gene Roddenberry Collection, Produced Episodes, The Galileo Seven, box 5, folder 4. Gene Roddenberry Papers, (Collection PASC 62), UCLA Library Special Collections, Charles E. Young Research Library, University of California, Los Angeles.

27. Bryce Russell to Gene Roddenberry, August 11, 1967, Gene Roddenberry Collection, Correspondence/General Files, Merchandising, box 29, folder 4, Gene Roddenberry Papers, (Collection PASC 62), UCLA Library Special Collections, Charles E. Young Research Library, University of California, Los Angeles.

28. Russell to Roddenberry (1967).

29. Matt Jefferies, https://memory-alpha.fandom.com/wiki/Matt_Jefferies.

30. Dorothy Fontana, "Behind the Camera: Walter M. Jefferies," *Inside Star Trek*, edited by Ruth Berman, *Inside Star Trek* 4 (October 1968): 3–4.

31. "Designing the Klingon Battle Cruiser," *Star Trek: The Magazine* 2, no. 9 (January 2002): 66–71.

32. "Designing the Klingon Battle Cruiser."

33. Stephen Whitfield to John Reynolds, January 2, 1968, Gene Roddenberry Collection, Correspondence/General Files, Merchandising, box 29, folder 4, Gene Roddenberry Papers, (Collection PASC 62), UCLA Library Special Collections, Charles E. Young Research Library, University of California, Los Angeles. As mentioned in note 24, the claim that nearly a million kits of the AMT *Enterprise* model were being sold during the first year of their production is not without merit. As noted in note 24, Bob Justman, in an October 19, 1967, memo to Roddenberry, stated that AMT had already sold over a million kits. Whitfield, in his letter that came out a little over two months later, on January 2, 1968, stated that demand for the kit "had only pushed sales toward

the 800,000 mark." Sometime near the end of 1967 or beginning of 1968, AMT was approaching the selling of one million kits of the *Enterprise*.

34. Marvin S. Katz to Stephen Whitfield, January 17, 1968, Gene Roddenberry Collection, Correspondence/General Files, Merchandising, box 29, folder 4, Gene Roddenberry Papers, (Collection PASC 62), UCLA Library Special Collections, Charles E. Young Research Library, University of California, Los Angeles.

35. Katz to Whitfield (1968).

36. Richard G. Van Treuren, "Star Trek Miniatures Part Two: The Other Space Ships," *Trek: The Magazine For Star Trek Fans*, no. 6, November 1976. According to their website, Smithsonian's filming model of the Klingon ship which is part of the collections at the National Air and Space Museum, measures 29¼ × 18½ × 7¼ inches in size. The second D7 model used to produce the AMT kits is a little shorter, measuring only 26 inches in length.

37. "Designing the Klingon Battle Cruiser," 66–71.

38. Dorothy Fontana, "The Klingons are Coming!" *Inside Star Trek*, edited by Ruth Berman, issue no. 2, August 1968, 2. Jefferies seems to have mistaken the overall length of the Klingon ship during the 1968 interview. It is not 747 feet as the detailed drawings he made and gave to AMT so they could make the two models clearly show the overall length of the ship to be 624 feet.

39. Dorothy Fontana, "Just Ask," *Inside Star Trek*, Edited by Ruth Berman, no. 9, March 1969, 2.

40. "Designing the Klingon Battle Cruiser," 66–71.

41. "Designing the Klingon Battle Cruiser," 66–71.

42. Frederick C. Durant III to Gene Roddenberry, letter, November 2, 1973, Smithsonian Institution Archives, Record Unit 000398, box 32, folder Gene Roddenberry Correspondence 1968–1979.

43. Richard G. Van Treuren, "*Star Trek* Miniatures" Part Two: The Other Space Ships," *Trek: The Magazine For Star Trek Fans*, 6 (November 1976).

44. According to the Smithsonian's own National Air and Space Museum's website, the model's actual measurements are 29¼ × 18½ × 7¼ inches. https://airandspace.si.edu/collection-objects/model-d7-klingon-battle-cruiser-star-trek/nasm_A19740482000.

45. Ron Barlow, George M. W. Snow, Anthony Frederickson, and Doug Drexler, eds., "Smithsonian Report," *Star Trek Giant Poster Book*, Voyage 10 (June 1977), (published by Paradise).

46. Barlow, Snow, Frederickson, and Drexler, EDS, "Smithsonian Report."

47. Gene Roddenberry to Donald J. Beebe, letter, April 24, 1967, Julien's Auction #3359, April 22–23, 2023. Beebe was division manager of AMT's Speed & Custom Equipment Division in Phoenix.

48. Bob Justman to Gene Roddenberry, memo, October 19, 1967, Gene Roddenberry Collection, Correspondence/General Files, Merchandising, box 29, folder 4, Gene Roddenberry Papers, (Collection PASC 62), UCLA Library Special Collections, Charles E. Young Research Library, University of California, Los Angeles.

49. Barry Rohan, "It's a Model Company, The No. 1 Vehicle Producer," *Detroit Free Press*, March 19, 1978. Labor and Urban Affairs Archives, Accession 512, 48-27 #824, AMT Corp., 1978, Wayne State University, Walter P. Reuther Library.

50. Glen E. Swanson email message to Jamie Hood, May 1, 2020.

51. Rohan, "It's a Model Company."

52. Amy Stamm, interview by Glen E. Swanson, Office of Communications, Smithsonian National Air and Space Museum, September 6, 2022.

53. Stephen Whitfield to John Reynolds, January 2, 1968, Gene Roddenberry Collection, Correspondence/General Files, Merchandising, box 29, folder 4, Gene Roddenberry Papers,

(Collection PASC 62), UCLA Library Special Collections, Charles E. Young Research Library, University of California, Los Angeles.

54. Roger Mitchell email messages to Glen E. Swanson, May 13, 2022, and December 27, 2022.

55. In *The World of Star Trek*, author David Gerrold gives the following definition of what the *Star Trek Guide* was. "The *Star Trek Guide* (in its earliest form) was a 20-page mimeographed book, distributed to all writers and prospective writers for the series. It contained descriptions of all the characters and sets as well as notes on the capabilities of the *Enterprise* and what kind of stories the series could use. It was revised twice as the show progressed. New material was added, and old material was updated. In the third edition, for instance, the *Enterprise* was upgraded from Cruiser Class to Starship Class—the feeling being that a 'starship' was a special kind of vessel with greater range, speed, power and other capabilities than other vessels in space." Source: David Gerrold, *The World of Star Trek* (Ballantine Books, 1973), 5.

56. Solow and Justman, *Inside Star Trek*, 401–402.

57. Gene Roddenberry to Alden Schwimmer, letter, Ashley Famous Agency, June 19, 1967, https://collectingtrek.ca/2019/01/15/gene-roddenberry-memo-endorsing-the-making-of-star-trek/.

58. Roddenberry to Schwimmer (1967).

59. Solow and Justman, *Inside Star Trek*, 401–402.

60. Whitfield and Roddenberry, *The Making of Star Trek* (1972), 12.

61. Whitfield and Roddenberry, *The Making of Star Trek* (1972), 13.

62. Whitfield and Roddenberry, *The Making of Star Trek* (1972), 13.

63. Bobbi Moore, interview by Glen E. Swanson, November 11, 2020.

64. Solow and Justman, *Inside Star Trek*, 402.

65. The photos and line drawings in the book were provided by Howard McClay and Frank Wright of the Desilu/Paramount Publicity Department. Whitfield began writing the manuscript of his book during the period of time when Desilu Studios was acquired by Paramount owner Gulf and Western and The Westheimer Company. See Whitfield and Roddenberry, *The Making of Star Trek* (1972), 15.

66. Susan Sackett, *Letters to Star Trek* (Ballantine Books, 1977), 17.

67. Whitfield and Roddenberry, *The Making of Star Trek* (1972).

68. Donald Freeman, "History of TV Series—Book on 'Star Trek' Makes Good Reading," *Shreveport (LA) Journal*, October 31, 1968.

69. Whitfield and Roddenberry, *The Making of Star Trek* (1972).

70. James W. Wright, "TV's Star Trek: How They Mix Science Fact with Fiction," *Popular Science*, 191, no. 6 (December 1967): 72–74.

71. G. Harry Stine, "To Make a 'Star Trek,'" *Analog* 80, no. 6 (February 1968): 70–85. Prior to publication of the article in *Analog*, G. Harry Stine wrote Roddenberry a letter on June 15, 1967, giving him a heads-up about the forthcoming piece. In this letter, Stine writes: "John [W. Campbell, editor of *Analog*] wants me to do an article for *Analog* entitled 'To Make a Star Trek' which would include the tremendous background material in the Guide [the *Star Trek Writers Guide*], much of which never is seen or mentioned in the show. This would also include the technical background on the ship and the technology as well as the background history…. Both John and I think this could be a very good bit. For example, John did not believe that all the action that is shown taking place aboard the USS *Enterprise* could actually happen there until I informed him of the location of the control room, what the various external pods were for, about the turbo elevators, etc. (You even had me fooled for a while. I built the AMT model of the *Enterprise* and had figured the location of the control room behind the dish on the lower hull.)" Source: G. Harry Stine to Gene Roddenberry, letter, June 15, 1967. Gene Roddenberry Collection, Correspondence/General Files, Publicity,

1967, box 30, folder 5, Gene Roddenberry Papers, (Collection PASC 62), UCLA Library Special Collections, Charles E. Young Research Library, University of California, Los Angeles.

72. "Out-of-this-World Special Effects for *Star Trek*, *American Cinematographer* 48. No. 10 (October 1967): 714–717. The following is an editor's note that appeared at the beginning of the original article: "In a unique parlay of highly technical skills and talents, three of the world's foremost studios of screen wizardry received a joint 1966 'Emmy' Award nomination in the Special Photographic Effects category in recognition of the spectacular illusions created for the Desilu-NBC outer space television series, STAR TREK. These studios are the Howard A. Anderson Company, Film Effects of Hollywood, Inc., and the Westheimer Co., all of which, headed by top ASC Special Effects experts, worked separately but in a spirit of cordial collaboration to produce some of the most stunning and literally 'far out' visual effects ever to be recorded on color film. At the request of *American Cinematographer*, the Presidents of the respective companies involved kindly consented to explain the technology involved in creating several of the aforementioned effects, and their individual comments follow."

73. Bettelou Peterson, "Bettelou Peterson Answers your TV Questions," *Detroit Free Press*, January 18, 1970, 2D.

74. "Son-in-Law Making Movie in Space," *Chillicothe (MO) Constitution-Tribune*, February 23, 1971.

75. "Action Line," *Miami Herald*, August 4, 1973.

76. "Sorry, Wrong Title," *The Record*, November 9, 1972.

77. Aaron Harvey and Richa Schepis, *Star Trek: The Official Guide to the Animated Series*, (Weldonowen, 2023), 108. On December 15, 1973, the fourteenth episode from the first season of *Star Trek: The Animated Series* aired on television. "The Slaver Weapon" came very close to having the Leif Ericson become part of *Star Trek*'s canon when animator Bob Kline used a sketch of the AMT model as a placeholder for the long-range shuttlecraft *Copernicus*. Sadly, the model was never shown in the final episode, only in pre-production artwork.

78. Frank Henriquez has an excellent webpage devoted to the history of the kit at http://frank.bol.ucla.edu/le.html?fbclid=IwAR0AkVGApFXtdsw0Bt4idJYQG5ePWBjPYdBiSFJZXRp4EO9o5WSccACIlmk.

79a. Whitfield, interview by Swanson, 2020.

79b. As a result of Whitfield's battle with leukemia, Whitfield sold his D7 model to a private collector in 1998 to help pay for medical expenses. The model then made its way to multiple public auctions. The first appearance was during the June 18, 1998, Christies Film and Television Auction 8115 where it sold as Lot 71 for $11,500. Six months later, the Hollywood: A Collectors Ransom 5 auction held on December 12, 1998, the model was up for auction again as Lot 245. This auction was run by Joseph Maddalena under the name "Pacific Design Center," the name of his auction house before he formally changed it to Profiles in History. Whitfield's D7 appears at auction a third time on March 31, 2004 at Profiles in History Auction No. 18 as Lot 211. It sold at this auction for $64,900 including buyer's premium. The model goes up for auction a fourth and final time during the March 31, 2006 Profiles in History Auction No. 24 as Lot 311. Memory Alpha states that it sold at this suction for $65,000 to Microsoft's co-founder Paul Allen. According to the Memory Alpha entry, the deal was closed before the auction started and was therefore not featured in the auction itself though it does appear in their catalog.

80a. Moore, interview by Swanson, 2020.

80b. Stephen Edward Poe, *A Vision of the Future: Star Trek Voyager*, (Pocket Books, 1998).

80c. Skorpus, " 'Vision' Quest: Reno Writer Publishes Second Behind the Scenes 'Star Trek' Book," 16.

81. Doug Shuit, "Star Trek: Still Luring a Galaxy of Aficionados," *Los Angeles Times*, June 27, 1972, 1, 3.

ENDNOTES

82. Sackett, Letters to Star Trek, 20.

83. José Moreno Hernández, *Reaching for the Stars*, (Center Street, 2012), 20–22, 179.

84. José Moreno Hernández, interview by Glen E. Swanson, February 10, 2023.

Chapter 4

1. The spelling of "Starfleet" evolved over time. In various first season scripts, it is listed as two words. Stephen Whitfield's *The Making of Star Trek* spells it as two words on multiple pages throughout. In a memo dated September 28, 1966, Roddenberry appears to lean toward a two-word preference for the term when he states: "It appears we may have a useful catchword deriving out of 'Star Trek' in our title. For example, STAR BASE, for command bases on various planets, STAR CUT describing the style of our pointed sideburn haircuts, STARSHIP, of course, is a generic term for the Enterprise, STAR COMMAND for fleet headquarters, and so on. Undoubtedly, other examples will occur to you. As with 'Batman,' 'Batmobile,' clever use of it may create a kind of terminology for our show." Source: Roddenberry memo to All Concerned, memo, September 28, 1966, Gene Roddenberry Collection, Correspondence/General Files, Publicity, up to and including 1966, box 30, folder 4, Gene Roddenberry Papers, (Collection PASC 62), UCLA Library Special Collections, Charles E. Young Research Library, University of California, Los Angeles. However, by the time the first season two-part episode "The Menagerie" aired at the end of November 1966, the Talos IV report that Commodore Mendez is shown giving to Kirk, viewers see "STARFLEET" clearly stamped in gold as one word on the report cover thereby establishing once and for all the correct spelling of the term. The third revision of the *Star Trek Writers Guide* also lists it as one word.

2. Stefan Rabitsch, *Star Trek and the British Age of Sail: The Maritime Influence Throughout the Series and Films* (McFarland, 2018), 81. Rabitsch postulates how *Star Trek* draws heavily from the British age of sail. Rabitsch writes how the franchise is "often draped with a symbolic veneer that is deceivingly indicative of the US Navy's post-war legacy. The resulting dense transatlantic mesh in nomenclature is easily obscured by the obvious prefix USS, the symbolic prominence of which might lead to premature conclusions about whose legacy Starfleet ships carry into space."

3. Roddenberry, *The Star Trek Writers Guide*, 27.

4. Whitfield and Roddenberry, *The Making of Star Trek* (1972), 25.

5. Roddenberry, *The Star Trek Writers Guide*, 10.

6. John E. Fahey, " 'A Tall Ship and A Star to Steer Her By'": *Star Trek and Naval History*, official blog of the Naval History and Heritage Command, April 5, 2023, https://usnhistory.navylive.dodlive.mil/Recent/Article-View/Article/3348268/a-tall-ship-and-a-star-to-steer-her-by-star-trek-and-naval-history/#_ftn18.

7. Initially the ship ran on "Lithium crystal circuits," which were mentioned in the show's second pilot "Where No Man Has Gone Before" as well as in the first season episode "Mudd's Women." In both episodes, the ship is crippled, and they have to go to a planet where lithium is mined. It was not clear how the ship "drew all her power" through the crystals. Perhaps Roddenberry thought crystals were capable of generating power (a New Age idea that became popular in the 1970s). In any case, the idea of matter/antimatter reaction was finally introduced later in the first season. The crystals, renamed "dilitihium," were still in use, but it was never explained exactly what they did.

8. Herb Solow interview by Ian Spelling, *Starlog* 241 (August 1997), 74–77.

9. Whitfield and Roddenberry, *The Making of Star Trek* (1972), 171. Here it is clarified that "the unit components [of the *Enterprise*] were built at the Star Fleet Division of what is still called the San Francisco Navy Yards, and the vessel was assembled in space. The *Enterprise* is not designed to enter the atmosphere of a planet and never lands on a planet surface." The term "Constitution-Class" does not appear in *The Making of Star Trek* and is not said or shown anywhere in the original series. Presumably, the term originated with Franz Joseph.

10. Whitfield and Roddenberry, *The Making of Star Trek* (1972), 203.

11. Film trims from Lincoln Enterprise showing the deleted scene of Kirk doing the military edge-of-hand-to-head salute were obtained by collector Christopher Beamish. This scene is also discussed in the book *Star Trek: Lost Scenes* by Curt McAloney and David Tilotta (Titan Books, 2018),148. Tilotta notes this scene appears in the May 2, 1966 outline of Paul Schneider's script and in the June 21, 1966 first draft, though the exact language is a bit different. Specifically, in the outline, Kirk salutes the Romulan commander, and the commander returns the salute. In the sequence that was filmed but then cut, Kirk salutes the commander and the commandeer responds "with a small, stiffly-precise bow."

12a. Okuda and Okuda, *The Star Trek Encyclopedia: A Reference Guide to the Future*, rev. and exp. ed., vol. 2 *M–Z* (Harper, 2016), 180. Spock first mentions "Our Prime Directive of non-interference" very late in the first season episode "Return of the Archons." But Kirk twists the term and uses it to goad Landreu into fulfilling its own "prime directive" of destroying evil. Landreu is evil, therefore it must destroy itself.

12b. "Rules of Engagement," in *Encyclopedia Britannica*, https://www.britannica.com/topic/rules-of-engagement-military-directives.

13. During World War II, Roddenberry served in the 394th Bomb Squadron, 5th Bombardment Group, Thirteenth Air Force, and flew bombing missions over Guadalcanal. Given this fact, it was ironic that George Takei served as the ship's helmsman, Lieutenant Sulu, in the series. See: A. J. Black, *Star Trek, History and Us: Reflections of the Present and Past Throughout the Franchise* (Jefferson, NC: McFarland, 2021), 6–9.

14. Gross and Altman, *The Fifty-Year Mission: The Complete, Uncensored, Unauthorized Oral History of Star Trek: The First 25 Years*, 69.

15. After graduating from high school in 1944, Robert Justman enlisted in the Navy. According to an interview conducted by the Television Academy Foundation, the recruiter told him he was not required to join up, but he insisted. During his first efforts to join, he flunked out due to poor eyesight. He protested their decision. The draft board performed another physical and eventually he got in. He served his time in the Navy overseas in destroyers and destroyer escorts. See Robert Justman, interview by Stephen J. Abramson, Television Academy Interviews, Television Academy Foundation, June 17, 2003, https://interviews.televisionacademy.com/interviews/robert-justman?clip=1#interview-clips.

16a. Whitfield and Roddenberry, *The Making of Star Trek* (1972).

16b. Wright, "TV's *Star Trek*," 74.

17. David Alexander, *Star Trek Creator: The Authorized Biography of Gene Roddenberry* (Penguin Books, 1994), 342–345. Alexander indicates that Roddenberry left on November 3, 1968, to begin a weeklong visit aboard the nuclear-powered attack carrier USS *Enterprise* (CVN-65). It is not clear if Roddenberry was actually on the carrier beginning November 3. The original CVN-65 Deck Logs which housed in the National Archives and Records Administration (NARA) in Suitland, Maryland, confirm the *Enterprise* was participating in combat readiness exercises off the Northern California operations area from November 2–10, 1968, and then again off Southern California from November 12–22. In addition, the Deck Logs did report the *Enterprise* left the Bay Area of San Francisco, clearing Pier 3 of Naval Air Station Alameda on Monday morning November 4, 1968. However, there is no specific mention of Roddenberry being onboard. This did not mean he wasn't, it just was not entered into the Deck Log. Most likely Roddenberry was aboard the *Enterprise* during the November 2–10 exercise. As Alexander indicates in his book, Roddenberry reported the results of his *Enterprise* visit on December 3, 1968. If he was aboard the *Enterprise* during the November 12–22 exercise, his report most likely would have included mention of the tragic accident that occurred during the exercise which cost the lives of two seamen. On November 16, 1968, Angel 73, a UH-2C (BuNo 150177) (073), flown by Lieutenant Ronald R. Bradley, pilot, Lieutenant (jg) George G. Kirsten, co-pilot, AMH2 Kenneth S. Carpenter, and Airman Brian S. Mullen,

crewmen, HC-1 Det 65, lifted off forward on the angled flight deck for plane guard, at 2232. Kirsten made the take-off and climbed straight ahead up to about 500 feet. Approximately one mile ahead and to port of *Enterprise*, Bradley took control, the tower instructing him to drop to 250 feet. As Angel 73 began a descending left turn, Kirsten noted he was unable to see the carrier, which was almost directly aft. Dropping rapidly, the helo impacted the water hard, but fixed wing recovery continued as the carrier attempted to regain radio communication with Angel 73. About 25 minutes after 73 launched, there was a single A-3 remaining to be recovered, and at that time it was determined that no one held visual or radar contact with the Seasprite. After three bolters, the A-3 arrested on deck, approximately 41 minutes after Angel 73 had launched. Search helo No. 83, Lieutenant (jg) Jack L. Berg, HC-1 Det 65, pilot, launched from Enterprise at 2255, shortly joined by No. 80, a SAR helo from Kitty Hawk, 45 minutes later. Surface fog and haze impeded rescue efforts to locate the survivors, who fired "numerous" flares. In addition, Berg's Doppler gear became inoperative, his radar altimeter failed in hover, and "gusty winds" and high swells complicating the rescue. At about 2355, No. 80 picked up Kirsten and Carpenter, 83 recovering Bradley, who was unconscious, a few minutes later. Bradley died of his injuries at 0055, and no trace of Mullen was ever found. Source: *CVN-65 USS Enterprise History 1961–2012*: https://www.seaforces.org/usnships/cvn/CVN-65-USS-Enterprise-history.htm.

18. Consolidated Film Industries to Gene Roddenberry, March 1, 1968, Gene Roddenberry Collection, Correspondence/General Files, Publicity, 1968 & Undated, box 30, folder 6, Gene Roddenberry Papers, (Collection PASC 62), UCLA Library Special Collections, Charles E. Young Research Library, University of California, Los Angeles.

19. Consolidated Film Industries to Roddenberry (1968).

20. Consolidated Film Industries to Roddenberry (1968).

21. Richard G. Van Treuren, interview by Glen E. Swanson, September 25, 2022.

22. J. M. Hession, USN commander to Gene Roddenberry, January 19, 1968, Gene Roddenberry Collection, Correspondence/General Files, Correspondence, Miscellaneous, box 27, folder 17, Gene Roddenberry Papers, (Collection PASC 62), UCLA Library Special Collections, Charles E. Young Research Library, University of California, Los Angeles.

23. Terry Rhodes, Facebook-directed message to author, June 18, 2019, https://m.facebook.com/dusty.rhodes.9847.

24. Jacques Le Roy Marcel Hartley was born in a hospital at Camp Knox, Kentucky, that later burned down. The Fort Knox depository was then built where the hospital was located. His mother, Helene Marie Boneufant, migrated from France and married Le Roy Poston Hartley, the thirteenth dentist to serve in the Army during World War I. Le Roy was originally "adopted" by an Ohio family with the last name of Poston, and he was one of eight children. They retired in Carmel, California.

25. Jack Hartley to Gene Roddenberry, Western Union Telegraph, September 9, 1966, Gene Roddenberry Collection, Correspondence/General Files, Correspondence, Miscellaneous, box 27, folder 17, Gene Roddenberry Papers, (Collection PASC 62), UCLA Library Special Collections, Charles E. Young Research Library, University of California, Los Angeles.

26. Alvin B. Webb Jr., "Some Spacemen May Be Dentists of Sorts," *Dental Times*, March 15, 1966.

27. January 10, 1966 episode of *To Tell The Truth*, https://www.youtube.com/watch?v=SS3upqk2mDY&fbclid=IwAR1XPuaZFaf5WX0wD7Cl8p-cVskMfTN9jQobw8Et1NwDUAVlmOtG6JqVUwA. According to Hartley's daughter Patricia, the hands shown in the *To Tell the Truth* episode clip demonstrating the astronaut dental kit are those of her father Jack.

28. Patricia Hartley, interview by Glen E. Swanson, November 8, 2020. Patricia, now retired, worked as an archivist at the Southwest Research Institute in San Antonio, Texas.

29. McCoy used three designs of medikit, first a large, commercially available black leather toiletries bag or "Dopp kit" carried in the hand, in the first season and occasionally thereafter. Early

in the second season, McCoy used a small leather pouch attached to his belt or pants with Velcro. Photos of the small pouch appear in *The Making of Star Trek*. After a few episodes, the small pouch was replaced with a long, tri-fold pouch, which was the one seen most often in the second and third seasons.

30. Jackie Wiley Hartley, interview by Glen E. Swanson, January 29, 2023. Jackie is the youngest of the two daughters of Jack Hartley.

31. As honorary chair of the American Dental Association's National Children's Dental Health Week in 1966, Jack Hartley appeared on television and radio shows to help promote the event. This included a January 10, 1966 appearance on the weekly CBS game show *To Tell the Truth*, where they exhibited his unique astronaut buddy-care dental hygiene kit. https://www.youtube.com/watch?v=SS3upqk2mDY&fbclid=IwAR1XPuaZFaf5WX0wD7Cl8p-cVskMfTN9jQobw8Et1NwDUAVlmOtG6JqVUwA. On February 7 of that same year, Hartley appeared on the *Tonight Show Starring Johnny Carson* along with Dr. Joyce Brothers and Clyde McCoy. He discussed dental care of the astronauts and demonstrated his buddy-care kit. He also presented Carson with a real human skull to show how to give injections to the upper and lower jaws to make them numb. Once Carson saw the skull, he couldn't resist using it for a gag. Johnny held it up and said: "Alas, poor Yorick!" "National Publicity for Children's Dental Health Week," *Journal of the American Dental Association* 72 (June 1966): 1,513–1,514. Newspaper clipping provided by Patricia Hartley, "Tonight Memories from S.A. Fan, Guest," by Jeanne Jakle, May 23, 1992. January 10, 1966, episode of *To Tell The Truth*.

32. Patricia Hartley, email message to author, March 8, 2023. Since the publication of this book, Jack Hartley's papers have been donated to the National Museum of Dentistry located on the campus of the University of Maryland in Baltimore.

33. Gene Roddenberry to Matt Jefferies, memo, June 18, 1965, Gene Roddenberry Collection, Correspondence/General Files, Sets, box 31, folder 1, Gene Roddenberry Papers, (Collection PASC 62), UCLA Library Special Collections, Charles E. Young Research Library, University of California, Los Angeles.

34. Gene Roddenberry to Howard McClay of McFadden, Strauss, Eddy, and Irwin (MSEI), the public relations firm for Desilu Productions, memo, January 17, 1968, Gene Roddenberry Collection, Correspondence/General Files, Kelly, Deforest, box 28, folder 15, Gene Roddenberry Papers, (Collection PASC 62), UCLA Library Special Collections, Charles E. Young Research Library, University of California, Los Angeles.

35. The Department of the Air Force to Gene Roddenberry, letter, January 11, 1968, Gene Roddenberry Collection, Correspondence/General Files, Publicity, 1968 & Undated, box 30, folder 6. Gene Roddenberry Papers, (Collection PASC 62), UCLA Library Special Collections, Charles E. Young Research Library, University of California, Los Angeles.

36. Sackett, *Letters to Star Trek*, 108–110.

37. Randy Roughton, "Squadron 19 Returns to the 'Starship,'" United States Air Force Academy Strategic Communications, https://www.usafa.edu/squadron-19-returns-to-the-starship/?fbclid=IwAR3TmTtDlBV47F-YRZVaFQXtyatpBCdwK3Mz3JRQjJ5fqZj0VAzoRcHR-Z4.

38. This intro wasn't written until the show went into production. The second pilot has a sort of an embryonic version with an opening "Enterprise log, Captain James Kirk commanding. We are leaving that vast cloud of stars and planets which we call our galaxy. Behind us, Earth, Mars, Venus, even our Sun, are specks of dust. The question: What is out there in the black void beyond? Until now our mission has been that of space law regulation, contact with Earth colonies and investigation of alien life. But now, a new task: A probe out into where no man has gone before." See original at https://www.youtube.com/watch?v=CMXVAbakvDQ Wikipedia states "The studio did not retain a print of this original version, and it was officially thought to be lost. In 2009, a German film collector discovered a print of it and brought it to the attention of CBS/Paramount, which then released it under the title 'Where No Fan Has Gone Before': The Restored, Unaired Alternate Pilot Episode as part of the TOS season 3 box set on Blu-ray."

ENDNOTES

39. Megan Shaw Prelinger, *Another Science Fiction: Advertising the Space Race 1957–1962* (Blast Books, 2010), 12–13.

40. Wayne R. Parrish. "A New Age Unfolds." *Missiles and Rockets* 1, no. 1, (October 1956): 5.

41. John Mack Faragher, ed., *Rereading Jackson Turner: "The Significance of the Frontier in American History" and Other Essays* (Yale University Press, 1998).

42. Faragher, *Rereading Jackson Turner*.

43. John Sisk, "The Six-Gun Galahad." *Time*, March 30, 1959, 52–60. https://time.com/archive/6827467/westerns-the-six-gun-galahad/

44. John F. Kennedy, *The New Frontier*, acceptance speech of Senator John F. Kennedy, Democratic National Convention, July 15, 1960. Papers of John F. Kennedy, John F. Kennedy Presidential Library and Museum.

45. John F. Kennedy, "If the Soviets Control Space… They Can Control Earth." *Missiles and Rockets* 7, no. 15 (October 10, 1960): 12–13.

46. John F. Kennedy, Address Before a Joint Session of Congress, May 25, 1961.

47. *Special Committee on Space Technology, Recommendations to the NASA Regarding A National Civil Space Program*, October 28, 1958, NASA Historical Reference Collection, History Office, NASA Headquarters, Washington, DC. https://www.nasa.gov/history/special-committee-on-space-technology-report-1958/

48. James Killian, *Introduction to Outer Space*, March 26, 1958, The White House.

49. Killian, *Introduction to Outer Space*.

50. Killian, *Introduction to Outer Space*.

51. Killian, *Introduction to Outer Space*.

52. Killian, *Introduction to Outer Space*.

53. Dwayne A. Day, "Boldly Going: Star Trek and Spaceflight." *Space Review*, November 28, 2005, https://www.thespacereview.com/article/506/1.

54. Valerie Fulton, "Another Frontier: Voyaging West with Mark Twain and *Star Trek*'s Imperialist Subject," *Postmodern Culture* 4, no. 3 (1994), 1–24.

Chapter 5

1. Robert W. Kamm, "Greetings from the NASA Western Operations Office," *Proceedings of the Science Education in the Space Age*, (NASA, November 1964), 3–4.

2. Hadley Meares, "How the aviation industry shaped Los Angeles," *Curbed Los Angeles*, July 8, 2019, https://la.curbed.com/2019/7/8/20684245/aerospace-southern-california-history-documentary-blue-sky.

3. Dwayne A. Day, "Blue skies on the West Coast: A History of the Aerospace Industry in Southern California," *Space Review*, August 20, 1007, https://www.thespacereview.com/article/938/1.

4. Kamm, "Greetings from the NASA Western Operations Office," 3–4.

5. Gene Roddenberry to North American Aviation, October 16, 1964, Gene Roddenberry Collection, Correspondence/General Files, Stock Footage, box 31, folder 7, Gene Roddenberry Papers, (Collection PASC 62), UCLA Library Special Collections, Charles E. Young Research Library, University of California, Los Angeles.

6. Norman Warren to Gene Roddenberry, October 20, 1964, Gene Roddenberry Collection, Correspondence/General Files, Stock Footage, box 31, folder 7, Gene Roddenberry Papers, (Collection PASC 62), UCLA Library Special Collections, Charles E. Young Research Library, University of California, Los Angeles.

7. Jim Mullins to Gene Roddenberry, November 20, 1964, Gene Roddenberry Collection, Correspondence/General Files, Stock Footage, box 31, folder 7, Gene Roddenberry Papers, (Collection PASC 62), UCLA Library Special Collections, Charles E. Young Research Library, University of California, Los Angeles.

8. Gene Roddenberry to "Those Concerned" titled "STAR TREK SPECIAL EFFECTS", memo, August 24, 1964, Gene Roddenberry Collection, Correspondence, General Files, Special Effects, box 31, folder 5, Gene Roddenberry Papers, (Collection PASC 62), UCLA Library Special Collections, Charles E. Young Research Library, University of California, Los Angeles.

9. During the re-editing of "Where No Man Has Gone Before," scenes were cut or reshot. These cut scenes remained officially unreleased until 2009, when the original pilot was released on Blu-ray.

10. H. L. Schwartzberg to Gene Roddenberry, January 8, 1968, Gene Roddenberry Collection, Correspondence/General Files, Kelley, DeForest, box 28, folder 15, Gene Roddenberry Papers, (Collection PASC 62), UCLA Library Special Collections, Charles E. Young Research Library, University of California, Los Angeles.

11. H. L. Schwartzberg to Gene Roddenberry, October 20, 1967, Gene Roddenberry Collection, Correspondence/General Files, Publicity, 1967, box 30 folder 5, Gene Roddenberry Papers, (Collection PASC 62), UCLA Library Special Collections, Charles E. Young Research Library, University of California, Los Angeles.

12. "Precision Radiation Model 111 and 111B Scintillator (ca. mid 1950s), ORAU, Museum of Radiation and Radiology, https://www.orau.org/health-physics-museum/collection/survey-instruments/1950s/precision-radiation-model-111-and-111b-scintillator.html?fbclid=IwAR3EWDpcu...%201/3.

13. http://national-radiation-instrument-catalog.com/new_page_42.htm.

14. Redshirtgal, tumblr, https://www.tumblr.com/redshirtgal/184327074794/what-a-cutie-pie-no-im-not-referring-to-captain.

15. "Props Re-used," The Trek BBS, https://www.trekbbs.com/threads/props-re-used.81174/page-7; "Flight Computer" https://en.wikipedia.org/wiki/Flight_computer.

16. The original series Astrogator became a shallow dome with a polar grid and stars on it ringed by green lamps in *Star Trek: The Motion Picture*. A big ceiling fixture showing another dome but with a flat picture of the *Enterprise* as viewed from the top is also present in the film, though it was never really clear what that was supposed to be. In *Star Trek: The Animated Series*, there is a dome, but it is the Bridge Defense Phaser Turret. Also, the Astrogator in the *Animated Series* became a flat screen and not just a round gadget.

17. David Tilotta and Curt McAloney "Promoting TOS," *Lost Star Trek History*, May 10, 2016, https://www.startrek.com/article/promoting-tos. In 2016, David Tilotta and Curt McAloney posted this article online. It was here I first noticed the early promotional efforts for *Star Trek* and a rather unusual item they created to promote the new series. Along with the article, Tilotta and McAloney posted a photo of a small cardboard standup of the *Enterprise*. Apparently, MSEI sent little cardboard standups to select NBC affiliates as part of an advance media campaign prior to the series's premiere. The fact that these kits featured a cutout of the *Enterprise* indicates how novel and unique this spaceship design was for the time and that their publicity department wanted to stress it. Most folks never saw anything like it before and would not know what it was. No other show featured such a huge ship that could travel through space. Its look was contrary to what television viewers were accustomed to. For many, this is not what a spaceship "should" look like. Digging deeper into the story on these little cardboard *Enterprise* promotional pieces, I learned Desilu, through MSEI, issued two different kinds to their NBC affiliate stations. The first had a premiere date and time printed on the base of "Thursday, September 15, 8:30–9:30 p.m." This date was moved up by one week when NBC did their "Sneak Peek," so the premiere of the series was Thursday September 8. Desilu then had MSEI re-issue the same cardboard cutouts but with newly printed

Because of the positive press that resulted when the Smithsonian secured a copy of the second *Star Trek* pilot for their archives, Gene Roddenberry sent this telegram to Fred Durant on March 8, 1968, shortly after learning of NBC's decision to grant the series a third season. *Photo courtesy of the Smithsonian Institution Archives, image SIA-2019-004908*

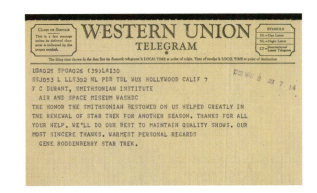

The Smithsonian is not the only place that the 11-foot *Enterprise* model first appeared on display. Prior to its 1974 arrival at the National Air and Space Museum, the original filming model was taken out of storage twice before for public viewing, both times at Golden West College in Huntington Beach, California. The first time was from April 9 to 15, 1972, when the model was exhibited in support of a space expo. The weeklong event featured a NASA traveling exhibit called "America in Space," which included hardware from the US space program along with a lunar sample from Apollo 11. In addition, a talk titled "Life in the Year 2001" was given by Arthur C. Clarke, plus lectures by film producer George Pal and aerospace engineer Krafft Ehricke. The *Star Trek* model came out once again the following year, when, after the success of the first space expo, a second one was held at the college from March 26 to April 7, 1973. Once again, the *Enterprise* was put on display. But this time it included the miniature filming model of the shuttlecraft from the television series along with miniatures from the MGM film *Earth II*. A highlight of this second space expo was seeing the flown Apollo 16 command module, which carried astronauts John Young, Charlie Duke, and Tom Mattingly to the moon and back. Apollo 16 was launched the day after the 1972 college space expo ended, so it was fitting to have that spacecraft on display. A moon rock from Apollo 16 was also shown. Both expos were the brainchild of former Paramount employee Craig O. Thompson, who worked on the original *Star Trek* television series before coming to Golden West College. *All photos courtesy of Ron Yungul*

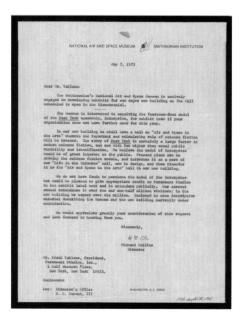 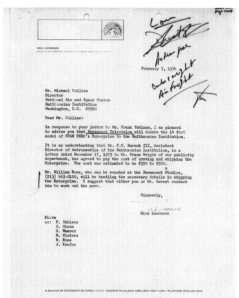

Left: May 7, 1973, letter from Michael Collins, the director of the National Air and Space Museum, formally requesting the original 11-foot production model *Enterprise* be given to the Smithsonian. Note the handwritten "FCD draft to MC" on the lower right of the letter. "FCD" is Frederick C. Durant, who drafted the letter for Michael Collins to sign. *Photo courtesy of the Smithsonian Institution Archives, image No. SIA2019-004907*

Right: February 7, 1974, letter from Dick Lawrence, executive vice president in charge of domestic television syndication for Paramount, agreeing to donate the production model *Enterprise* to the Smithsonian. The text of this letter first appeared in the *Star Trek Giant Poster Book No. 10*, which was published in 1977. *Photo courtesy of the Smithsonian National Air and Space Museum, Washington, DC, copy in author's files, original in registrar's records*

Curators at the Smithsonian's National Air and Space Museum took these two black-and-white Polaroid photographs to document the original 11-foot *Enterprise* model's condition when it arrived from Paramount on February 28, 1974. It was disassembled and in two separate shipping crates, and the declared insurance value of the production model was $5,000. *Photos A19740668000CP01 and A19740668000CP02, courtesy of the Smithsonian National Air and Space Museum*

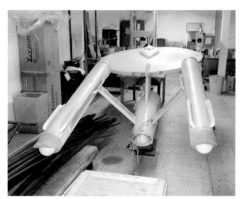

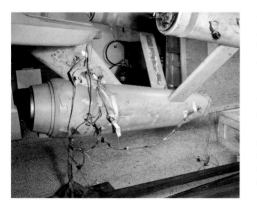

The original 11-foot production filming model of the *Enterprise* is shown being assembled in this series of photos assembled after its arrival at the Smithsonian Institution's Arts and Industries building from Paramount on February 28, 1974. The model is missing its two warp engine domes, along with the main deflector dish. These would later be fabricated by the Smithsonian. *Photos by Richard B. Farrar; courtesy of the Smithsonian National Air and Space Museum, NASM 74-3977, 3978, 3979, 3981; photo courtesy of the Smithsonian Institution Archives, image no. 743980*

Above: Guests at an invitation-only black-tie gala given for the first *Life in the Universe?* exhibit, held on September 19, 1974, at the Smithsonian's Arts and Industries building. Shown to the left are Smithsonian curator Tom Crouch and his wife. The three individuals to the right of the photo are NASA deputy administrator George Low and his wife, Mary, and, to the right of Low, is exhibit curator Alexis "Dusty" Doster, III. *Photo courtesy of the Smithsonian Institution Archives, image SIA-74-9715-03*

Left: On April 12, 1971, former astronaut Michael Collins became the director of the Smithsonian's National Air and Space Museum. This photo was taken of him attending an invitation-only black-tie gala given for the first *Life in the Universe?* exhibit, held on September 19, 1974, at the Smithsonian's Arts and Industries building. *Photo courtesy of the Smithsonian Institution Archives, image SIA-74-9715-34*

Guests at an invitation-only black-tie gala given for the first *Life in the Universe?* exhibit, held on September 19, 1974, at the Smithsonian's Arts and Industries building. Shown are NASA deputy administrator George Low and his wife, Mary, while exhibit curator Alexis "Dusty" Doster, III explains part of the exhibit. *Photo courtesy of the Smithsonian Institution Archives, image SIA-74-9715-18*

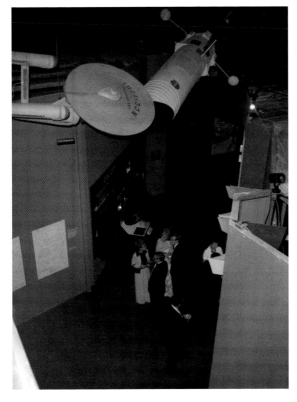

Hanging from the ceiling in the Smithsonian's first *Life in the Universe?* exhibit is shown the original 11-foot production filming model of the starship *Enterprise*, next to a proposed model of what eventually became the Hubble Space Telescope. Beneath the *Enterprise* model can be seen framed and mounted on the wall a set of original blueprints of the *Enterprise* drawn by Franz Joseph Schnaubelt. The blueprints were untrimmed versions that Joseph made prior to the commercial sets that eventually would be available for purchase from Ballantine Books beginning in the spring of 1975. This photo was taken during an invitation-only black-tie gala given for the new *Life in the Universe?* exhibit, held on September 19, 1974, at the Smithsonian's Arts and Industries building. *Photo courtesy of the Smithsonian Institution Archives, image SIA-74-9715-11*

Young *Star Trek* fans shown during an invitation-only black-tie gala given for the first *Life in the Universe?* exhibit, held on September 19, 1974, at the Smithsonian's Arts and Industries building. *Photo courtesy of the Smithsonian Institution Archives, image SIA- 74-9715-27*

A 1976 photo of Franz Joseph Schnaubelt at the Marine Corps Air Station Miramar in San Diego, California. They were showing the play *Pardon Me, Is This Planet Taken?* and included a set of Franz Joseph's blueprints displayed in the auditorium lobby. Note the plans are stamped "CLASSIFIED" in red. *Photo courtesy of Charles G. Weir and Karen Schnaubelt*

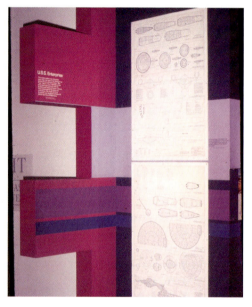

Left: The exit to the new *Life in the Universe?* exhibit that opened in the new National Air and Space Museum building in 1976 features the 11-foot *Enterprise* model. *Photo courtesy of the Smithsonian Institution Archives, image SIA 76-17086*. *Right*: Photo of Franz Joseph's blueprints mounted on the wall beneath the 11-foot *Enterprise* model as they appeared in 1977 at the Smithsonian's "new" *Life in the Universe?* exhibit, which opened in 1976 as part of the newly built National Air and Space Museum. The text for the *Enterprise* model reads: "This studio model of an interstellar space ship was used in the filming of the science fiction television series, 'Star Trek.' Many of the 79 episodes of the series dealt specifically with the problems and results of human interaction with extraterrestrial lifeforms and civilizations. The model of the USS *Enterprise* was designed by Walter M. Jefferies and Gene Roddenberry." The wording of the text next to Franz Joseph's blueprints on the right reads: "These drawings of the details and interior arrangements of 'Star Trek's' USS *Enterprise* were created by aerospace designer, Franz Joseph Schnaubelt." *Photo courtesy of Charles G. Weir and Karen Schnaubelt*

Because of the resulting public interest in Franz Joseph's blueprints being displayed at the Smithsonian, Ballantine Books signed a contract to print sets of his drawings for commercial resale. The resulting *Star Trek Blueprints* became a global phenomenon shortly after they appeared in early 1975. This was soon followed later that same year with the publication by Ballantine of Franz Joseph's *Star Fleet Technical Manual*. To celebrate the *Technical Manual* becoming the number 1 *New York Times* bestseller for trade paperbacks, Judy-Lynn del Rey, Joseph's editor at Ballantine, sent him a bottle of champagne. The successful publication of Franz Joseph's *Star Trek Blueprints and Technical Manual* allowed Ballantine Books to lead in the production and sale of *Star Trek* books in the 1970s. *Photo courtesy of the author, Karen Schnaubelt, Ballantine Books, and the Jay Kay Klein Collection, Special Collections & University Archives, University of California, Riverside*

Dane Boles (*left*), director of the Rocketeer communications department at Estes Industries, is shown holding an Estes flying model rocket version of the starship *Enterprise* from *Star Trek*. The model kit became a big seller after first being made available to the public in April 1975. Michael Collins, director of the Smithsonian's National Air and Space Museum, is shown holding the model. Estes donated their 1:668 scale assembled model kit to the museum during a presentation by Estes held at the Smithsonian on February 20, 1976. *Photo courtesy of the Smithsonian Institution Archives, image No. 76-3194-32A*

Boles is shown holding an Estes flying model rocket Klingon ship from *Star Trek*. The model kit was released at the same time as the abovementioned *Enterprise* model in 1975. For comparison, Collins is shown holding the actual model used in the filming of the original television series. The 29-inch filming model was donated to the Smithsonian on October 26, 1973. Like they did with their *Enterprise* model rocket, Estes also gave their 1:483 scale assembled Klingon model kit to the museum during a presentation by Estes held at the Smithsonian on February 20, 1976. *Photo courtesy of the Smithsonian Institution Archives, image No. 76-3194-36A*

During the February 20, 1976, meeting, representatives from Estes Industries of Penrose, Colorado, makers of flying model rocket kits, met with Smithsonian officials in Washington, DC, to present copies of their new line of flying model rocket kits of the starship *Enterprise* and Klingon battle cruiser from *Star Trek*. Assembled versions of both Estes kits can be seen on the center coffee table alongside the original 29-inch filming Klingon model that was used in the television series. Shown in this photo, *left to right*, are Fred Durant, assistant director of astronautics, NASM; Meredith Marsh of Daniel J. Edelman Company, the PR firm for NASM; Elizabeth Linderman from the Damon Corporation, owners of Estes Industries; Dane Boles, director of the Rocketeer communications department at Estes Industries; and Michael Collins, director of the Smithsonian's National Air and Space Museum. *Photo courtesy of the Smithsonian Institution Archives, image no. 76-3194-28A*

Left: During its stay at the Smithsonian, the original 11-foot production filming model *Enterprise* moved around. In 1979 after the close of the *Life in the Universe?* exhibit, the model moved across the main floor to the History of Rocketry and Spaceflight Gallery. In 1984, the model moved upstairs to be showcased in the museum's yearlong space art exhibit that featured the work of artist Bob McCall. Here is *Star Trek* fan Clint Young posing underneath the *Enterprise* while it was being displayed in the McCall art exhibit. *Photo courtesy of Clint Young, www.cyproductions.com*

Views of the 11-foot filming model *Enterprise* shown suspended for photography at the Smithsonian's National Air and Space Museum Garber Facility, Silver Hill in Suitland, Maryland, on October 8, 1991. NASM exhibits designer Barbara Brennan stands off to the lower right, steadying the model for the photograph. The model was being moved from display to begin restoration by Ed Miarecki in preparation for inclusion at the National Air and Space Museum's *Star Trek* exhibit. The popular exhibit ran from February 28, 1992, through January 31, 1993. The exhibit then moved to New York City, where it was shown at the Hayden Planetarium from 1993 to 1994. *Photos by Mark Avino, Smithsonian National Air and Space Museum, NASM2023-05094, NASM2023-05095, NASM2023-05096*

Beginning in late 1991, the 11-foot *Enterprise* model underwent an extensive restoration by Ed Miarecki. The model was displayed as the centerpiece of a *Star Trek* exhibit hosted at the Smithsonian from 1992–93 in conjunction with the 25th anniversary of the original television series. The original *Enterprise* joined other filming models from the franchise in the popular exhibit including original filming models of the Botany Bay from the episode "Space Seed," and the *Enterprise* refit as seen in the first six *Star Trek* theatrical films. On February 26, 1992, Star Trek cast members attended an advance VIP tour of the exhibit. Shown *left to right* are: Lieutenant Hikaru Sulu (George Takei), Pavel Chekov (Walter Koenig), Spock (Leonard Nimoy), Lieutenant Nyota Uhura (Nichelle Nichols), Captain James T. Kirk (William Shatner), Dr. Leonard "Bones" McCoy (DeForest Kelley), Nurse Christine Chapel (Majel Barrett-Roddenberry) and Montgomery Scott (James Doohan). *Photos courtesy of the Smithsonian Institution Archives, Image 92-1441-9, and the author*

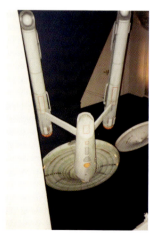

To help educate visitors as to why the 11-foot production model *Enterprise* was at the National Air and Space Museum, Smithsonian officials made a point whenever they could to acknowledge the role that *Star Trek* played in promoting public interest in space travel and exploration. Examples of this message could be found on souvenir floaty pens that were first sold at the museum in 1974. This message continued to be included on other *Star Trek* items that were later sold to the public at the museum, including a special line of "Smithsonian Museum Shops" toys that were licensed to tie in with the museum's popular 1992 *Star Trek* exhibit. *Photos courtesy of the author, the Smithsonian, and Ertl*

2024 marked the fiftieth anniversary of the 11-foot production model *Enterprise* arriving at the Smithsonian and being displayed. *Photos courtesy of Karen Schnaubelt, CBS/Paramount, Ballantine Books, the Smithsonian Institution Archives, the National Air and Space Museum, and the author*

JOURNEYS OF THE STARSHIP ENTERPRISE
TO THE SMITHSONIAN AND BEYOND

1964-1968
Construction, use on the production of Star Trek

1969-1971
After being filmed for "The Trouble With Tribbles," the filming model was put into storage at Paramount

Photo Courtesy AMPAS Margaret Herrick Library Linwood G. Dunn Papers 70118127

April 9-April 15, 1972
Model was displayed publicly for the first time at Golden West College in Huntington Beach, California

Mar. 26-Apr. 7, 1973
Model was again displayed at Golden West College in Huntington Beach, California

Photo Courtesy Ron Yungul

Feb. 28, 1974
The Smithsonian National Air and Space Museum (NASM) in Washington, D.C. takes delivery of the model

Photo Courtesy National Air and Space Museum A1974066800CP01

Sept. 23, 1974- Jan. 1, 1975
After the initial "Rogay" restoration, the model is displayed in the Arts & Industries Building *Life in the Universe?* exhibit

Photo Courtesy National Air and Space Museum A1974066800CP02

Photo Courtesy Smithsonian Institution Archives SIA-74-9716-11

July 1, 1976- Mar. 1, 1979
After "Phase 2" restoration, hung in new National Air and Space Museum building *Life in the Universe?* exhibit

Photo Courtesy Smithsonian Institution Archives SIA-76-17086

1979
Moved to History of Rocketry and Spaceflight Gallery

Photo Courtesy Smithsonian Institution Archives SIA-87-14651CT

1984
Large sections of the secondary hull, the connector between the secondary hull and the primary hull, and the two nacelles were all repainted white

Photo Courtesy National Air and Space Museum NASM-2023-05096

1984-1985
Yearlong display in Robert McCall Space Art Exhibit Flight and Arts Gallery

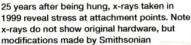

Photo Courtesy of the Author

1992-1993
25th anniversary restoration, display in the NASM Star Trek Exhibit and later in NYC's Hayden Planetarium. Afterward the model was taken off display and was put into storage

1999
25 years after being hung, x-rays taken in 1999 reveal stress at attachment points. Note x-rays do not show original hardware, but modifications made by Smithsonian

X-ray Courtesy of the Smithsonian Institution with graphic overlay by Karl Tate

Photo Courtesy Clint Young www.cyproductions.com

2000
Moved to basement gift shop

Photo Courtesy Karl Tate

2015-2016
Conservation member Gary Kerr is shown installing a replacement for a missing navigation beacon on the *Enterprise's* starboard nacelle during the 2015-16 NASM restoration.

Photo Courtesy Gary Kerr

2016
The model rests today in the Independence Avenue Entrance lobby opposite the Robert McCall mural

Photo Courtesy Karl Tate

A promotional item issued by Paramount Television circa 1969. Note that this item includes mentions of *Star Trek*'s 1966 World Science Fiction convention distinguished contribution award as well as Leonard Nimoy's visit to NASA's Goddard Spaceflight Center the following year to attend the Goddard Memorial Dinner. It also mentions the show's second pilot being added to the archives of the Smithsonian Institution. Finally, the ad states that some of the show's optical screen effects were shown in New York's Museum of Modern Art. It turns out that some of the opticals used in *Star Trek* were included in a special film shown at the museum. The film, called *Special Screen Effects*, showed excerpts from well-known movies, artistic experimentations, television series, and commercials that employed special effects. The film was shown at the museum during the spring of 1969 and proved to be so popular that repeat screenings were scheduled. *Ad courtesy of Paramount and David Arland*

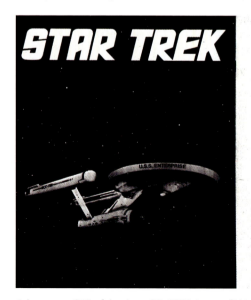

Ad on page 845 of the June 23, 1976, issue of *Weekly Variety*

Syndication ad from the August 4, 1969, issue of *Broadcasting* magazine.

Ad from the September 21, 1976 issue of the *New York Times*

Nichelle Nichols and *Star Trek* creator Gene Roddenberry at NASA's Jet Propulsion Laboratory in Pasadena, California, during the Viking 1 landing on Mars in July 1976. *Photo courtesy of Don Davis*

September 17, 1976: the rollout of the Space Shuttle orbiter OV-101 *Enterprise*. *Left to right*: NASA administrator, Dr. James D. Fletcher; DeForest Kelley, who portrayed Dr. "Bones" McCoy on the series; George Takei (Mr. Sulu); James Doohan (Chief Engineer Montgomery "Scotty" Scott); Nichelle Nichols (Lt. Uhura); Leonard Nimoy (Mr. Spock); series creator Gene Roddenberry; US representative Don Fuqua (D-FL), chairman of the House Science and Technology Subcommittee; Walter Koenig (Ensign Pavel Chekov); NASA deputy administrator George M. Low (*head turned*); and Jesco von Puttkammer (*partly obscured by Low*), NASA engineer of the Space Shuttle Project, and later a consultant and technical advisor for *Star Trek: The Motion Picture*. Upon invitation by NASA, Leonard Nimoy attended the landing of the first space shuttle, *Columbia*, at Edwards Air Force Base, California, on April 14, 1981. Source: *Daily Variety* 191, no. 29 (April 15, 1981). *Photo courtesy of NASA*

ENDNOTES

stickers stuck over the original printing. The base of the cutouts now read "TIME: Thursdays, 8:30 p.m. (7:30 p.m. Central)." What is not known for certain is if Desilu shipped out the revised sticker versions of this promotional item because of the premiere date switch or simply issued them after as a second promotion with the more generic time. Both items are extremely rare and sought-after collectibles. I located a copy of the sticker version of the promotional piece in the archives at the University of Wyoming's American Heritage Center. Source: *Star Trek* Cardboard Enterprise model, box 83, Forrest J. Ackerman collection, collection number 2358, American Heritage Center, University of Wyoming.

18. News release included with Tektites issued by McFadden, Strauss, Eddy & Irwin Public Relations (MSEI), Gene Roddenberry Collection, Correspondence/General Files, Publicity, 1968 & Undated, box 30 folder 6, Gene Roddenberry Papers, (Collection PASC 62), UCLA Library Special Collections, Charles E. Young Research Library, University of California, Los Angeles.

19. Brother Guy Consolmagno and Glen E. Swanson, email correspondence, March 22, 2023.

20. *Douglas Progress* Fourth Quarter 1966, Gene Roddenberry Collection, Correspondence/General Files, Miscellaneous, box 29 folder 5, Gene Roddenberry Papers, (Collection PASC 62), UCLA Library Special Collections, Charles E. Young Research Library, University of California, Los Angeles.

21. Gene Roddenberry to Superintendent of Documents, May 13, 1966. Gene Roddenberry Collection, Correspondence, General Files, Correspondence, Miscellaneous, box 27, folder 17, Gene Roddenberry Papers, (Collection PASC 62), UCLA Library Special Collections, Charles E. Young Research Library, University of California, Los Angeles.

22. Whitfield and Roddenberry, *The Making of Star Trek* (1972), 79–84.

23. Edward Gross and Mark A. Altman, *Captain's Logs: The Unauthorized Complete Trek Voyages,* (Little, Brown, 1995), 9.

24a. Leo R. Lunine, Manager Support Section Technical Information & Documentation Division, Jet Propulsion Laboratory to Gene Roddenberry, September 22, 1967, Gene Roddenberry Collection, Correspondence, General Files, Publicity, 1967, box 30, folder 5, Gene Roddenberry Papers, (Collection PASC 62), UCLA Library Special Collections, Charles E. Young Research Library, University of California, Los Angeles.

24b. William Frech, General Electric Company, Defense Programs Operation, to Gene Roddenberry, October 21, 1964, Gene Roddenberry Collection, Correspondence, General Files, Stock Footage, box 31, folder 7, Gene Roddenberry Papers, (Collection PASC 62), UCLA Library Special Collections, Charles E. Young Research Library, University of California, Los Angeles.

25. Gene Roddenberry "U.S.S. Enterprise Controls & Instrumentation" to Pato Guzman, A. Nelson, J. Paisly, R. Brooks, memo, August 25, 1964, Gene Roddenberry Collection, Correspondence/General Files, Art Direction, box 27, folder 4, Gene Roddenberry Papers, (Collection PASC 62), UCLA Library Special Collections, Charles E. Young Research Library, University of California, Los Angeles.

26. Herb Solow to Gene Roddenberry, memo, August 13, 1965, Gene Roddenberry Collection, Correspondence/General Files, Special Effects, box 31, folder 5, Gene Roddenberry Papers, (Collection PASC 62), UCLA Library Special Collections, Charles E. Young Research Library, University of California, Los Angeles.

27. Alexander, *Star Trek Creator*, 202. Patricio "Pato" Guzman was a native of Chile who worked in the United States as an art director, production designer, and producer of television and film. He served as production designer for the first *Star Trek* pilot and worked directly with Matt Jefferies and Roddenberry in the design of the early sets for the series as well as the *Enterprise*.

28. Gene Roddenberry to Oscar Katz, memo, July 21, 1964. Gene Roddenberry Collection, Correspondence/General Files, Miscellaneous, box 29, bolder 5, Gene Roddenberry Papers, (Collection PASC 62), UCLA Library Special Collections, Charles E. Young Research Library, University of California, Los Angeles.

29. Gene Roddenberry to Herb Solow, memo, August 3, 1964, Gene Roddenberry Collection, Correspondence/General Files, Miscellaneous, box 29, folder 5, Gene Roddenberry Papers, (Collection PASC 62), UCLA Library Special Collections, Charles E. Young Research Library, University of California, Los Angeles.

30. Gene Roddenberry to Kerwin Coughlin, memo, October 14, 1964, Gene Roddenberry Collection, Correspondence/General Files, Casting, box 35, folder 5, Gene Roddenberry Papers, (Collection PASC 62), UCLA Library Special Collections, Charles E. Young Research Library, University of California, Los Angeles.

31. Alan Gess, *Googie: Ultramodern Roadside Architecture*, (Chronicle Books, 2004), 66–68. The origin of the name Googie dates to 1949, when architect John Lautner designed the Googies Coffee Shop in Hollywood, which had distinct architectural characteristics. "Googie" had been a family nickname of Lillian K. Burton, the wife of the restaurant's original owner, Mortimer C. Burton. Googie became a rubric for the architectural style when editor Douglas Haskell of *House and Home* magazine and architectural photographer Julius Shulman were driving through Los Angeles one day. Haskell insisted on stopping the car upon seeing Googies and proclaimed, "This is Googie architecture." He popularized the name after an article he wrote appeared in a 1952 edition of *House and Home* magazine. Haskel wrote in the article "In Googie whenever possible the building must hang from the sky. Where nature and engineering can't accomplish this, art must help… nothing need appear to rest on anything else, least of all on the earth; in Googie architecture both the glass and the stone are conceived to float."

32. Stuart W. Leslie, "Spaces of the Space Age," *Smithsonian Air & Space Magazine*, August 2013, https://www.smithsonianmag.com/air-space-magazine/spaces-of-the-space-age-332231/

33. The Los Angeles Conservancy, https://www.laconservancy.org/locations/northrop-grumman?fbclid=IwAR1zBY8rBoxM-DA60a50JaUTJBxoFMPPYAflvHMLaBCr-PZl-ClFXmYygz8

34. Beaming down to Deneva's main city, Kirk, Spock. McCoy, and Scotty, along with a security detail, are confronted by an armed group of men who warn them to get away prior to attacking them. After stunning the group, Kirk heads to the home where his brother Sam Kirk lives with his wife and son. Sam is found dead while Aurelan, Sam's wife, acts irrationally, then passes out. Her son Peter is found nearby in a comatose state. Kirk and McCoy beam back to the *Enterprise* with Aurelan and Peter to treat their condition. During this time, Aurelan briefly tells Kirk about creatures that arrived eight months earlier on a ship whose crew was forced to travel to Deneva. Kirk learns the creatures use the Denevans as their bodies, forcing them with pain to build ships so they can spread to other systems. Kirk beams back to the surface to rejoin Spock and the landing party. Here they find some squeaking, jelly-looking creatures clinging to the side of a shadowed wall. They stun one of the creatures. As they leave, the stunned creature flies up and attaches itself to Spock. Kirk rips the creature off, but not until leaving a painful puncture wound on Spock's back. Back in sickbay, McCoy learns the creatures attack their victims by stinging them, leaving behind a strand of tissue that affects the host's nervous system. Spock is in great pain but, because of his Vulcan physiology, can overcome it enough to resume his duties. He beams down to the planet to capture one of the creatures and bring it back to the ship for study, where the *Enterprise* crew learns the creatures are part of a single entity. Kirk orders McCoy to study it further to learn how to destroy it. It is during this time they discover ultraviolet light kills the creatures. The *Enterprise* then deploys a ring of satellites into orbit around Deneva that emit ultraviolet light. The creatures are killed, and the infected Denevans are cured.

35. Vela was a group of satellites first developed in the early 1960s as part of Project Vela. The program was developed by the US to detect nuclear detonations to monitor compliance with the 1963 Partial Test Ban Treaty by the Soviet Union. All the spacecraft were manufactured by TRW and launched in pairs, either on an Atlas-Agena or Titan IIIC.

36. Schoenberg Hall opened on the UCLA campus in the fall of 1955. The building, of modernist architecture style, was named after Arnold Schoenberg, the twentieth-century composer who was a member of UCLA's music faculty during the 1940s. It was one of the "most modern

music education facilities of the time." The building is equipped with fiberglass insulation to keep musician rehearsals contained in the 66 basement practice rooms. It also features a library and performance spaces. Unique to its exterior is a 164-foot mosaic mural above the foyer made by artist Richard Haines that tells the story, in sixteen panels, of the history of music from a global perspective. Growing up on a farm in Iowa, Haines began his artistic career as a designer at a greeting-card company and then at a calendar firm before attending the Minneapolis School of Art and the École des Beaux-Arts in France. Returning to the US, Haines worked on several New Deal mural projects and eventually made his way to Los Angeles, where he worked for Douglas Aircraft during WWII. After the war, Haines was commissioned to design, animate, and illustrate a huge world map for Lockheed's Super Constellation passenger aircraft. Called *The Super Constellation World*, portions of the map were used in advertising by Lockheed and various airlines, such as TWA and Qantas, to help promote their global flights. Source: Kristen Hardy, "Look Up and Learn," *UCLA Magazine*, October 2017, https://newsroom.ucla.edu/magazine/ucla-building-mosaics.

37. *Star Trek* was not the only television show to make use of the Space Park campus. "Cold Hands, Warm Heart," the second season episode of the original *The Outer Limits* television series was one of a host of space-themed TV shows and films that utilized the space-age architecture. It helped that the sprawling complex was located relatively close to Hollywood, which made on-location shooting more convenient and less expensive. William Shatner starred in the 1964 *The Outer Limits* episode about the first astronaut to visit Venus. Upon returning to Earth, Shatner contracts an ailment that alters his hands, changes his blood chemistry, and reduces his body temperature. Shatner comes back to the TRW campus three years later in "Operation: Annihilate!" The first season *Star Trek* episode aired in 1967, the same year that the big-screen film *Countdown* starring James Caan premiered in theaters, which also featured scenes shot in Space Park.

38. M. W. Root, "Structural Concept Satellite Space Station" Douglas Aircraft Company, Inc., SM-35661, March 20, 1959; also see M. W. Root "Structure Design Criteria for Manned Satellites," Douglas Engineer Paper 908, in preparation.

39. Cover story on George Akimoto, CA, *The Journal of Commercial Art and Design* 3, no. 10 (October 1961). The claim that George Akimoto was the artist responsible for illustrating the 1959 Douglas design that eventually became *Star Trek*'s K-7 space station was first made by Megan Prelinger. Prelinger is author of the excellent book *Another Science Fiction: Advertising the Space Race 1957–1962* (Blast Books, 2010). In her book she first states that Akimoto did the artwork in a caption that she wrote beneath a vintage Douglas Missiles and Space System ad that features the space station design (see p. 169 of *Another Science Fiction*). The ad originally appeared on p. 21 of the December 18, 1959, issue of *Missiles and Rockets*.

40. Eugene B. Konecci and Neal E. Wood, "Design of an Operational Ecological System," in *Proceedings of the Manned Space Stations Symposium, Los Angeles, California, April 20–22, 1960* (Institute of the Aeronautical Sciences, 1960), 137–50.

41. *Douglas Annual Report 1959* published February 1959 by Douglas Aircraft Company, Inc. Artwork depicting elements of what would become the K-7 space station in *Star Trek* appear both on the cover and throughout the report.

42. The original Root paper is referenced in another paper "Design of an Operational Ecological System" by Eugene B. Konecci and Neal E. Wood, two engineers with the Missiles and Space Systems Engineering Department of Douglas Aircraft Company in Santa Monica, published in the *Proceedings of a Manned Space Stations Symposium*, including papers presented at a symposium held in Los Angeles from April 20 to 22, 1960, and sponsored by NASA and the RAND Corporation.

43. Noel Datin McDonald and Richard C. Datin. Jr., *The Enterprise NCC 1701 and The Model Maker*, (CreateSpace, 2016), 90.

44. McDonald and Datin. Jr., *The Enterprise NCC 1701*, 91–92.

45. McDonald and Datin. Jr., *The Enterprise NCC 1701*, 91.

46. McDonald and Datin. Jr., *The Enterprise NCC 1701*, 90–91.

47. McDonald and Datin. Jr., *The Enterprise NCC 1701*, 92.

48. Day, "Boldly Going: *Star Trek* and Spaceflight," *Space Review*, November 8, 2005, https://www.thespacereview.com/article/506/1.

49a. Peter E. Glaser, "A Structural Approach to the Thermal Control of Space Vehicles," in *Proceedings of the Manned Space Stations Symposium: Los Angeles, California, April 20–22, 1960* (Institute of the Aeronautical Sciences, 1960), 202–06.

49b. Dorothy Fontana, "Behind the Camera: Walter M. Jefferies," *Inside Star Trek 4*, October 1968, 4.

49c. Cover photo, *Space World: The Magazine of Space News*, A-13, (November 1964). The cover photo depicts a Bell Aerosystem's Remora capsule (foreground) which enabled individuals to assemble, inspect, service, and maintain satellites and space stations or shuttle crews between space vehicles while being protected from the hazards of meteorites and radiation. Taking its name from the fish that attaches itself to sharks, Remora is equipped with mechanical grappling arms by which it can attach itself to space stations and satellites. Inside the capsule astronauts would have the freedom to manipulate the arms that would allow them to engage in assembly and maintenance activities. In the background is a reentry vehicle like the design depicted in Peter E. Glaser's "A Structural Approach to the Thermal Control of Space Vehicles."

Chapter 6

Portions of this chapter were published by Glen E. Swanson, "How *Star Trek* Helped NASA Dream Big and How NASA Helped *Star Trek* Stick Around," *Air & Space Smithsonian*, 35 no 7 (February/March 2021) and Glen E. Swanson, "The Making of an Enterprise: How NASA, the Smithsonian and the Aerospace Industry Helped Create *Star Trek*." *Space Review*, September 7, 2021, https://www.thespacereview.com/article/4240/1?fbclid= IwAR2mPm7I3LteQuOF7SzisOA_6VPLP4YJXt4RKJtTsRwLF9jmj924C3Ery8A.

1. According to Marc Cushman, the ratings the night *Star Trek* premiered show "The Man Trap," which was the fifth episode filmed and the first episode to air, won its time slot "hands down." However, Cushman's assessment of the show's ratings has been a source of debate. A more detailed analysis of the ratings conducted by other *Star Trek* scholars supports an alternate claim that "The Man Trap" did indeed premiere to many viewers, but with the added caveat that viewership trailed off during the first half of the hour-long episode. This was mainly the result of television viewers flipping channels, a common occurrence, since people wanted to see what else was being shown. Despite the declining numbers during the first episode airing, *Star Trek* still did very well for a premiere. Source: Marc Cushman with Osborn, *These Are The Voyages: TOS; Season One*, 202–203; and Michael Kmet, "The Truth about *Star Trek* and the Ratings," *Star Trek* Fact Check, July 5, 2014, https://startrekfactcheck.blogspot.com/2014/07/the-truth-about-star-trek-and-ratings.html

2. "CBS LANDS SPACE FOOTAGE FROM USSR," *Weekly Variety*, 244, no. 3, (September 7, 1966): 22. Back in 1966 during the Gemini 11 mission, CBS obtained exclusive footage from the Russian Embassy they could intersperse with their live network coverage of the flight. This was good because the launch ended up being delayed several days. The delay allowed CBS to make full use of their rare Russian footage to show millions of Americans their first views of the Soviet Union's equivalent of Cape Kennedy. The footage included launch procedures of a Russian spacecraft and its recovery, as well as close-ups of the spacecraft controls and interiors of their launch control room.

3. "NBC Wins Gemini," *Weekly Variety*, 244, no. 4, (September 14, 1966): 32.

4. The Jupiter 2 from CBS's *Lost in Space* televisions series, a competitor with *Star Trek* at the time, was originally called "Gemini 12." An undocumented story that has circulated among fans is that series creator Irwin Allen wanted "to avoid confusion with NASA's actual Gemini space program." Viewers were watching missions from NASA's Gemini Program unfold on their televisions

◀ ENDNOTES ▶

at the same time *Lost in Space* was being shown, so there was not only the potential for confusion but, if Allen stuck with the name "Gemini 12," this would also give the appearance of dating the show to the present rather than in the show's projected year of 1997. For these reasons, Allen therefore changed the name of his spacecraft to the "Jupiter 2." Source: Marc Cushman, *Irwin Allen's* Lost in Space: *The Authorized Biography of a Classic Sci-Fi Series*, (Jacobs/Brown, 2016), 199.

5. Wright, "TV's *Star Trek*," 72–74.

6. Mariner V was not the first spacecraft to visit Venus. Mariner 2 performed the first successful flyby of another planet when it reached Venus five years earlier. The Soviet Union's Venera 4 then followed to become the first spacecraft to survive entry into another planet's atmosphere when it deployed a probe that returned atmospheric data during its October 18, 1967 descent to the planet's surface. The Venera 4 probe did not survive landing. A replica of Mariner 10 hung in the Smithsonian's *Life in the Universe?* exhibit near the production model *Enterprise* when the new National Air and Space Museum opened in 1976. The model was sent back to JPL in 1989. The Smithsonian's Mariner 10 they have on display now is a flight spare that JPL transferred to the Museum in 1981. It is not 100% complete as it is missing some scientific instruments and a few subsystems, but it remains one of the Smithsonian's most complete robotic spacecraft in their collection. It now hangs in the Exploring the Planets gallery along with Voyager. Source: Glen Swanson email exchange with Matt Shindell, Smithsonian National Air and Space Museum, April 25, 2024; "Spacecraft to Smithsonian," NASM5225, 5.

7a. *Star Trek Initial Pitch,* dated March 11, 1964.

7b. Star Trek Voyage One "The Cage" notes dated September 15, 1964, Gene Roddenberry Collection, Produced Episodes, The Menagerie & Materials From "The Cage," box 7, folder 1, Gene Roddenberry Papers, (Collection PASC 62), UCLA Library Special Collections, Charles E. Young Research Library, University of California, Los Angeles.

8. Alexander, *Star Trek Creator*, 212.

9. Gene Roddenberry to Ed Perlstein, memo, March 31, 1967, Gene Roddenberry Collection, Correspondence/General Files, DeForest Kelley, box 28, folder 15, Gene Roddenberry Papers (Collection PASC 62), UCLA Library Special Collections, Charles E. Young Research Library, University of California, Los Angeles.

10. "BONES SENDS WIRE TO ASTRONAUTS," Hollywood (November 21, 1968), NBC Feature Typescript. DeForest Kelley Papers, Star Trek Publicity, Academy of Motion Pictures Arts and Sciences Margaret Herrick Library. Also referenced in Terry Lee Rioux, *From Sawdust to Stardust: The Biography of DeForest Kelley, Star Trek's Dr. McCoy* (Pocket Books, 2005), 192. Rioux indicates the source for this is the Roddenberry Papers at UCLA. I have seen the notice in UCLA's Margaret Herrick Library but have not been able to verify if and where it was published.

11. Department of the Air Force to Gene Roddenberry, May 5, 1967, in Whitfield and Roddenberry, *The Making of Star Trek*, (1972), 382.

12. NASA's Dryden Flight Research Center, now Armstrong Flight Research Center, has in their archives a series of photos taken of the 1967 visit by the *Star Trek* cast and crew. The NASA files indicate these photos were taken on April 13, 1967, a little less than a month before the May 5, 1967, invitation from Buchanan, as shown on p. 382 of Whitfield and Roddenberry, *The Making of Star Trek*. Also, in Gross Altman's, *Captain's Logs*, (Little, Brown, 1995), 20, Marc Daniels mentions the visit during an interview.

13. A month after Scotty's meeting, Peterson was flying the M2-F2 when he lost control shortly before landing. The vehicle crashed and somersaulted across a dry lakebed at Edwards before finally coming to a stop. Amazingly, Peterson survived, though he lost an eye. Footage of the crash was used at the beginning of every episode of the popular 1970s TV series *The Six Million Dollar Man*. The M2-F2 was rebuilt and modified with an additional third vertical fin—centered between the tip fins—to improve control. The new and improved lifting body was renamed the M2-F3 and flew again until it retired from service at the end of 1972. It can be seen on display at the Smithsonian's National Air and Space Museum in Washington, DC.

14. Gene Roddenberry to Robert Justman, memo, June 15, 1966, Gene Roddenberry Collection, Correspondence/General Files, Publicity, up to and including 1966, box 30, folder 4, Gene Roddenberry Papers (Collection PASC 62), UCLA Library Special Collections, Charles E. Young Research Library, University of California, Los Angeles.

15. Art Wilcox to Scott Carpenter, July 12, 1966, Gene Roddenberry Collection, Box 30, Folder 4. Correspondence/General Files, Publicity, up to and including 1966, fox 30, folder 4, Gene Roddenberry Papers (Collection PASC 62), UCLA Library Special Collections, Charles E. Young Research Library, University of California, Los Angeles.

16. Harvey Pack, "Producer of *Star Trek* Struggling to Keep Space Ship From Crashing," *The Wichita (KS) Eagle*, November 26, 1967, 76. Pack was a syndicated writer based in New York who went on to become a noted personality in the field of horse racing. After being hired by the New York Racing Association in 1974, he became the host of *Thoroughbred Action*, a thirty-minute nightly recap show of racing from New York Racing Association tracks that aired on cable. Source: David Grening, "Harvey Pack, Legendary On-Air Racing Personality, Dead at 94," DRF, July 6, 2021, https://www.drf.com/news/harvey-pack-legendary-air-racing-personality-dead-94.

17. *Science Education in the Space Age*, proceedings of a National Conference Held in Los Angeles, June 1–4, 1964 (NASA and US Government Printing Office, November 1964), 3–4.

18. Gene Roddenberry to Bill Heath, memo, March 29, 1966, Gene Roddenberry Collection, Correspondence/General Files, Stock Footage, box 31, folder 7, Gene Roddenberry Papers (Collection PASC 62), UCLA Library Special Collections, Charles E. Young Research Library, University of California, Los Angeles.

19. Gene Roddenberry to Byron Haskin, memo, November 23, 1964, Gene Roddenberry Collection, Produced Episodes, The Menagerie & Materials from "The Cage," box 7, folder 6, Gene Roddenberry Papers (Collection PASC 62), UCLA Library Special Collections, Charles E. Young Research Library, University of California, Los Angeles.

20. SUBLIMINAL, KNOWLEDGE MONTAGE, memo, December 21, 1964, Gene Roddenberry Collection, Correspondence/General Files, Special Effects, box 31, folder 5, Gene Roddenberry Papers (Collection PASC 62), UCLA Library Special Collections, Charles E. Young Research Library, University of California, Los Angeles.

21. *Space: The New Frontier*, publication 637081 (US Government Printing Office, 1962). The first edition of this publication was published in 1959 and did not include the photos used in "The Cage." The issue that *Star Trek* used for "The Cage" was published by NASA in 1962. NASA continued to publish this title for many years, updating it to reflect the times. The last issue NASA published using that name appeared in 1966.

22. In addition to the Talosian slide show, use of original NASA photos as more permanent set dressing can be seen in the first season episode of "Court Martial." In this episode, Kirk is in an office on a Federation Starbase and viewers can clearly see NASA photos from the Gemini Program hanging on the walls.

23. Gene Roddenberry to Ed Orzechowski, February 14, 1968, Gene Roddenberry Collection, Correspondence/General Files, N.A.S.A., box 29, folder 6, Gene Roddenberry Papers (Collection PASC 62), UCLA Library Special Collections, Charles E. Young Research Library, University of California, Los Angeles.

24. *Earth Photographs from Gemini III, IV, and V, NASA SP-129* (Scientific and Technical Information Division, Office of Technology Utilization, NASA, 1967), 20. The idea of a book devoted to publishing photos of the Earth as taken from the Gemini spacecraft originated with Dr. Robert R. Gilruth, the first director of NASA's Manned Spacecraft Center (now the Johnson Space Center), in Houston, Texas. Gilruth was very interested in the early photos taken of the planet by the Gemini astronauts. As a result, NASA produced a three-volume set of hardbound books. The first volume covered selected images taken during the first three Gemini missions, Gemini III–V. Two other volumes were subsequently published that contained images taken from the Gemini VI-A through Gemini XII missions. Volume I contains 3 Gemini III, 96 Gemini IV, and 145

◀ ENDNOTES ▶

Gemini V high quality color lithographs, each seven inches square, or about three magnifications of the original Hasselblad photos. Jointly sponsored by NASA's Office of Manned Space Flight and MSC's director, Vol. I of the series was edited by Dr. Jocelyn R. Gill, the Gemini Sciences Manager. The informative captions accompanying each photograph was a joint effort of Dr. Paul Lowman of NASA Goddard, Ken Nagler and Stan Soules of the US Weather Bureau, Art Alexiou of the Naval Oceanographic Office, and Dick Underwood and Herb Tiedemann of MSC's Photographic Technology Laboratory. Sandra Scaffidi of the Scientific and Technical Information Division of Washington organized the material for publication to ensure the reproductions would be of first-rate quality. (Source: NASA Manned Spacecraft Center *Space News Roundup*, June 23, 1967, 6).

25. Earlier versions of the script for "Assignment: Earth" dated December 4, 1967, had the episode opening with the crew on the bridge of the *Enterprise* watching a scene on the ship's view screen from NBC's hit western *Bonanza*. NBC loved the idea, but it would have been cost prohibitive, as all the *Bonanza* people would have had to be paid along with securing all the required releases from the writers, directors, and producers of the episode. The red tape and added cost were simply not in the show's budget, even though it would have been a most memorable scene.

26. "Assignment: Earth" Spin-Off Pilot Synopsis Created by Gene Roddenberry and Art Wallace, Paramount Studios, December 5, 1967, 2.

27 "Assignment: Earth" Spin Off (1967).

28 Apollo 4, whose footage was used by *Star Trek* in this episode, was an "all-up" test, meaning all rocket stages and spacecraft were fully functional on the initial flight of this new launch vehicle. This mission marked many firsts, including the first time the Saturn V's S-IC first stage and S-II second stage flew. The flight was designated Apollo 4 because there had been three previous Apollo/Saturn flights, all using the Saturn IB launch vehicle but with no crew onboard. After Apollo 4's successful launch on November 9, 1967, the command module splashed down in the Pacific Ocean nine hours later, achieving all its mission goals. NASA deemed Apollo 4 a complete success, proving the Saturn V worked, an important step toward achieving Apollo's objective of landing astronauts on the Moon and bringing them back to the Earth before the end of the decade.

29. Onboard staging footage used in "Assignment: Earth" showed dramatic views of the Sinai Peninsula that includes the Suez Canal and the Great Bitter Lake in the lower center of the screen, with the Red Sea to the left of center and the Mediterranean Sea at the bottom. The Sinai Peninsula dominates the center of the screen, while the rest of Egypt can be seen on the lower left side.

30. "Assignment: Earth" also makes use of footage of another Saturn V launch vehicle. This second spacecraft, the Saturn V Facilities Integration Vehicle, also known as the SA-500F (alternately SA500F, 500F), was the first Saturn V rocket built. Unlike its predecessors, it was never designed to go to the moon, let alone leave the earth. SA-500F was a full-sized flight configured Saturn V in every way, except that as a facilities demonstrator it was never meant to fly. NASA needed dummy Saturn Vs to use as structural test vehicles or as dynamic test articles, and the 500F was one of five non-flight configurations built but never flown. The 500F made sure that, among other things, all the equipment designed to assemble, transport and prepare the real launch vehicle at NASA's Kennedy Space Center would work correctly. The 500F is easy to spot because of its peculiar paint scheme. The intertank of the S-IC first stage has a unique black circumferential band. However, this band was removed on all subsequent Saturn Vs, starting with Apollo 4, because engineers found the black paint created excessive heat, generated by the absorption of sunlight, that affected the accuracy of the strain gauge readings in the intertank. After painting the area white instead, engineers learned it would reflect heat more evenly, resulting in fewer temperature differences in the stage's interior. Additional unique visual features of the 500F are its distinctive black-and-white roll patterns plus "USA" emblazoned on the side in big, bold, red letters on the S-IVB third stage. Since this was the first fully assembled Saturn V, model makers duplicated its unique paint scheme, preserving an appearance we can still see on many scale replicas. Source: Milton Alberstadt, "New Paint Job," *Space World*, E-4-52 (April 1968) 28.

31. "Assignment: Earth (Episode)," Memory Alpha, https://memory-alpha.fandom.com/wiki/Assignment:_Earth_(episode).

32. Launch Complex 13 was constructed in 1956 and launched the first of two Atlas B rockets two years later. Later, the pad was converted to support launches of the Atlas Agena. Three Vela spacecraft (two satellites each) were launched from the launch complex between 1963–1965. In 1966, the launch complex was turned over to NASA to support civilian Atlas Agena D launches. From 1966 to 1968, six NASA Atlas Agena D missions launched, which included five Lunar Orbiter missions. In 2005, the Mobile Service Tower came down by a controlled explosion, and in 2012 the Blockhouse was demolished. Beginning in 2015, SpaceX leased the site and renamed it Landing Zone One (LZ-1). It has since been used as a landing platform for the Falcon 9 first stage.

33. Marc Cushman, memo to Gene Roddenberry from Kellam De Forest, December 22, 1967; and Cushman with Osborn, *These Are The Voyages TOS: Season Two*, 588.

34. Gene Roddenberry to James E. Webb, March 16, 1968. Gene Roddenberry Collection, Correspondence/General Files, N.A.S.A., box 29, folder 6, Gene Roddenberry Papers (Collection PASC 62), UCLA Library Special Collections, Charles E. Young Research Library, University of California, Los Angeles. In the Gene Roddenberry Papers at UCLA, there is a NASA document containing twenty-five pages of stock footage requested from NASA by *Star Trek*. Descriptions contained of the footage in this document include scenes of a car driving out to the Saturn V launch complex. This same car closely matched the set scenes that show Gary Seven emerging from the trunk. Source: Gene Roddenberry Collection, Produced Episodes, Assignment: Earth, box 32, folder 23, Gene Roddenberry Papers (Collection PASC 62), UCLA Library Special Collections, Charles E. Young Research Library, University of California, Los Angeles.

35. Letter from Walter E. Whitaker, NASA Audio-Visual Officer, Educational Programs Division, Office of Public Affairs to Edward Milkis dated February 27, 1968, NASA History Office, Washington, DC, (courtesy Dwayne A. Day). Letter from Walter E. Whitaker, NASA Audio-Visual Officer, Educational Programs Division, Office of Public Affairs to Frank Capra dated May 9, 1968, Martin Caidin Papers, Collection 02946, Box 11, American Heritage Center, University of Wyoming.

36. Even though NASA could not be called out by name, there are several episodes in the original series in which other branches of the federal government are mentioned. Both the Air Force and the Navy are called out, for example. In the episode "Tomorrow is Yesterday," after the pilot of the F-104 chasing the *Enterprise* is beamed aboard from his disintegrating aircraft, he states he is "Captain John Christopher, United States Air Force. Serial number 4857932." Later, when commenting about how big the *Enterprise* is, Christopher asks "Did the Navy…" Kirk then cuts him off by saying "we're a combined service, Captain. Our authority is the United Earth Space Probe Agency." The term "Starfleet" had not yet been invented. The Navy is also mentioned by Edith Keeler in the episode "City on the Edge of Forever." When she sees McCoy's Starfleet uniform and after Bones explains he is a surgeon aboard the *Enterprise*, Edith comments, "I don't mean to disbelieve you, but that's hardly a Navy uniform." In "Assignment: Earth" Roberta Lincoln asks if Gary Seven is with the CIA. Seven confirms this while showing her a fake ID, one of several he creates as part of his cover. The National Security Agency (NSA) is also shown in one of Seven's fake IDs but is not mentioned by name in the episode.

37. *Star Trek* "Assignment: Earth," (pilot spin-off), teleplay by Art Wallace, story by Gene Roddenberry and Art Wallace, revised first draft, December 20, 1967. Production no, 60355, p. 34, scenes 83–85. Also on p. 36, scene 90, Seven is described pulling out the plans to the rocket base. He then spots a tall ladder running up the side of a tall building marked "ROCKET ASSEMBLY—ORBITAL STAGE. USAF." In scene 129 of this same version of the script, Seven takes a gantry elevator that takes him to the top of the rocket, "with the vast panorama of Cape Kennedy behind it."

38. Cushman with Osborn, *These Are the Voyages: TOS; Season Two*, 589, 590, 634.

39a. *The Lieutenant* was a television show created by Gene Roddenberry that aired on NBC from 1963 to 1964. The series was based at Camp Pendleton in Southern California, the West

ENDNOTES

Coast base of the US Marines. The show was set against a Cold War backdrop and focused on the men of the corps during peacetime. The title character was a rifle platoon leader and training instructor at Camp Pendleton. Second Lieutenant William Tiberious Rice, played by Gary Lockwood, would later be cast as Lieutenant Commander Gary Mitchell in *Star Trek*'s second pilot, "Where No Man Has Gone Before." *The Lieutenant* was an hour-long dramatic series that featured guest stars such as Majel Barrett, Leonard Nimoy, and Walter Koenig, all of whom would later appear as regulars in *Star Trek*. Roddenberry enjoyed the support of the Marines since they allowed the series to be filmed at Camp Pendleton, which included access to the camp's troops, uniforms, buildings, vehicles, and weapons. The military endorsement was a nice benefit, but the hospitality would soon come to an end. "To Set It Right" was an episode of *The Lieutenant* that dealt with the military and racial prejudice. The story centers on Rice, who seeks to resolve a racial dispute between two members of his platoon, characters played by Dennis Hopper and Don Marshall. Marshall would later play Lieutenant Boma in the *Star Trek* episode "The Galileo Seven." Hopper and Marshall are enemies from their civilians days, and this feud is carried over when they meet up again in the Marines. The white Hopper is openly prejudiced toward Black people, and Marshall is equally prejudiced against White people. Nichelle Nichols, who portrayed *Star Trek*'s Lieutenant Uhura, played the Black soldier's fiancée in what was her television debut. The Marine Corps had issues with the content of the episode and a Department of Defense spokesperson told the studio "it didn't want to be involved in this type of story." They saw the subject matter as being too controversial and objected to its being shown. As a result, they withdrew their seal of approval from the episode "on grounds the story dealing with segregation doesn't present either white or black race in a good light." Seeking to appease the Pentagon's concerns and to maintain their ongoing support for the series, NBC decided not to air it or pay the licensing fee. To change the network's mind, Roddenberry brought the issue to the attention of the NAACP (National Association for the Advancement of Colored People). What happened to the controversial episode has been the topic of some debate. Some sources indicate it never aired, while others, including multiple newspapers from the time, seem to suggest it was shown in some markets, including ones in the South. A *Variety* article from early 1964 indicated MGM-TV had plans to go ahead despite opposition and air the episode on Saturday, February 22, 1964. The entertainment organization even ran a review of it in their Monday February 24 issue of *Daily Variety*. What is known for sure is that the Marine Corps withdrew all support for the show. This meant Camp Pendleton was no longer available for use by Roddenberry for the series. Despite decent ratings, losing the Marine's seal of approval forced NBC to cancel *The Lieutenant* after twenty-nine episodes. The Paley Center for Media in New York City has a videotape of the episode. This episode was eventually broadcast on cable channel TNT in the early 1990s. Sources: "'Lieutenant' Rubs Defense Dept. But Seg Will be Aired," *Variety*, no. 13, (February 19, 1964,): 22; and James Van Hise, *The Man Who Created Star Trek: Gene Roddenberry* (Pioneer Books, 1992), 16, Roddenberry is quoted as saying, "I had only one thing I could do… I went out to [the] NAACP, and an organization named CORE, and they lowered the boom on NBC. They said, 'Prejudice is prejudice, whatever the color.' And so we were able to show the show." Sources: *Cincinnati (OH) Post*, February 15, 1965, 36; *Tulsa (OK) Tribune*, February 22, 1964, 20; *The State*, February 22, 1964, 4; and "Reaction, Telepix Followup, THE LIEUTENANT ("To Set it Right")," *Variety*, 122, no. 55 (February 24, 1964): 8.

    39b. Edward Gross, "Art Wallace: An 'Obsession' On 'Assignment: Earth,'" *Starlog* 112 (November 1986): 37.

    39c. Teri Garr with Henriette Mantel, *Speedbumps: Flooring It Through Hollywood*, (Hudson Street, 2005), 63–64. After Garr secured her role in *Star Trek*, she did what many new actors did at the time: she took out an ad in *Variety* to announce her major television debut. As Garr recalls in her autobiography, "It was shameless self-promotion, but it was the best way to let the industry know you were working and earning enough to afford the ad." As the airing date of "Assignment: Earth" approached, Garr tried to think of some attention-getting ad to place. "I went to the dentist (thank you, SAG, for insurance) and had my annual X-rays taken. And I thought, I know! I'll put X-rays of my teeth in *Variety*." Garr claims the *Variety* ad said, "Watch Terry Garr smile on *Star*

Trek." Garr said that as a result, "For the next few weeks all I heard were recommendations for dentists" (ba-dum-bump)." In the March 29, 1968, issue of *Daily Variety,* there does appear a smiling Garr on p. 24, along with a promotion for the new *Star Trek* episode that would air that same evening. The text Garr claimed to have been included is not shown and neither are her dental X-rays. There is, however, a small graphic in the lower right corner of the ad, which is the logo for the William Morris Agency.

40. Gene Roddenberry to James Webb, March 16, 1968, Gene Rodenberry Collection, Correspondence/General Files, N.A.S.A., box 29, folder 6, Gene Roddenberry Papers (Collection PASC 62), UCLA Library Special Collections, Charles E. Young Research Library, University of California, Los Angeles.

41. Gene Roddenberry memo, September 15, 1966, Gene Rodenberry Collection, Correspondence/General Files, Publicity, up to and including 1966, box 30, folder 4, Gene Roddenberry Papers (Collection PASC 62), UCLA Library Special Collections, Charles E. Young Research Library, University of California, Los Angeles.

42. Robert H. Hood Jr. joined Douglas Aircraft in 1965 as a government sales representative. In 1971, he went to work as assistant deputy administrator of NASA. From 1973 until 1981, Hood worked at Pullman Inc. as vice president. He rejoined McDonnell Douglas in 1982 as vice president for program development and marketing. He transferred to California in 1984 to become vice president and deputy general manager of Astronautics Co.'s Huntington Beach Division. In 1986, he moved back to St. Louis as vice president for aerospace group business development, and in 1989, he was named president of the newly formed McDonnell Douglas Missile Systems Company.

43. Alberta Moran, interview by Glen E. Swanson, December 9, 2019.

44. Pamela "Penny" Jeanne Moran, interview by Glen E. Swanson, September 29, 2022.

45a. Alberta Moran to Gene Roddenberry, February 21, 1967. Gene Rodenberry Collection, Correspondence/General Files, N.A.S.A., box 29, folder 6, Gene Roddenberry Papers (Collection PASC 62), UCLA Library Special Collections, Charles E. Young Research Library, University of California, Los Angeles.

45b. Gene Roddenberry to Alberta Moran, February 28, 1967, Gene Rodenberry Collection, Correspondence/General Files, N.A.S.A., box 29, folder 6, Gene Roddenberry Papers (Collection PASC 62), UCLA Library Special Collections, Charles E. Young Research Library, University of California, Los Angeles.

46. Moran, interview by Swanson, 2019.

47. *Independent Star-News*, August 13, 1967, 62.

48. Leonard Nimoy to Gene Roddenberry, dated March 28, 1967.

49. "Guests of Honor," Tenth Annual Goddard Memorial Dinner, March 15, 1967.

50. Text of speech "Progress in Peace and Technology," by Vice President Hubert H. Humphrey before the Tenth Annual Goddard Memorial Dinner, Washington, D.C., Wednesday March 15, 1967.

51. Leonard Nimoy to Gene Roddenberry, March 28, 1967.

52. Moran, interview by Swanson, 2019.

53. Moran, interview by Swanson, 2022.

54. Leonard Nimoy to Gene Roddenberry, March 28, 1967.

55. *Daily Variety* 135, no. 31, (April 17, 1967): 28–29. The Emmys are to network television what the Oscars are to film. For both, there is great prestige not only in being awarded one but in being nominated for one. Before winning an Emmy, shows first submit their entries for consideration for nomination by the Academy of Television Arts & Sciences. As a result, prior to being nominated, studios will promote their Emmy submissions through ads in various trade publications, hoping it will help get their shows noticed and nominated. In anticipation of the 1967 Emmy Awards, Desilu began placing ads that spring in *Variety*, the leading trade publication that covered the film and

ENDNOTES

television industry. About a week after NBC formally announced it had renewed *Star Trek* for a second season, Desilu issued a full-page ad in the March 14, 1967, issue of *Daily Variety* to not only congratulate the show for being picked up again by NBC but also with an eye toward helping promote the series for Emmy consideration. Later, on April 17, just a few short weeks before Emmy nominations were formally announced, the ad ran again, but it appeared next to a full-page ad for Desilu's *Mission: Impossible*. Together, the ads made an eye-catching two-page spread that included the following notification: "Desilu Productions, Inc. respectfully submits these series for consideration by the Members of the National Academy of Television Arts and Sciences." Two ads for two different shows both having the same goal: to win a coveted Emmy. Some argue the *Star Trek* ad contains material solicited or contrived by Roddenberry, who was well known to stretch the truth, if not outright lie for the sake of his show. As a result, it is worth examining the individuals quoted in the ad. **Alberta Moran (1926–2022)**: It is fitting the first person examined is someone who worked for NASA. This was not an accident as Roddenberry courted the nation's space agency from *Star Trek*'s inception and had a very good working relationship with them through all three seasons of the show. Most likely, this quote was taken from correspondence that Moran had with Roddenberry as part of her efforts to secure a speaker to attend the National Space Club's annual Goddard Memorial Dinner in March 1967. **Richard C. Hoagland (1945–)**: Hoagland was a Curator of Astronomy and Space Sciences at the Springfield, Massachusetts Science Museum from 1964 to 1967, and Assistant Director at the Gengras Science Center in West Hartford, Connecticut from 1967 to 1968. Hoagland's interest in space is clear from an early age, beginning with his time as a planetarium lecturer. In 1965 he produced a program called "Mars: Infinity to 1965" that coincided with the Mariner 3 and 4 missions to Mars. G. Harry Stine, one of the founding figures of model rocketry as well as a science and technology writer, and (under the name Lee Correy) an author of science fiction, mentions Hoagland in a letter to Roddenberry. Hoagland was active in early *Star Trek* fandom and participated, along with Stine, in a letter-writing campaign that helped secure a third season of the show. In a January 7, 1968 letter to Roddenberry, Stine writes, "Hoagland called me from Hartford last Friday night and informed me of the current crisis for *Star Trek*. I'd like to report to you that I've done a few things that may help a little bit." In 1976, Hoagland initiated a letter-writing campaign that was then picked up by John and Bjo Trimble that successfully lead to President Gerald R. Ford naming the first Space Shuttle *Enterprise* after the famous starship. Today, Hoagland is a promoter of various conspiracy theories. **Isaac Asimov (1920–1992)**: Of all the individuals listed in the ad, Isaac Asimov is the closest to a celebrity and therefore most likely the only one the public may have known. In addition to being a prolific author, Asimov was one of the first authorities within the science fiction community to write about *Star Trek*. Such support for the show was evident in a *Star Trek* article Asimov published in the April 29, 1967 issue of *TV Guide*. The article contains language that most likely was exchanged in earlier letters between Roddenberry and Asimov, language that could have been co-opted for use in the *Variety* ad. In particular, Asimov mentions in the April 29 article (my emphasis in *italic*), "A revolution of incalculable importance may be sweeping America, thanks to *television*. And thanks particularly to *Star Trek*, which, in its noble and successful effort to present *good science fiction* to the American public, has also presented everyone with an *astonishing revelation*." The language is not out of character for Asimov, and I don't think Asimov would have denied what was quoted by him in the *Variety* ad were his genuine thoughts about the show. **Philip Cohn (1928–2017)**: Of the six endorsements given in the *Variety* ad, this one is the most obscure. Philip L. Cohn was born in the Bronx and after a stint in the Merchant Marines, he moved to Florida, where he and his father opened a real estate company. After doing well selling land in Miami, they both began selling land in Brazil. In addition to selling real estate, Philip Cohn created several schools in the Miami area. Called "tutorial schools," these schools catered to students who ran into trouble at more conventional private or public schools. One of Cohn's schools was called Yale. "My purpose is to get these kids through high school," said Cohn in an interview about his schools and their programs for troubled youth. "If it means them looking out the window, smoking a cigarette or listening to music here at school, that's OK." Because of the special nature of his students, Cohn may have encouraged his teachers to have their students watch *Star Trek* so they could incorporate the series into the classroom. Such innovation may have especially worked at Yale to help get through to students who faced many personal problems, including drug use. "I think kids are getting the message that you can die if you stick a needle in your arm," Cohn said of his students at Yale. *Star Trek* may have helped them cope. Even though Cohn may have written to Roddenberry telling him how much he liked the show and how he used the series to reach through to the kids at his school, Roddenberry may have also simply thought readers might see Yale and associate the name with the more prestigious university and not some obscure private school for troubled youth. I'm sure the kids often told how their school got confused with Yale University.

Jack Hartley (1921–2000): Even though Hartley's last name is spelled "Hackley" in the ad, there is no doubt it is talking about Dr. Jack L. Hartley the Lt. Colonel in the United States Air Force and an aerospace dental scientist discussed in Chapter 4.

56. Letter by Paul L. Klein, vice president of NBC, Audience Measurement, published in *TV Guide*, 15, no. 21, issue #739 (May 27, 1967): A-2.

57. NBC Television Network, 1967–68 2nd YEAR, *Star Trek* (in color), promotional item issued by NBC, June 20, 1967, written by Chuck Appel and styled by Herb Cytryn of Sales Planning Art. Copy provided to the author by Bill Kobylak from his personal collection. This was one in a series of promotional items printed in the same style and format that NBC issued for other programs that year. For example, they also printed one that same season that promoted the sixth season of *The Virginian*.

58. Moran, interview by Swanson, 2022.

59. Alberta Moran to Leonard Nimoy, letter, March 22, 1967, Gene Roddenberry Collection, Correspondence/General Files, N.A.S.A., box 29, folder 6, Gene Roddenberry Papers (Collection PASC 62), UCLA Library Special Collections, Charles E. Young Research Library, University of California, Los Angeles.

60. Moran, interview by Swanson, 2022.

61. Moran, interview by Swanson, 2022.

62. Moran, interview by Swanson, 2019.

63. Moran, interview by Swanson, 2022.

64. Moran, interview by Swanson, 2022.

65. Gene Roddenberry to Alberta Moran, dated March 31, 1967, Gene Rodenberry Collection, Correspondence/General Files, N.A.S.A., box 29, folder 6, Gene Roddenberry Papers (Collection PASC 62), UCLA Library Special Collections, Charles E. Young Research Library, University of California, Los Angeles.

66. Roddenberry to Moran (March 31, 1967).

Chapter 7

1. The often-told stories by Roddenberry that *Star Trek*'s first pilot was rejected because network executives thought it was "too cerebral" and that *Star Trek* was the first television series to have had a network order a second pilot are well known. But are these stories true? Stephen Whitfield, in his 1968 book *The Making of Star Trek,* states, "The overall reason given for the rejection was that the pilot was just 'too cerebral.' NBC felt the show would go over the heads of most of the viewers, that it required too much thought on the part of the viewer in order to understand it." (p. 124). Oscar Katz, who was the vice president in charge of production at Desilu Studios at the time *of Star Trek*, spoke at the first major *Star Trek* convention, held in New York City on January 21–23, 1972, only two and a half years after NBC aired the show's last episode. At the convention, Katz gave a more detailed explanation of why the network rejected the first pilot. "My version of what happened is somewhat different. Nobody blamed Gene, neither NBC nor Desilu. The only one who blamed Gene, I guess, was Gene. What NBC said—I believe it was a conversation between myself and the vice president of programs at NBC in Hollywood, a fellow named Grant Tinker, Mary Tyler Moore's husband—he said, 'You delivered everything you promised us, but we can't

schedule it and we can't sell it because we don't think we can sell this show from this episode to advertisers.' And I said, 'Why not?' And he said, 'It's the wrong story. We see this as not typical of the series. We see it as the kind of show that you would schedule once every thirteen weeks to give the series a hypo or a shot in the arm. And we feel that if we go to market with this, go to advertisers, we have to have a pilot that's more typical.' And I said, 'Hey, fellas, you picked the story. Now you're telling us that we delivered a perfect pilot and you can't go to market with it because the story is wrong, and you've cost the studio $360,000. This is surely cavalier.' And Grant, who is quite a gentleman and a very honorable guy, said, 'It's our fault. We take the full blame. And we'd like to order another pilot for next season.'" Source: "Oscar Where Are You? Part 1," Fact Trek, .https://www.facttrek.com/blog/oscarkatz?fbclid=IwAR1BYtB6fnThS43JiibvD6tr4HPlJbBjaA1-Av4bGkP1RYPQFJQ5wTcQjDE Katz's account seems to confirm that even though the word "cerebral" may have not been used, the characterization was present. From this we can conclude Roddenberry did not invent the story; he simply embellished it. In examining the claim that *Star Trek* was the first television series to have had a network order a second pilot, let's look to Stephen Whitfield. In his book *The Making of Star Trek,* Whitfield states, "NBC shattered all television precedents and asked for a second pilot. This caused quite a stir within the industry, because up until that time no network had ever asked for a second pilot." Source: Whitfield and Roddenberry, *The Making of Star Trek*, (July 1972), 126. Since Whitfield's first mention, the second-pilot story claim has grown in stature to become one of the most popular myths to emerge from the series. Unfortunately, it is a myth. NBC did order a second pilot for the series, but this was not without precedent. In Michael Kmet's blog *Star Trek Fact Check*, the predecessor to his current "Fact Trek" (https://startrekfactcheck.blogspot.com/search?updated-max=2016-0), his posting "Second Pilot Episodes Before *Star Trek*?" from April 6, 2018, does an excellent job at debunking *Star Trek*'s second-pilot claim. He lists ten shows prior *to Star Trek* in which the initial pilot episodes were rejected but followed by a second pilot (and sometimes even a third). Of course, not all these second pilots became a series, since many simply did not sell, but all of them were produced prior *to Star Trek* and clearly debunk the myth that Roddenberry's show was the first to have resulted from a second pilot.

2. As of this writing, I have yet to find evidence disputing the claim that *Star Trek* was the first network television series to have the Smithsonian request a copy of its pilot for its archives. During research at both the Smithsonian Institution Archives (SIA) and the Smithsonian Archives of the National Air and Space Museum (NASM), I have not found any evidence to dispute this claim. I did, however, find a letter in which NASM director Michael Collins wrote MGM asking for a print copy of the film *2001: A Space Odyssey* for permanent retention in their archives. The letter dated January 24, 1974, was sent to Charles Powell, director of the MGM's advertising and publicity. Collins stated in his letter that "MGM has previously donated to the Museum the four fine paintings by Robert McCall depicting scenes from the film. We hope that it may be feasible for MGM to add further a print of the film to the national collection in the interest of historical preservation." Karla Davidson of MGM's legal department, responded in a letter dated February 25, 1974, stating "Although we are honored by your request that we donate a print of the film to the National Air & Space Museum, we regret that we must turn it down at this time." By 1974, six years after the film's first release, the film still enjoyed theatrical and non-theatrical showings. As a result, prints remained a valuable commodity. Davidson closed her letter by stating, "We are cognizant of the great historical and cultural value of this film and, following our policy throughout the years, we plan to preserve it. Additionally, the Library of Congress has been obtaining prints of all important films being registered currently for copyright to preserve the same for archival purposes, and it already has a print of *2001: A Space Odyssey*." Source: Michael Collins to Charles Powell, letter, January 24, 1974, and Karla Davidson to Michael Collins, letter, February 25, 1974. Smithsonian Institution Archives, National Air and Space Museum, Office of the Director, NASM Files, 1971–1978 and undated, Record Unit 000306, box 2, folder: Astronautics 1974.

3. Richard K. Preston Sr. "Security Investigation Data for Sensitive Position" provided to the author by Richard K. Preston, II. The museum was called the National Air Museum prior to 1966,

the year President Lyndon Johnson signed a law that changed the name of the National Air Museum to the National Air and Space Museum to recognize significant developments in both aviation and spaceflight.

4. Larkin Preston, interview by Glen E. Swanson, May 7, 2023.

5. Richard K. Preston II, interview by Glen E. Swanson, February 21, 2023.

6. Preston II, interview by Swanson (2023).

7. Preston II, interview by Swanson (2023).

8. Bannon Preston, interview by Glen E. Swanson, February 14, 2023.

9. "DURANT APPOINTED TO NAM POST," *Smithsonian Torch*, 1, (February 1965): 3.

10. Steve Durant, interview by Glen E. Swanson, April 13, 2023. Norman Rockwell did many paintings dealing with the space program, which is most likely how Durant came to know him.

11. Durant, interview by Swanson, (2023).

12. Randy Lieberman, "Frederick C. Durant (1916–2015)," *Space Review*, November 2, 2015, https://www.thespacereview.com/article/2856/1

13. Durant, interview by Swanson, (2023).

14. Durant, interview by Swanson, (2023).

15. Alberta Moran to Leonard Nimoy, March 22, 1967, Gene Roddenberry Collection, Correspondence/General Files, N.A.S.A., box 29, folder 6: Gene Roddenberry Papers, (Collection PASC 62), UCLA Library Special Collections, Charles E. Young Research Library, University of California, Los Angeles.

16. Gene Roddenberry to Richard K. Preston, April 19, 1967, Gene Roddenberry Collection, Correspondence/General Files, Fan Letters, 1967, box 28, folder 2: Gene Roddenberry Papers, (Collection PASC 62), UCLA Library Special Collections, Charles E. Young Research Library, University of California, Los Angeles.

17. Richard K. Preston to Gene Roddenberry, June 1, 1967, Gene Roddenberry Collection, Correspondence/General Files, Publicity, 1967, box 30, folder 5, Gene Roddenberry Papers, (Collection PASC 62), UCLA Library Special Collections, Charles E. Young Research Library, University of California, Los Angeles.

18. "Smithsonian Seeks TV Pilot," *Los Angeles Times*, June 13, 1967, 63.

19. Gene Roddenberry to Richard K. Preston, August 16, 1967, Gene Roddenberry Collection, Correspondence/General Files, Publicity, 1967, box 30, folder 5: Gene Roddenberry Papers, (Collection PASC 62), UCLA Library Special Collections, Charles E. Young Research Library, University of California, Los Angeles.

20a. A series of press photos at the news event on August 28, 1967 depict Gene Roddenberry and his first wife Eileen-Anita Rexroat, presenting the second *Star Trek* pilot to S. Paul Johnston, then director of the National Air and Space Museum. Johnston served as the director of the National Air and Space Museum from 1964 to 1969. Johnston graduated from the Massachusetts Institute of Technology in 1921, where he studied mechanical engineering and aeronautics. In 1941 he served on the National Advisory Committee for Aeronautics (NACA), NASA's predecessor, where he was coordinator for research. From 1942 to 1944, he served as manager of the Washington office of the Curtiss-Wright Corporation. During that time, he authored several works on the history of aviation, including *Wings After War*, a book that forecast the postwar development of aviation. In 1944, he became staff engineering officer for the Naval Air Transport Service in Hawaii. Johnston was also director of the Institute of Aeronautical Sciences, now the American Institute of Aeronautics and Astronautics. During this time, he was an advisor to three US presidents as well as the Navy and Central Intelligence Agency. Known for his ability to present complex topics clearly, Johnston served as editor of *Aviation Magazine,* as well as the author of numerous articles on aviation and airpower.

20b. Gene Roddenberry to Richard Preston, August 16, 1967, Gene Roddenberry Collection, Correspondence/General Files, Publicity, 1967, box 30, folder 5: Gene Roddenberry Papers,

ENDNOTES

(Collection PASC 62), UCLA Library Special Collections, Charles E. Young Research Library, University of California, Los Angeles.

21a. Motion Picture Pilot film "*Star Trek*" TV Serial, 16 mm Color, 1,080 feet, Title: "Where No Man Has Gone Before," Accession Worksheet, Luis R. Purnell curator, Received August 28, 1967, Smithsonian Institution Archives, National Air and Space Museum, Office of the Director, NASM Files, 1971–1978 and undated, Record Unit 000306, box 2, folder: Astronautics 1974.

21b. Richard G. Van Treuren to Fred Durant, April 27, 1974, Smithsonian Institution Archives, National Air and Space Museum, Dept. of Astronautics, Correspondence, 1965–1984, with materials dating from c. 1953, Record Unit 000398, box 33, folder: Gene Roddenberry, Richard G. Van Treuren Correspondence, 1973–1977.

21c. Fred Durant to Richard G. Van Treuren, May 30, 1974, Smithsonian Institution Archives, National Air and Space Museum, Dept. of Astronautics, Correspondence, 1965–1984, with materials dating from c. 1953, Record Unit 000398, box 33, folder: Gene Roddenberry, Richard G. Van Treuren Correspondence, 1973–1977.

21d. According to Greg Kennedy, who used to work at the Smithsonian and remembers visiting the Arts and Industries Building around the time Roddenberry formally presented a copy of the pilot to Preston in 1967, NASA had a traveling exhibit temporarily set up in the building. The exhibit included a mock Apollo command model. Source: Greg Kennedy, interview by Glen E. Swanson, November 15, 2022.

21e. "Star Trek' Pilot Film to Smithsonian," *Santa Maria (CA) Times*, September 16, 1967, 22.

21f. "Telle-Talk," *Independent Star-News*, September 17, 1967, 93.

21g. "Smithsonian Seeks RV Pilot, *Los Angeles Times*, June 13, 1967, 63.

21h. Collage showing assorted news articles announcing the *Star Trek* Pilot request by the Smithsonian. Newspapers shown are: *Los Angeles Times*, June 13, 1967; *Santa Maria Times*, September 16, 1967; and *Independent Star-News*, September 17, 1967.

22. Gene Roddenberry to Fred Durant, telegram, March 8, 1968, Gene Roddenberry Correspondence 1968–1979, Smithsonian Institution Archives, Record Unit 000398, Box 32.

23. "In Hollywood… Gene Roddenberry, Producer of 'Star Trek,' Makes the Proud Claim That Pilot of His show Is the Only Episode Ever to Be Requested by Smithsonian Institution…" *Weekly Variety*, 248, no. 5, (September 20, 1967): 42. In addition, an August 13, 1967 article in The *Detroit Free Press* entitled "Star Trek's Upward Flight," Roddenberry mentions "We're up for an award this year from the World Science Fiction Assn. when it holds its convention. The only other two nominees are movies," said Roddenberry. "And the Smithsonian Institution has asked for our pilot film for its archives." Source: *Detroit Free Press*, Sunday August 13, 1967, p. 70.

24. Fred Durant, memo, March 31, 1971, Smithsonian Institution Archives, National Air and Space Museum, Dept. of Space Science and Exploration, Subject Files, c. 1960–1986, Record Unit 000348, box 6, folder: Gene Roddenberry Correspondence 1968–1979, NASM Admin General.

25. Fred Durant, memo, April 27, 1971, Smithsonian Institution Archives, National Air and Space Museum, Office of the Director NASM Files 1971–1978 and undated, Record Unit 000306, box 8, folder: Exhibits General, 1971–1973.

26. Fred Durant to Gene Roddenberry, September 14, 1970. Smithsonian Institution Archives, National Air and Space Museum, Dept. of Astronautics Records, 1965–1980, with materials dating from c. 1953, Record Unit 000398, box 32, folder: Gene Roddenberry Correspondence 1968–1979.

27. There remains some confusion regarding the fate of the original production series bridge from *Star Trek*. Bjo Trimble recalls in her 1983 book, *On the Good Ship Enterprise*, a time in 1974, "When we were still putting on Equicon, the West Coast *Star Trek* Convention, some of the committee decided it would be fun to build a bridge set for the con. After much planning and paperwork, we heard that a UCLA electronics class had built an *Enterprise* bridge set as a class

project." Trimble explains "we were to find out later that many of the people involved with this set were also busy rebuilding parts of the real *Enterprise* bridge set, which had been donated to UCLA." Source: Bjo Trimble, *On the Good Ship Enterprise: My 15 Years with Star Trek* (Donning, 1982), 93. Housed in the archives of the Smithsonian Institution is a collection of documents and photos pertaining to the original bridge set pieces. Among these records are a series of letters sent in late 1974 by Stephanie F. Lawrence of Burbank, California. In these letters, Lawrence explains "my brother and I, along with a friend of ours, possess the original helm from the television science fiction series *Star Trek*." She asks, "in light of the Institution's continuing interest in the space program, and its concern for preserving the varied cultural aspects of the American society, we would like at this time to offer this piece of equipment to the Smithsonian Institution." Fred Durant, then the Assistant Director of Astronautics at the Smithsonian's National Air and Space Museum, replied to Lawrence's letter which thereby began a series of correspondence in efforts to confirm the authenticity of the prop. The Smithsonian received photos of the helm along with additional documents in efforts to prove its provenance. "The set is being stored with a friend at UCLA," explained Lawrence in a March 3, 1975, letter to Durant. "My friend has told me that although the electrical wiring has been redone, it can easily be restored to the original design." Source: Letter from Stephanie Lawrence to the Smithsonian dated December 20, 1974, follow-up letter to Lawrence from Durant dated March 14, 1975, Smithsonian Institution Archives, Record Unit 000398, box 33, folder: Gene Roddenberry Inquiries by fans, 1972–1977. This correspondence with Durant helps to confirm Trimble's account of what happened to the original bridge set. Though the Smithsonian expressed initial interest, Lawrence's offer to come to "terms which would be most acceptable to you" for the helm's acquisition by the Smithsonian never materialized. As a result, the 160-lb. prop was never acquired. The helm did make an appearance years later as a featured artifact in a popular *Star Trek* exhibit hosted by the Smithsonian from 1992 to 1993. After that exhibit closed, the helm was sold at auction to Planet Hollywood, where it became part of their corporate collection. Ten years later, in 2003, the helm was sold once again during a Profiles in History auction where Microsoft co-founder Paul Allen purchased it. Source: *Profiles in History: "The Ultimate Sci-Fi Auction"* catalog, August 26, 2003, 42–43. More recently, the helm was seen as part of a popular traveling *Star Trek* exhibit initially developed by Seattle's Museum of Pop Culture (MoPOP), which originally went by the awkward-sounding name of the Experience Music Project and Science Fiction Museum and Hall of Fame (EMPISFM). In 2016, the museum changed its name to the EMP Museum before finally settling on MoPOP. Called *Star Trek: Exploring New Worlds*, the exhibit showed a fully restored helm along with other original set pieces not only from the original television series but from the various spinoffs. In addition to the helm, the Smithsonian also received an urgent request they take the original full-scale version of the Galileo shuttlecraft. This request was made by Roger Heisman of Palos Verdes, California, who came into possession of the prop. In his letter dated February 18, 1977, he writes: "It was originally built for Paramount Studios at a cost of $60,000.00 When I got the shuttlecraft it had fallen into disrepair and I haven't been able to afford the time or money to restore it. I have had legal action against me by the Los Angeles County Zoning Board to have the Shuttlecraft moved from my property by March 17, 1977. In review of my alternatives of actions I felt that this historical item would find its best possible new home in the Smithsonian Museum. If you are interested, as I hope you are, please immediately contact me." He goes on to describe its dimensions: "For further consideration, its dimensions are 24 feet long, 7½ feet high, and wings and nacelles attached 14 feet wide, without them attached it is 11 feet wide. Its damage is cosmetic in nature as it is primarily fiberglass and Masonite over a steel frame. It basically simply needs repainting and some minor body work and a new roof. It weighs approximately two to three thousand pounds." Source: Letter to Durant from Heisman dated February 18, 1977, Smithsonian Institution Archives, Record Unit 000398, box 33, folder: Gene Roddenberry inquiries by fans 1972–1977.

    28. Fred Durant to Roddenberry, June 5, 1972, Smithsonian Institution Archives, National Air and Space Museum, Dept. of Astronautics Records, 1965–1980, with materials dating from c. 1953, Record Unit 000398, box 32, folder: Gene Roddenberry Correspondence 1968–1979.

ENDNOTES

29a. Undated document from Roddenberry to Durant regarding "Collection of *Star Trek* Material." Even though the document is undated, it most likely was issued in the weeks after "The Nature of Scientific Discovery" Copernicus symposium held at the Baird Auditorium in the Smithsonian's National Museum of Natural History on April 24, 1973. This event was held in honor of the 500th anniversary of the birth of Nicolaus Copernicus (1473–1543). Smithsonian Institution Archives, National Air and Space Museum, Dept. of Astronautics Records, 1965–1980, with materials dating from c. 1953, Record Unit 000398, box 32, folder: Gene Roddenberry Correspondence 1968–1979. It should also be pointed out that the same reasons the Smithsonian sought materials from *Star Trek* were also published in the book *Letters to Star Trek* by Susan Sackett (Ballantine, 1977) 196–198. Sackett was Roddenberry's personal secretary after the original series ended. In this book, Durant summarizes the history of the Smithsonian's interest in *Star Trek* in a letter dated September 24, 1975 which reads: "In response to your letter of 14 September [1975], we are pleased to tell you of our interest in *Star Trek*. It all started in August 1967 when Gene Roddenberry offered, and the museum accepted, a print of the *Star Trek* pilot film, #2, 'Where No Man Has Gone Before.' This film is in a protected archive and not available for public screening. We believe that science fiction can play a 'mind-stretching' role in the minds of creative, scientific and technically inclined persons. All three of the acknowledged rocket pioneers, Tsiolkovsky, Goddard and Oberth, acknowledged the influence of Jules Verne. In addition, Goddard had high interest in H. G. Wells' *First Men on the Moon*. The large number of scientific and technical professionals who indulge in speculative fiction today reinforces this view: motion pictures, e.g., '2001: A Space Odyssey,' and the *Star Trek* series represent the same kind of invitation to imaginative thinking as books but in different genre. As plans for our new museum building developed, we inquired of possible artifacts available from the *Star Trek* production sets, the program having been terminated. Matt Jefferies has donated the Klingon battle cruiser model (about 30 inches long) and a 4-inch block of plastic containing a tiny model of the *Enterprise* used in the episode 'Catspaw.' David Gerrold has given us a few tribbles. Paramount has donated the large (11½-foot long) studio model of the *Enterprise* to the museum. Received in early 1974, it was in several pieces and in deteriorated condition. A few parts were missing. We performed some cosmetic restoration and displayed the model from mid-September 1974 to January this year in our 'Life in the Universe' exhibit. Franz Joseph's blueprints of the interior were exhibited nearby. We have moved to our new building and are developing exhibits preparatory to our opening 4 July 1976. The new museum is a major structure on the Mall here in Washington. It is located at 600 Independence Avenue, S.W. The sweeping story of flight from earliest times to possible space exploration will be shown in twenty-four exhibit galleries. When we open, the *Enterprise* will be fully refurbished and re-wired for lighting. It will be exhibited in the 'Life in the Universe' gallery. Besides Gene Roddenberry and Matt Jefferies, we have enjoyed the assistance and cooperation of Dorothy Fontana and Bjo Trimble as well as may others in developing a reference file of scripts, fan magazines, etc. on the *Star Trek* phenomenon. This is a quick view and report on the Smithsonian's interest in the subject. We are limited in the time we can give to the subject but recognize high interest on the part of a sizable number of our visiting public." Source: Susan Sackett to Fred Durant, letter, September 14, 1975, Smithsonian Institution Archives, National Air and Space Museum, Dept. of Astronautics Records 1965–1980, with materials dating from c. 1953, Record Unit 000398, box 32, folder: Gene Roddenberry Correspondence, 1968–1979.

29b. Fred Durant to Jacqueline Lichtenberg, July 31, 1973, Smithsonian Institution Archives.

30. Fred Durant to Michael McMaster, October 30, 1973, Smithsonian Institution Archives, National Air and Space Museum, Dept. of Astronautics Records 1965–1984, with materials dating from c. 1953, Record Unit 000398, box 33, folder: Roddenberry, Gene, Correspondence, 1968–1979. Also, Robert Greenberger, *Star Trek: The Complete Unauthorized History* (Minneapolis, MN: Voyageur Press, 2012), 69.

31. Whitfield and Roddenberry, *The Making of Star Trek* (1972).

32. Fred Durant to Russell A. Kirby, March 26, 1973, Smithsonian Institution Archives, National Air and Space Museum, NASM Office of the Director, NASM Files 1971–1978 and

undated, Record Unit 000306, box 2, folder: Astronautics 1973. During this time, the Smithsonian had an interactive Gemini display provided by IBM. Visitors who completed the display received a wallet-sized card as a souvenir. Greg Kennedy, who worked for the Smithsonian and remembers the exhibit, recalls "whenever there was a problem with it, we had to call a local IBM rep to take care of it. The problem also could have been something as simple as the wallet card dispenser being empty." Source: Greg Kennedy email message to Glen E. Swanson, April 22, 2023.

33a. "Life on Other Planets Studied at Smithsonian," *Spokesman-Review* (Spokane, WA), September 22, 1974, 19; Smithsonian Institution Archives, National Air and Space Museum, Office of the Director, NASM Files 1971–1978 and undated, Record Unit 000306, box 8, folder: "Exhibit: Life in the Universe." This memorandum is the final approved listing of the exhibit hall names for the new NASM Building. This listing includes the title, along with a general synopsis of the scope of each hall.

33b. Undated document entitled "Studio Model of Starship Enterprise," Smithsonian Institution Archives, National Air and Space Museum, Office of the Director Dept. of Astronautics Records 1965–1980, with materials dating from c. 1953, Record Unit 000398, box 32, folder: Roddenberry, Gene, 'Enterprise' Model Restoration, 1973–1975. This document gives a date of January 6, 1975, as the last date of the *Life in the Universe?* exhibit.

33c. James M. Murphy to Michael Collins, memo, July 29, 1974, Smithsonian Institution Archives, National Air and Space Museum, Office of the Director, NASM Files 1971–1978 and undated, Record Unit 000306, box 8, folder: "Exhibit: Life in the Universe" The memo pertains to the Arts and Industry's Building termination date for the *Life in the Universe?* exhibit which was originally scheduled for the end of 1974 to allow for the removal of the Douglas World Cruiser and provide access for the cooling towers.

33d. Michael Collins to James M. Murphy, memo, August 5, 1974, Smithsonian Institution Archives, National Air and Space Museum, Office of the Director, NASM Files 1971–1978 and undated, Record Unit 000306, box 8, folder: "Exhibit: Life in the Universe" Memo states the decision was made to retain the *Life in the Universe?* exhibit until January 1, 1975.

34. "Life on Other Planets Studied at Smithsonian," 19.

35. Michael Collins to Gene Roddenberry, June 1, 1973, Smithsonian Institution Archives, National Air and Space Museum, Office of the Director, NASM Files 1971–1978 and undated, Record Unit 000306, box 8, folder, "Exhibit: Life in the Universe." Michael Collins to Gene Roddenberry, letter, June 1, 1973, Smithsonian Institution Archives, National Air and Space Museum, Dept. of Astronautics Records 1965–1980 with materials dating from c. 1953, Record Unit 000398, box 32, folder: Roddenberry, Gene, Correspondence, 1968–1979. Collins also mentioned in this letter that he was intrigued with a "reverse Pioneer 10 plaque" idea Roddenberry had that would "constitute an extraterrestrial 'message'" museum visitors would have to decipher. The Pioneer plaques Collins was referring to are a pair of gold-anodized aluminum plaques mounted onboard the twin Pioneer 10 and 11 spacecraft launched to Jupiter. Pioneer 10 launched on March 2, 1972, and Pioneer 11 on April 5, 1973. The twin plaques feature nude human figures, one male and one female, along with symbols designed to show information on where the spacecraft originated. The Pioneer 10 and 11 plaques were the first human-built objects to achieve escape velocity from our solar system. Both plaques stemmed from an idea by Eric Burgess. Burgess approached Carl Sagan, who was enthused about the idea. NASA gave them three weeks to build them. Together with Frank Drake, they designed the plaque using artwork prepared by Linda Salzman, who was Sagan's wife at the time.

36. There were two Powers of Ten documentary films made. The first film, *A Rough Sketch for a Proposed Film Dealing with the Powers of Ten and the Relative Size of Things in the Universe*, was a prototype done in 1968. This was the version shown in the Smithsonian's *Life in the Universe?* exhibit beginning with the first 1974–75 exhibit and also in the new building. The second film, *Powers of Ten: A Film Dealing with the Relative Size of Things in the Universe and the Effect of Adding Another Zero*, was completed in 1977. The 1977 film had several changes from the original, including

being shot entirely in color, having the starting location be Chicago instead of Miami, removing the relativistic (time) dimension, introducing an additional two powers of ten at each extreme, a change in narrator from Judith Bronowski to Philip Morrison and better graphics. The 1977 version of the film was selected for preservation by the Library of Congress as being "culturally, historically, or aesthetically significant."

37. "Life on Other Planets Studied at Smithsonian," 19.

38. Tom Shales, "What's the Smithsonian Trying to Tell Us?" *Honolulu Advertiser*, November 1, 1974, 63.

39. Karen Peterson and Fan Service, "Artist Creates 'Life' Forms," *The Lima (OH) News*, January 30, 1976, 7.

40. Shales, "What's the Smithsonian Trying to Tell Us?," 63.

41. The twenty-five exhibit hall names in the new National Air and Space Museum building that opened in 1976 and their respective Hall Codes are: Vertical Flight (1AN), Air Transportation (1BN), Earliest Flight (1CN), Milestones of Flight (1DN), Stacking Area for People Waiting to Enter the Theater (1EN), Space Hall (1GN), Benefits from Flight (1GS), Satellites (1FS), Life in the Universe (1ES), Lobby (1DS), Air Traffic Control (1GS), Exhibition Flight (1BS), General Aviation (1AS), Sea-Air Operations (2AN), Spacearium (2CN), Theater (2EN), Flight Technology (2GN), Flight and the Arts (2GS), Apollo to the Moon (2FS), The X Airplanes (2ES), Special Exhibits (2DS), World War I Aviation (2CS), Balloon and Airships (2BS), and World War II Aviation (2AS). Memo from Melvin B. Zisfein dated February 26, 1974. Smithsonian Institution Archives, National Air and Space Museum, Office of the Director, NASM Files 1971–1978 and undated, Record Unit 000306, box 8, folder: Exhibits: General 1974–1975.

42. Smithsonian Curator Emeritus Tom Crouch, interview by Glen E. Swanson, July 19, 2023.

43. "A Pilot Study of Visitors to the 'Life in the Universe' Exhibit," February 1975, p. 1, Smithsonian Institution Archives, National Air and Space Museum, Dept. of Space Science and Exploration, Subject Files, c. 1960–1986, Record Unit 000348, box 14, folder: A Pilot Study of Visitors to the "Life in the Universe" Exhibit.

44. Crouch, interview by Swanson, (2023).

45. "A Pilot Study of Visitors to the 'Life in the Universe' Exhibit," February 1975, p. 1, Smithsonian Institution Archives, National Air and Space Museum, Dept. of Space Science and Exploration, Subject Files, c. 1960–1986, Record Unit 000348, box 14, folder: A Pilot Study of Visitors to the "Life in the Universe" Exhibit.

46. I always thought photos from the new building showing the 11-foot production model *Enterprise* hanging in the center of the pink-painted interconnected walls of the *Life in the Universe?* exhibit was the entrance. After obtaining a copy of the original architectural drawings of the exhibit, I learned this was not the case. The exhibit had a separate entrance where visitors would first enter to see the beginnings of the Universe. Then, as they wandered through the display, they would eventually come to an area to see how life formed on Earth. Finally, near the end where the *Enterprise* hung, visitors engaged with various interactive exhibits designed to help them speculate about the possibility of life elsewhere while looking up at the *Enterprise*.

47. Karen Dick Schnaubelt, interview by Greg Taylor conducted via email from June 1999 to July 1999, Trekplace, *Uncharted Content from the Final Frontier Since 1999*, http://www.trekplace.com/fj-kdint01.html.

48. Karen Schnaubelt, interview by Glen E. Swanson, September 22, 2023.

49. Schnaubelt, interview by Swanson, (2023).

50. Paul Newitt, "An Interview with Franz Joseph," 1982. The interview was originally published in the June 1984 "Spotlight on the Technical Side" special edition of James van Hise's *Enterprise Incidents*. The interview was conducted face to face and was transcribed from an audio recording. www.trekplace.com/fj-fjnewittint02.html.

51. Newitt, "An interview with Franz Joseph."

52. Franz Joseph to Gene Roddenberry, May 14, 1973, Roddenberry.com

53. Gene Roddenberry to Fred Durant, March 29, 1974, Smithsonian Institution Archives, National Air and Space Museum, Dept. of Astronautics, Records 1965–1980, with materials dating from c. 1953, Record Unit 000398, box 32, folder: Roddenberry, Gene, Correspondence, 1968–1979.

54. The filming model of the *Enterprise* is not 14 feet long, which Roddenberry was often fond of stating. The model is closer to 11 feet in length. To be more precise, the model is exactly 135 inches from stem to stern or just 3 inches over 11 feet.

55. Fred Durant to Franz Joseph, April 17, 1974, Smithsonian Institution Archives, National Air and Space Museum, Dept. of Astronautics, Records 1965–1980, with materials dating from c. 1953, Record Unit 000398, box 32, folder: Roddenberry, Gene, Franz Joseph Designers, 1974–1976.

56. Franz Joseph to Fred Durant, April 20, 1974, Smithsonian Institution Archives, National Air and Space Museum, Dept. of Astronautics, Records 1965–1980, with materials dating from c. 1953, Record Unit 000398, box 32, folder: Roddenberry, Gene, Franz Joseph Designers, 1974–1976.

57. Fred Durant to Franz Joseph, May 6, 1974, Smithsonian Institution Archives, National Air and Space Museum, Dept. of Astronautics, Records 1965–1980, with materials dating from c. 1953, Record Unit 000398, box 32, folder: Roddenberry, Gene, Franz Joseph Designers, 1974–1976.

58a. Franz Joseph to Fred Durant, August 15, 1974, Smithsonian Institution Archives, National Air and Space Museum, Dept. of Astronautics, Records 1965–1980, with materials dating from c. 1953, Record Unit 000398, box 32, folder: Roddenberry, Gene, Franz Joseph Designers, 1974–1976.

58b. Fred Durant to Franz Joseph, September 17, 1974, Smithsonian Institution Archives, National Air and Space Museum, Dept. of Astronautics, Records 1965–1980, with materials dating from c. 1953, Record Unit 000398, box 32, folder: Roddenberry, Gene, Franz Joseph Designers, 1974–1976.

59. Fred Durant to Franz Joseph, November 22, 1974, Smithsonian Institution Archives, National Air and Space Museum, Dept. of Astronautics, Records 1965–1980, with materials dating from c. 1953, Record Unit 000398, box 32, folder: Roddenberry, Gene, Franz Joseph Designers, 1974–1976. The stamp shows the *Enterprise* orbiting the planet Gamma Taurus IV, a completely fictious world created by Franz Joseph and is not mentioned in any of the original series episodes.

60. Fred Durant to Gene Roddenberry, September 6, 1974, Smithsonian Institution Archives, National Air and Space Museum, Dept. of Astronautics, Records 1965–1980, with materials dating from c. 1953, Record Unit 000398, box 32, folder: Roddenberry, Gene, Correspondence, 1968–1979.

61. Photos taken by Dwight R. Bowman of the NASM *Life in the Universe?* black-tie opening on September 19, 1974. General shots of guests at the Smithsonian's Arts and Industries Building. Smithsonian Institution Archives, National Air and Space Museum, Dept. of Space Science and Exploration., Subject Files, c. 1960–1986, Record Unit 000348, box 11, folder: Life in the Universe? Folder 1 of 3.

62. Glen E. Swanson, ed., *"Before This Decade Is Out…" Personal Reflections on the Apollo Program*, NASA SP-4223, (National Aeronautics and Space Administration, History Office, Office of Policy and Plans, NASA, 1999), 307–309. George W. Low served in a variety of administrative positions within NASA during the Apollo program including deputy center director of the Manned Spacecraft Center (now the Johnson Space Center) and as manager of the Apollo Spacecraft Program Office (ASPO) where he was responsible for directing the numerous redesign changes made to the Apollo spacecraft to get the manned lunar landing program on back on track after setbacks resulting from the Apollo 1 fire. Low is also credited with conceiving the idea of the Apollo 8 lunar orbital flight and proving its technical soundness to others. On December 3, 1969, Low became NASA's deputy administrator, serving with administrators Thomas O. Paine and James C. Fletcher. After Paine's resignation, Low served as acting NASA administrator from September 1970 to May 1971.

ENDNOTES

Low played a significant role in the development of the Skylab program as well as the Apollo-Soyuz Test Project (ASTP) and the Space Shuttle program. Low retired from NASA on June 5, 1976, to become president of Rensselaer Polytechnic Institute, where he remained in that position until his death on July 17, 1984. Low received many awards and honors during his aerospace career, including the NASA Distinguished Service Medal (twice awarded), NASA Outstanding Leadership Award, Honorary Doctor of Engineering from RPI, and Honorary Doctor of Science from the University of Florida.

63a. Franz Joseph to Fred Durant and Holly B. Laffon, Secretary to Durant, September 27, 1974, Smithsonian Institution Archives, National Air and Space Museum, Dept. of Astronautics, Records 1965–1980, with materials dating from c. 1953, Record Unit 000398, box 32, folder: Roddenberry, Gene, Franz Joseph Designers, 1974–1976.

63b. Holly Laffon to Franz Joseph, September 23, 1974, Smithsonian Institution Archives, National Air and Space Museum, Dept. of Astronautics, Records 1965–1980, with materials dating from c. 1953, Record Unit 000398, box 32, folder: Roddenberry, Gene, Franz Joseph Designers, 1974–1976.

63c. Holly Laffon to Franz Joseph, October 1, 1974, Smithsonian Institution Archives, National Air and Space Museum, Dept. of Astronautics, Records 1965–1980, with materials dating from c. 1953, Record Unit 000398, box 32, folder: Roddenberry, Gene, Franz Joseph Designers, 1974–1976.

63d. Franz Joseph to Fred Durant, October 7, 1974, Smithsonian Institution Archives, National Air and Space Museum, Dept. of Astronautics, Records 1965–1980, with materials dating from c. 1953, Record Unit 000398, box 32, folder: Roddenberry, Gene, Franz Joseph Designers, 1974–1976.

64. Franz Joseph to Fred Durant, November 18, 1974, Smithsonian Institution Archives, National Air and Space Museum, Dept. of Astronautics, Records 1965–1980, with materials dating from c. 1953, Record Unit 000398, box 32, folder: Roddenberry, Gene, Franz Joseph Designers, 1974–1976.

65. Karen Dick, "Franz Joseph Timeline," with annotations by Greg Tyler. Entries are condensed from FJ's "Writing Activities Log" (April 14, 1973–April 14, 1988) and FL's "Design Work, Script, and Correspondence—Planet Earth" notebook (1974). http://www.trekplace.com/fj-timeline.html.

66. Steve Lohr, "Promotors Hitchhike on *Star Trek*," *Press and Sun-Bulletin*, January 27, 1976, 46.

67. Newitt, "An Interview with Franz Joseph."

68. "It Lives!" from "Paper Back Talk," *New York Times Book Review*, July 13, 1975, 39.

69. Bestseller Lists, July 9, 1975, indicating the *Star Trek* blueprints were no. 10 at B. Dalton. From the files of Karen Schnaubelt.

70. "Paper Back Talk," *New York Times Book Review*, December 21, 1975, p. 19.

71. Christmas 1976 letter written by Franz Joseph and mailed to friends. Fred Durant at the Smithsonian received a copy on December 13, 1976. Smithsonian Institution Archives, National Air and Space Museum, Dept. of Astronautics, Records 1965–1980, with materials dating from c. 1953, Record Unit 000398, box 32, folder: Roddenberry, Gene, Franz Joseph Designers, 1974–1976.

72. Dick, "Franz Joseph Timeline," with annotations by Greg Tyler. Entries here are condensed from FJ's "Writing Activities Log" (April 14, 1973–April 14, 1988) and FL's "Design Work, Script, and Correspondence—Planet Earth" notebook (1974). http://www.trekplace.com/fj-timeline.html.

73. *New York Times Book Review*, January 4, 1976, 23.

74. Christmas 1976 letter written by Franz Joseph.

75. "Paper Back Talk," *New York Times Book Review*, p. 19.

76. Newitt, "An Interview with Franz Joseph," 1982.

77. Schnaubelt, interview by Swanson, (2023).

78. First published in a Ziplock baggie as a pocket game called *Task Force Games #4*, co-founder of Task Force Games, Steve Cole recalls the original creation of *Star Fleet Battles*. "The design of *Star Fleet Battles* began during 1975… Jim Brown… and I were playing a lot of *Jutland*… One afternoon I was studying the *Jutland* battle that was in progress on my floor (left from the previous evening) when the [*Star Trek*] re-run of the day came on. I began to consider the possibility of doing a space game on the *Jutland* system. *Jagdpanther* was in operation at the time, and I had vague thoughts that I could somehow get a license for the game. By the time Jim came by to collect me for dinner, I had a Federation CA and a Klingon D7 fighting it out. In the brief space of an hour-long re-run, I had two SSD's, the proportional movement system, and the charts for phasers and disruptor bolts. All were to change drastically within a week and were to continue evolving for five years, but the start was made." Source: Steve Cole "Retrospect: Star Fleet Battles," *Space Gamer*, Number 42 (August 1981), 4–5.

79. Fred Durant to Gene Roddenberry, September 14, 1970, Smithsonian Institution Archives, National Air and Space Museum, Dept. of Astronautics, Records 1965–1980, with materials dating from c. 1953, Record Unit 000398, box 32, folder: Roddenberry, Gene, Correspondence, 1968–1979.

80. Gene Roddenberry to Fred Durant, May 2, 1973, Smithsonian Institution Archives, National Air and Space Museum, Dept. of Astronautics, Records 1965–1980, with materials dating from c. 1953, Record Unit 000398, box 32, folder: Roddenberry, Gene, NASM/Washington Visit, May 24–28, 1973.

81. Michael Collins to Michael Yablena, May 7, 1973, Smithsonian Institution Archives, National Air and Space Museum, Dept. of Astronautics, Records 1965–1980, with materials dating from c. 1953, Record Unit 000398, box 32, folder: Roddenberry, Gene, Correspondence, 1968–1979.

82. Gene Roddenberry to Fred Durant, August 22, 1973, Smithsonian Institution Archives, National Air and Space Museum, Dept. of Astronautics, Records 1965–1980, with materials dating from c. 1953, Record Unit 000398, box 32, folder: Roddenberry, Gene, Correspondence, 1968–1979.

83. Fred Durant memo regarding the *Enterprise* model, January 24, 1974, Smithsonian Institution Archives, National Air and Space Museum, Dept. of Astronautics, Records 1965–1980, with materials dating from c. 1953, Record Unit 000398, box 32, folder: Roddenberry, Gene, "Enterprise" model restoration, 1973–1975.

84a. Dick Lawrence to Michael Collins, February 7, 1974, Smithsonian's National Air and Space Museum, copy in author's files, original in registrar's records.

84b. Fred Durant to Frank Wright, December 17, 1973, Smithsonian Institution Archives, National Air and Space Museum, Dept. of Astronautics, Records 1965–1980, with materials dating from c. 1953, Record Unit 000398, box 32, folder: Roddenberry, Gene, "Enterprise" model restoration, 1973–1975.

85. Airbill #LAX 12001, consignee's Copy, Emery Air Freight Corporation, February 26, 1974, Smithsonian Registrar's Records, A19740668000, NASM 4037, National Air and Space Museum, Washington, DC.

86. When shipped from Paramount, two crates contained the *Enterprise* model while a third stored the electronic control panel or brains that lit up the ship's portholes, powered its running lights, and controlled the spinning "pod motors" of the warp engines. We know what this device looked like as there are studio photos showing a simple black box labeled "SPACE – CONTROL," with switches labeled on top that operators activated to "fly" the starship. In a 2016 article entitled "Two Enterprises: *Star Trek*'s Iconic Starship as Studio Model and Celebrity" in *Journal of Popular Film and Television*, 44, no. 1 (2016): 2–13, Margaret Weitekamp, a curator of the Smithsonian's National Air and Space Museum who oversaw the 2016 restoration of the original eleven-foot production model, acknowledges there were three crates delivered to the Smithsonian by Paramount.

ENDNOTES

Included in the article are two printed black and white Polaroid snapshots showing the crated model shortly after arriving at the Smithsonian, which she indicates were part of the paperwork used by museum officials to document the model after it was received. Weitekamp states "museum staff took only three Polaroid pictures, one of each shipping crate." Absent from her article is a photo showing the third crate that contained the control panel. A careful examination of subsequent photos taken by the Smithsonian chronicling the unpacking and assembly of the eleven-foot model shows the third crate in the foreground of the partially assembled *Enterprise*. The photo is clearly labeled "1974-668 (3) ELECTRONIC CONTROL PANEL." In addition, William S. McCullars wrote an article entitled "Starship '72 The Forgotten Detour of the NCC-1702" (*Star Trek Communicator: The Magazine of the Official Star Trek Club*, No. 120, December 1998/January 1999), 18–19, 77. This article is about the eleven-foot filming *Enterprise* model which was on display at Golden West College in Huntington Beach, California. Craig Thompson, who once worked for Desilu/Paramount Television as an office manager for postproduction on the lot during the original filming of the *Star Trek* television series, borrowed the filming model for two space expos he organized at the college. After leaving Desilu/Paramount Television, Thompson completed graduate work in education and became the director of student services at Golden West College. The article talks about the model during the two space expos. The first was held in 1972 and the second in 1973. In the article, Thompson confirms when the *Enterprise* model was on display at Golden West College during the space expos, it had the control panel.

87a. Lecture by Arthur C. Clarke, *Los Angeles Times*, April 14, 1972, 23.

87b. "Space Week Set For Blast Off At Golden West," *Daily Pilot*, April 9, 1972.

87c. "'2001' Author to Talk at Exposition," *Los Angeles Times*, April 9, 1972, Golden West College Archives, Golden West Library.

88. "Moon Rock Guarded at College in HB," *Register*, April 12, 1972, Golden West College Archives, Golden West Library.

89a. "Space Display At Golden West," *Orange County Evening News*, April 12, 1972, Golden West College Archives, Golden West Library.

89b. "NASA Exhibit Opens at JC," *News-Tribune*, April 12, 1972, Golden West College Archives, Golden West Library.

89c. "NASA Brings Exhibit to GW College," *Independent*, April 13, 1972, Golden West College Archives, Golden West Library.

90. "Space Exposition Showing Moon Rock at Golden West," *Daily Pilot*, April 11, 1972, Golden West College Archives, Golden West Library.

91. "Apollo 16 Goes On Display At Space 'Expo,'" *Register*, March 22, 1973, 36, Golden West College Archives, Golden West Library.

92a. "Space Exposition Will Open Monday, *Los Angeles Times*, March 25, 1973, 285.

92b. "Apollo 16 Goes On Display At Space Expo," *Register*, March 22, 1973, Golden West College Archives, Golden West Library.

92c. "GWC's Space Exposition," *Branding Iron*, March 30, 1973, Golden West College Archives, Golden West Library.

93. "Space Exhibits at GW College," *News Enterprise*, April 5, 1973, Golden West College Archives, Golden West Library.

94. William S. McCullars, "Starship '72: The Forgotten Detour of the NCC-1701," *Star Trek Communicator: The Magazine of the Official Star Trek Fan Club*, No. 120, December 1998–January 1999, 18–19, 77. This article includes an interview with Craig O. Thompson about the first appearances of the 11-foot production model at Thompson's two space expos held at Golden West College in Huntington Beach California during 1972 and 1973. Craig Thompson, who once worked for Desilu/Paramount Television as an office manager for postproduction on the lot during the original filming of the *Star Trek*, television series, was able to borrow the filming model for two

space expos that he organized at the college. After leaving Desilu/Paramount Television, Thompson completed graduate work in education and became the director of student services at Golden West College.

95. Gary Kerr, "Restoring a Legend—Part 2, A History of the Starship Enterprise, From Its Construction in 1964 to the Restoration of 2016," *Sci-fi & Fantasy Modeller* 45 (2017): 73–92.

96. Scott Steidinger interview by Glen E. Swanson, January 10, 2024. Scott Steidinger was a student of Craig Thompson's attending Golden West College during the years they had the space expos. During this interview, he talks about how he replaced some of the Christmas tree lightbulbs burned out inside the nacelle domes of the warp engines on the eleven-foot production model *Enterprise* when it was displayed during the second space expo held in 1973. Steidinger also mentioned that during the first space expo in 1972, the space station model from the 1971 made-for-television pilot *Earth II* was loaned to them for display from MGM studios. "It was so large they had to put it on display outside because it would not fit through any of the building doors," said Steidinger. During the 1973 space expo, Steidinger said Craig also got the loan of some of George Pal's "Puppet Toons" stop motion animation characters as well as the 22-inch filming miniature of the shuttlecraft from *Star Trek*. You can just see the starboard side of the shuttlecraft model in the shadows behind the *Enterprise* in one of Ron Yungul's photos. Steidinger mentioned the first year they did the space expo in 1972, they did not have many people attending. "During the second year, they had many school field trips from the local schools. It could be rather crowded at times," he said. Steidinger also confirmed they only had the two space expos, one in 1972 and the second in 1973. After that, Craig left Golden West College.

97a. Fred Durant to Walter M. Jefferies, September 18, 1974, Smithsonian Institution Archives, National Air and Space Museum, Dept. of Astronautics, Records 1965–1980, with materials dating from c. 1953, Record Unit 000398, box 32, folder: Roddenberry, Gene, Jefferies, Walter M.,1974–1975.

97b. Kerr, "Restoring a Legend–Part 2," 73–92. A reference photo taken by the Smithsonian during this period shows a pile of rags and a can of "Sail" brand scouring cleanser next to the eleven-foot model. All the weathering on the model, except for the upper saucer, was scrubbed off, possibly in the belief that it was dirt and grime.

98. Fred Durant to Francis Baby, memo, April 30, 1974, Smithsonian Institution Archives, National Air and Space Museum, Dept. of Astronautics, Records 1965–1980, with materials dating from c. 1953, Record Unit 000398, box 32, folder: Roddenberry, Gene, "Enterprise" model restoration, 1973–1975.

99. "Star Trek 'Enterprise' Spacecraft Model" dated April 30, 1974. Smithsonian Institution Archives, National Air and Space Museum, Dept. of Astronautics, Records 1965–1980, with materials dating from c. 1953, Record Unit 000398, box 32, folder: Roddenberry, Gene, Correspondence, 1968–1979.

100a. "Paramount TV Donates *Star Trek's* USS *Enterprise* to the Smithsonian Institution," Paramount Television News, Smithsonian Institution Archives, National Air and Space Museum, Dept. of Astronautics, Records 1965–1980, with materials dating from c. 1953, Record Unit 000398, box 32, folder: Roddenberry, Gene, "Enterprise" model restoration, 1973–1975.

100b. "*Star Trek* Ship Goes to Museum," *Paducah* (Paducah, Kentucky), November 8, 1974, 33.

101. Jacqueline Lichtenberg, Sondra Marshak and Joan Winston, *Star Trek Lives!* (1975), 20.

102a. Charles Cushing, July 20, 1979, Smithsonian Institution Archives, National Air and Space Museum, Dept. of Astronautics, Records 1965–1980, with materials dating from c. 1953, Record Unit 000398, box 32, folder: Roddenberry, Gene, Correspondence, 1968–1979. Fred Durant to Charles Cushing, July 23, 1979, Smithsonian Institution Archives, National Air and Space Museum, Dept. of Astronautics, Records 1965–1980, with materials dating from c. 1953, Record Unit 000398, box 32, folder: Roddenberry, Gene, Correspondence, 1968–1979.

◀ ENDNOTES ▶

102b. Astronautics Department Status Report, Gallery No. 107, *Life in the Universe?* December 15, 1975. Report states that "*Enterprise* model to be refurbished by Otano. Mars globe to be returned by NASA in April. Mariner 10 installed. Viking Mars Lander to be delivered first week in June 1976." Smithsonian Institution Archives, National Air and Space Museum, Dept. of Space Science and Exploration., Subject Files, c. 1960–1986, Record Unit 000348, box 11, folder: *Life in the Universe?* (107).

102c. Gallery No. 107, Astronautics Department Status Report No. 2, *Life in the Universe?* December 23, 1975. Report states that "The USS *Enterprise* model is to be restored by NASM Audio/Visual personnel." Smithsonian Institution Archives, National Air and Space Museum, Dept. of Space Science and Exploration., Subject Files, c. 1960–1986, Record Unit 000348, box 11, folder: Life in the Universe (107).

102d. Astronautics Department Status Report, Gallery No. 107, *Life in the Universe?* February 18, 1976. Report states that "Restoration of USS *Enterprise* model is proceeding well." Smithsonian Institution Archives, National Air and Space Museum, Dept. of Space Science and Exploration., Subject Files, c. 1960–1986, Record Unit 000348, box 11, folder: Life in the Universe (107).

102e. Astronautics Department Status Report, Gallery No. 107, *Life in the Universe?* March 10, 1976. Report states that "Fabrication: USS *Enterprise* model restoration complete." Smithsonian Institution Archives, National Air and Space Museum, Dept. of Space Science and Exploration., Subject Files, c. 1960–1986, Record Unit 000348, box 11, folder: Life in the Universe (107).

102f. Astronautics Department Status Report, Gallery No. 107, *Life in the Universe?* April 7, 1976. Report states that "Artifacts and Specimens: The USS *Enterprise* has been hung." Smithsonian Institution Archives, National Air and Space Museum, Dept. of Space Science and Exploration., Subject Files, c. 1960–1986, Record Unit 000348, box 11, folder: Life in the Universe (107).

102g. Astronautics Department Status Report, Gallery No. 107, *Life in the Universe?* May 18, 1976. Report states that "Artifacts: The plastic hemisphere which the painters knocked off the *Enterprise* has been glued together. Fitzpatrick will find out if he can repaint it to hide the cracks. If not, a new one will be ordered from Rogay." Smithsonian Institution Archives, National Air and Space Museum, Dept. of Space Science and Exploration., Subject Files, c. 1960–1986, Record Unit 000348, box 11, folder: *Life in the Universe?* (107).

103. "Smithsonian Sets Space Art Exhibit," *Durham Morning Herald*, August 26, 1984, 53.

104. Text from the signage that appeared beneath the *Enterprise* model when it was displayed at the second *Life in the Universe?* exhibit which opened in the new building in 1976. Floaty Pens remain a popular souvenir novelty made by Eskesen in Denmark. They feature a pen with an object moving back-and-forth inside a liquid-filled barrel. The pen was named "The Original Floating Action Pen" and is still manufactured today, over seventy years since they were first invented. The *Star Trek Enterprise* pen has its liquid-filled barrel encapsulating a small model of the starship *Enterprise* that moves back-and-forth in front of a stylized galactic backdrop. The reverse side of the encased backdrop reads "Smithsonian Institution USS ENTERPRISE National Air and Space Museum." The printed text on the outside body of the pen reads: "U.S.S. 'Enterprise' from 'Star Trek' television series spurred interest in space travel." The "*Star Trek* USS *Enterprise*" pen was first sold in gift shops by the National Air and Space Museum during the original *Life in the Universe?* exhibit when it was housed at the Arts and Industries building in 1974. The pens continued to be sold at the new museum building when it opened to the public in 1976.

105. Text from blister packaging of licensed Ertl 4.5-inch metal refit *Enterprise* toy that was sold under the series "Smithsonian Museum Shops" product line ca. 1991, SKU 1244409, 1372-7HFW. This was a repackaging of a series of *Star Trek* die-cast metal miniatures that originally included not only the refit *Enterprise* but also the *Excelsior* and the Klingon *Bird of Prey* that were first released by Ertl with the 1984 premiere of *Star Trek III: The Search for Spock*. Between February 28, 1992, and January 31, 1993, the National Air and Space Museum (NASM) presented "*Star Trek*: A Retrospective Exhibition." Mary S. Henderson, Curator of Art at NASM, was the chief curator. Despite drawing record crowds (close to 1 million visits were made by the public during the 11 months of the

exhibition), there was a great deal of internal opposition by NASM officials to the *Star Trek* exhibit. Many people at the museum, including Martin Harwit, NASM's director, opposed the idea arguing that exhibits which focus on popular culture did not belong in the National Air and Space Museum. After the close of the *Star Trek* exhibit, Henderson proved her critics wrong by following up with the even more popular "*Star Wars*: The Magic of Myth" exhibit which premiered at NASM in 1997. Henderson also wrote a best-selling book that accompanied the exhibit. Even though the *Star Wars* exhibit proved immensely popular, Lucasfilm objected to its academic tone, arguing that both the exhibit and the book compared Luke Skywalker's trials to heroes in Greek mythology. Lucas preferred to have a chronological exhibit, highlighting his *Star Wars* films and culminating in the promotion of *The Phantom Menace* which was then currently in production. To help convince management of the appeal of popular culture exhibits at NASM and their educational value, Henderson suggested that a study be conducted. As a result, the Smithsonian's Institution Studies Office did a survey during the last few months of the *Star Trek* exhibit. In November 1994, they published their findings in a formal report entitled "Space Fantasy and Social Reality: A Study of the *Star Trek* Exhibition at the National Air and Space Museum." As part of the study, some 1,365 visitors responded to survey questions and offered comments. Among the survey questions asked was whether respondents "thought that it was appropriate to present a *Star Trek* exhibit alongside artifacts of flight and space exploration at NASM." The report stated that 92 percent thought that it was, with the most common reason given that *Star Trek* was "about space," followed by statements that *Star Trek* was part of our culture, that both NASM and *Star Trek* were about "the future," and that exhibits like *Star Trek* helped interest people in science and space exploration n general. As a result of the popularity of these two exhibits and the findings of the Smithsonian report, the museum eventually created a new position that was filled by Margaret Weitekamp who, in 2004, became NASM's first curator of Social and Cultural History of Spaceflight. Twelve years later, Weitekamp would lead a major restoration by NASM of the 11-foot production model *Enterprise* to prepare it for permanent display in conjunction with the 50th anniversary of *Star Trek* in 2016. Today, millions of visitors can see the original filming model prominently displayed on the museum's main floor across from the Independence Avenue entrance and in front of a giant spaceflight mural painted by artist Bob McCall. Sources: January 30, 2005, interview with Barbara Brennan; "Space Fantasy and Social Reality: A Study of the *Star Trek* Exhibition at the National Air and Space Museum," by Adam Bickford, Zahava D. Doering, and Andrew Pekarik, November 1994, Institutional Studies Office, Smithsonian Institution. https://repository.si.edu/bitstream/handle/10088/17193/opanda_94-5-StarTrek.pdf?sequence=1&isAllowed=y&fbclid=IwY2xjawlJwLVleHRuA2FlbQIxMAABHYnZIONptomMJfFy7I-w7Pvl_dqZPonn9IKsD-mmWmj7NmijDVEPWw3Yjw_aem_WOpyT9LQqsz19WkYNgVIBw

106. Crouch, interview by Swanson, (2023).

107. The first time I saw the *Enterprise* hanging in the Smithsonian was during the summer of 1978 when our family rented an RV to do a road trip to Washington, DC. My brother was getting ready to enter his senior year in high school that fall, and my mom figured it would probably be the last opportunity we had as a family to all go together on an extended vacation. My mom scrimped until she had enough funds saved to rent a motorhome (my father would never have indulged in such an extravagance). My dad did, however, purchase a brand-new Nikon FE 35mm single lens reflex camera for the trip that he let me use. It turned out to be a great trip, despite my father worrying we would break something in the RV along the way. I was thrilled to finally see the *Enterprise* up close and in person. Then, as a fifteen-year-old-expert on all things having to do with *Star Trek*, I expressed dismay at what I saw. "I saw the USS *Enterprise* and the way that they have it is an insult to *Star Trek*!" I wrote on postcards sent to my friends after making the pilgrimage to see the holy relic. "They have it terribly lit, and the way that they have it displayed is awful!" Apparently, I was not too traumatized by it all since I took photos with my dad's camera and purchased postcards and posters from the giftshop that depicted the model to give to envious friends back home. Source: Author postcard depicting the studio model *Enterprise* purchased at the Smithsonian and sent to Jerry Fellows on August 18, 1978. These postcards could be purchased at

the museum gift shop for ten cents each, along with lithograph posters depicting the same postcard image for $2. Source: Fred Durant to Robert Craft, March 22, 1977, Smithsonian Institution Archives, National Air and Space Museum, Dept. of Space Science and Exploration, Subject Files c. 1960–1986, Record Unit 000348, box 6, folder: Correspondence – General, 1977.

108. This story is mentioned in the fan-produced publication *Star Trektennial News*. In issue No. 17 (September/October 1976) there appears on p. 3 a group photo taken on July 10, 1976 showing Majel Barrett, Gene Roddenberry, James Doohan, and George Takei all standing under the original studio model as it hung in the Smithsonian's *Life in the Universe?* exhibit. The photo was taken just nine days after the new museum first opened to the public on July 1, 1976. The text next to the photo states that Fred Durant is the one who originated the story. When I asked Steve Durant, Fred's son, about this, he mentioned that his father often mingled with museum visitors. "He enjoyed talking with people about the museum's collections," said Steve. "In talking about the Smithsonian and *Star Trek* he would often share that story with the public."

Chapter 8

1. John D. F. Black, interview by Marc Cushman, 2013, *These Are the Voyages TOS: Season One*, 25.

2. Charles Witbeck, "TV Topics: New Series Pits 400-Man Craft Against Space," *The Buffalo (NY) Evening News*, September 3, 1966.

3. "*Star Trek's* Upward Flight," *Detroit Free Press*, August 13, 1967, 70.

4. Solow and Justman, *Inside Star Trek: The Real Story*, 351.

5. Michelle Hilmes, ed., *NBC: America's Network*, (University of California, 2007), 218.

6. Gerrold, *The World of Star Trek*, 166–67.

7. "Star Trekkers Are Restored," *Hartford Courant*, March 17, 1968, 168.

8. Trimble, *On the Good Ship Enterprise*, 36.

9. Bob Thomas, "'Eggheads' Like TV's *Star Trek*," *Lancaster (PA) New Era*, May 6, 1967, 9.

10. Many stories have been told about *Star Trek*, making it a challenge to distinguish fact from fiction. A prime example is the story about the infamous letter-writing campaign launched by fans of the series in late 1967, petitioning NBC to renew *Star Trek* for a third season. It is true NBC received tens of thousands of pieces of mail asking the network to keep *Star Trek* on the air. What is less clear is whether *Star Trek* was ever in any real danger of being canceled prior to the third season. It certainly did not hurt for fans to show their support, but it was not unprecedented for television viewers to campaign for their favorite TV shows. It happened at least six times before *Star Trek*, with the first being in 1951, when viewer feedback helped convince CBS to renew *Mr. I. Magination*. After the Friday, March 1, 1968, episode showing of "The Omega Glory," NBC told its viewers on the air that *Star Trek* would be renewed for a third season. The subtext of such a move was clear: Stop writing letters. NBC's corporate policy at the time was to reply, by return mail, to each viewer letter received. Answering every *Star Trek* letter was costing the network money. The network issued a press release about the on-air announcement, the text of which was reprinted in Stephen Whitfield's *The Making of Star Trek*, 394–395: "MARCH 4TH, 1968 UNPRECEDENTED VIEWER REACTION IN SUPPORT OF 'STAR TREK' LEADS TO ON-AIR ANNOUNCEMENT OF SERIES' SCHEDULING FOR 1968–69. In response to unprecedented viewer reaction in support of the continuation of the NBC Television Network's STAR TREK series, plans for continuing the series in the Fall were announced on NBC-TV immediately following last Friday night's (March 1) episode of the space adventure series. The announcement will be repeated following next Friday's (March 8) program. From early December to date, NBC has received 114,667 pieces of mail in support of STAR TREK, and 52,151 in the month of February alone. Immediately after last Friday night's program, the following announcement was made: 'And now an announcement of interest to all viewers of STAR TREK. We are pleased to tell you that STAR TREK will continue to be seen on NBC Television. We know you will be looking forward to seeing the weekly adventure in space on STAR TREK.'"

11. John Trimble, interview by Marc Cushman, 2012, *These Are The Voyages TOS: Season Two*, 445.

12. Solow and Justman, *Inside Star Trek: The Real Story*, 402.

13. Mark Altman interview with Richard Block, vice president and general manager, Kaiser Broadcasting Corp., in Gross and Altman, *The Fifty-Year Mission: The Complete, Uncensored, Unauthorized Oral History of Star Trek: The First 25 Years*, 232–233.

14. Altman, interview with Block

15. KBHK is now called KBYX and is owned by CBS/Paramount.

16. *Broadcasting*, February 10, 1975, 43.

17. *Broadcasting*, March 24, 1969, 87; *Broadcasting*, August 4, 1969, 31; *Broadcasting*, February 16, 1970, 43; *Weekly Variety* 283, no. 7 (June 23, 1976): 44–45.

18. Solow and Justman, *Inside Star Trek: The Real Story*, 402.

19. White House memo to the president from William F. Gorog, titled "Naming the Space Shuttle," dated September 3, 1976. Before coming to the White House, Gorog was founder and CEO of the Data Corporation, which helped create LexisNexis, the computerized information retrieval system.

20. The mention of CB or citizens band radio in the White House memo is in reference to First Lady Betty Ford's talking on CB. During the mid 1970s, the use of CB radio was common among truckers. It grew in popularity among the public due in part because of the novelty song "Convoy" performed by C. W. McCall. The song became a number 1 hit on both the country and pop charts and is listed ninety-eighth among *Rolling Stone* magazine's 100 Greatest Country Songs of All Time. America's fascination with CB radio was seen in the television series *Movin' On*, which ran for two seasons on NBC from 1974 to 1976, and in the theater through the films *Smokey and the Bandit* (1977) and *Convoy* (1978). Mrs. Ford received a CB radio in 1976 as a gift from her daughter Susan. After the FCC quickly granted her a license, she took to the airwaves during the campaign trail for her husband. She picked the handle "First Mama" at the suggestion of comedian Flip Wilson; a handle is an identifying name used in CB conversations. Among the CB operators who called the first lady were two who identified themselves as "Starship Enterprise," and "Peg Leg Charlie." Source: "'Smokies' Abound Texas: 'First Mama' Betty Ford Makes Debut on CB Radio," *Pittsburgh Post-Gazette, Sun-Telegraph*, April 21, 1976, 2.

21. Transcription of September 8, 1976, speech given by President Ford, naming the first space shuttle OV-101 "Enterprise."

22. White House memo to the president from Jim Connor, "Naming the Space Shuttle," dated September 7, 1976.

23. "Naming the Space Shuttle."

24. "And Now, A Real Space Shuttle Named 'Enterprise,'" NASA Current News, *Washington Post*, September 9, 1976.

25. That the rollout of the world's first reusable spacecraft OV-101 occurred on September 17 was by design. Originally, the space shuttle orbiter was going to be named *Constitution,* and the rollout was planned to occur on September 17, which is Constitution Day. The year 1976 was the nation's bicentennial, so it would have been fitting to name the first orbiter in recognition of the two hundredth anniversary. NASA specifically scheduled the rollout to occur on the 189th anniversary of the day the Constitutional Convention adopted the document. According to Dennis Jenkins, author of the massive three-volume history *Space Shuttle Developing an Icon, 1972–2013* (Specialty Press, 2016)), "Most NASA documentation claims the name was in honor of the bicentennial of the US Constitution. However, since the Constitution was not completed until 17 September 1787, this is obviously in error. Given that 1976 was the US Bicentennial (signing of the Declaration of Independence) we will assume this is what NASA intended" (p. I-422). The rollout occurred on September 17, 1976. In one of the photos taken during the rollout ceremony at Rockwell International Space Division's assembly plant in Palmdale, one can see a large banner calling the orbiter a "new

era in transportation." Suspiciously absent from the banner is the name of the orbiter *Enterprise*, most likely because of the last-minute name change. Even though President Ford changed the name to *Enterprise*, the last-minute name change was not universally well received. In an October 6, 1976, Newhouse News Service article by Peter Cobun titled "NASA Oldsters Bristle at *Star Trek* Influence," Cobun spoke with Jesco von Puttkamer, a writer and senior manager at NASA headquarters. "There may have been politics involved," admits von Puttkamer. "There's always something political about everything that comes out of the White House." Von Puttkamer has had many close connections to *Star Trek* and was a strong supporter of the series. About the name change, he told Cobun that some NASA officials "were kind of caught by surprise and didn't quite see the connection between us and *Star Trek*. The connection is what I call the 'space connection.'" Von Puttkamer said the *Star Trek* world "is helping to establish a better awareness of the space program in the public mind." For those old-timers at NASA who find the pairing of science and show biz "demeaning," Von Puttkamer told Cobun "I would say to them that after all, we are using the public money, and why shouldn't we establish a dialogue with the public to find out what the public is thinking." Source: Peter Cobun, "NASA Oldsters Bristle at *Star Trek* Influence," Newhouse News Service, *Houston Chronicle*, October 6, 1976.

26. NASA originally intended to have OV-101 become a fully functional space shuttle orbiter that would bring astronauts into space and back. NASA documents stated that while OV-101 was being built, it was "designed for operational missions" just like all the others (Jenkins, *Space Shuttle*, I-422). However, once completed, both OV-101 *Enterprise* and OV-102 *Columbia* proved to be heavier than expected and could not carry national security payloads to their desired orbits. In addition, OV-101 would require a teardown to make structural modifications and other changes the approach and landing tests (ALT) revealed would be necessary to make it into a fully operational space vehicle. Taking STS-099 and converting it into a space-flown orbiter would require only a buildup. Simply put, engineers were more efficient in building orbiters than in taking them apart. As a result, *Enterprise* remained "grounded" to be used only as a test vehicle on Earth and not to be flown in space. STS-099 was then built up to join the fleet and become a space shuttle *Challenger*.

27. Gene Roddenberry to Fred Durant, September 16, 1976. Smithsonian Institution Archives, Record Unit 000398, Box 32, National Air and Space Museum, Dept. of Astronautics, Records 1965–1980, with materials dating from c. 1953, folder: Roddenberry, Gene, Correspondence, 1968–1979. Surprisingly, Roddenberry did not approve of the name change of the space shuttle from *Constitution* to *Enterprise*. In a letter dated September 16, 1976, sent to Fred Durant, Roddenberry voices his displeasure: "I have been quite upset over the change of the Shuttlecraft name to 'Enterprise.' I may have told you at one time I had heard of a letter-writing campaign on the subject (Hoagland and Preston) and had completely disassociated myself from it. Frankly, I thought nothing would ever come of it anyway. (If I had my own choice, I would have wanted the Shuttlecraft named 'Goddard' or at least something completely unmilitary). What troubles me is that I know all of us get stereotyped one way or another. It even happens to museum curators. Unfortunately, we in the entertainment industry often are thought of materialistic creatures who will do literally anything for publicity. You know that this isn't true but some don't. I just wanted to let you know personally that I not only would not toy with the space program but have also strongly recommended to Paramount that they seek no publicity advantage in the naming of the Shuttlecraft. Whether the front office here needs my advice or not may be another thing. I will be present at the rollout since I was long ago invited there and consider it a momentous occasion. But I certainly will not be making speeches about our starship or movie. If anyone asks, I'll tell them approximately what I have just written you."

28. On January 8, 2024, a new rocket called Vulcan and built by United Launch Alliance, launched on its first-ever flight, called Cert-1, from Space Force Station in Cape Canaveral, Florida. Its primary payload was the privately built Peregrine lunar lander made by the Pittsburgh-based company Astrobotic. In addition to the lander, other payloads carried aboard the rocket include two memorial spaceflights designed by the space memorial company Celestis, which sends customer DNA and cremated remains into space. The private company offers a range of missions for flying

customer remains, which include suborbital flights to Earth orbit, lunar orbit, the lunar surface and to a permanent orbit around the Sun. Celestis had sent payloads to the lunar surface once before, when some of Eugene Shoemaker's remains were carried aboard NASA's successful Lunar Prospector mission that was launched in 1998. At the conclusion of that mission, the spacecraft impacted the lunar surface inside a permanently shadowed crater near the south lunar pole. Now Celestis sought to do this again with a memorial flight called Luna Tranquility. Tranquility includes the DNA of 66 mission "participants," whose remains would be permanently placed on the moon's surface following Peregrine's successful landing. Tranquility was not the only Celestis payload launched on this flight. The company also has a payload called "Enterprise" that was carried aboard the rocket's Centaur upper stage. Named after the starship *Enterprise* from *Star Trek*, the Enterprise memorial flight contains cremated remains and DNA from *Star Trek's* creator Gene Roddenberry and his wife, Majel Barrett-Roddenberry, as well as the remains from several actors from the original TV series, including Nichelle Nichols, James Doohan and De Forest Kelley, who played Lieutenant Uhura, Chief Engineer "Scotty," and Chief Medical Officer Leonard "Bones" McCoy, respectively. The DNA from some of the living relatives of the *Star Trek* cast is also flying on the Enterprise flight, including Eugene "Rod" Roddenberry, the son of Gene and Majel Roddenberry, and Wende Doohan, the widow of James. Shortly after launch, the Peregrine lunar lander ran into difficulties, which included a fuel leak that jeopardized its mission. Officials soon determined there would be no opportunity for the spacecraft to land on the lunar surface. As a result, Peregrine re-entered the Earth's atmosphere on January 18, 2024, where it broke apart in the skies above the South Pacific, 1,500 miles off the eastern coast of Australia. As of this writing, the Enterprise portion of the flight is doing well, with that payload headed toward the sun where it will enter a heliocentric orbit. When it reaches its final orbit, it will be renamed "Enterprise Station." Source: https://www.space.com/ula-vulcan-centaur-first-launch-peregrine-celestis-moon-mission.

Afterword

1. Andrew Chaikin, *A Man on the Moon: The Voyages of the Apollo Astronauts* (Blackstone Publishing Audiobooks, 2015) transcribed Afterword for the 50th Anniversary of the Space Age.

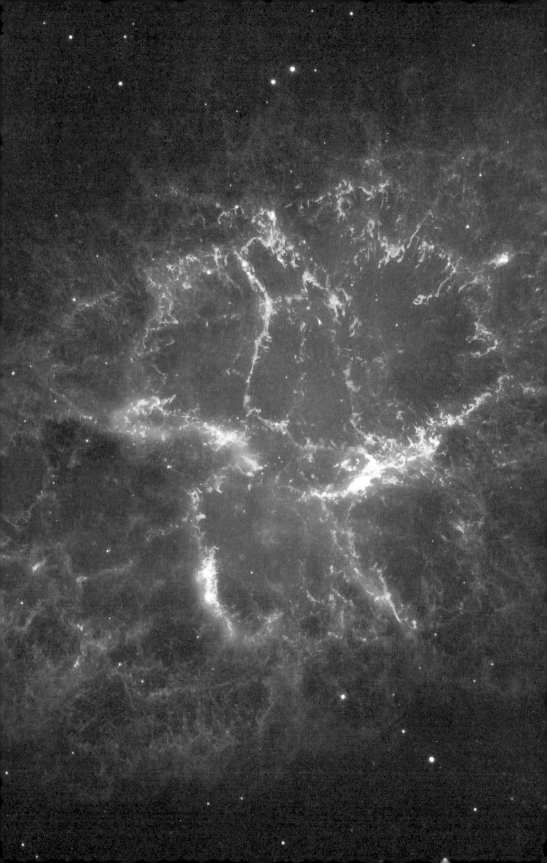

BIBLIOGRAPHY

Adamson, Joe. *Byron Haskin*. Directors Guild of American Oral History. Scarecrow, 1984.

Addey, Dave. *Typeset in the Future: Typography and Design in Science Fiction Movies*. Abrams, 2018.

Alberstadt, Milton. "New Paint Job," *Space World*, E-4-52 (April 1968).

Alexander, David. *Star Trek Creator: The Authorized Biography of Gene Roddenberry*. Penguin Books, 1994.

Asherman, Allan. *The Star Trek Interview Book*. Pocket Books, 1988.

Asherman, Allan. *The Star Trek Compendium*. Simon and Schuster, 1981.

Barlow, Ron, George M.W. Snow, Anthony Frederickson, and Doug Drexler, eds. "Smithsonian Report." *Star Trek Giant Poster Book*, Voyage 10 (June 1977).

Beard, Jim, ed. *Galloping Around the Cosmos: Memories of TV's Wagon Train to the Stars from Today's Grown-Up Kids*. Becky's Books, 2023.

Bernardi, Daniel Leonard. *Star Trek and History: Race-ing Toward a White Future*. Rutgers University Press, 1998.

Black, A. J. *Star Trek, History and Us: Reflections of the Present and Past Throughout the Franchise*. McFarland & Company, 2021.

Bond, Jeff. *The Fantasy Worlds of Irwin Allen*. Creature Features, 2019.

Bond, Jeff and Gene Kozicki. *Star Trek: The Motion Picture: Inside the Art & Visual Effects*. Titan Books, 2020.

Britt, Ryan. *Phasers on Stun! How the Making (and Remaking) of Star Trek Changed the World*. Plume, 2022.

Buck, Jerry. "Around the Dial." *The San Bernardino County Sun*, October 7, 1984.

Carrington, Andre M. *Speculative Blackness: The Future of Race in Science Fiction*. University of Minnesota Press, 2016.

Chaikin, Andrew. *A Man on the Moon: The Voyages of the Apollo Astronauts,* Audiobooks, 2017. Includes transcribed special afterword for the fiftieth anniversary of the space age.

Chavkin, Dan and Brian McGuire. *Star Trek Designing the Final Frontier: How Midcentury Modernism Shaped Our View of the Future*. Weldon Owen, 2021.

Clarke, Arthur C. *The Exploration of Space*. Harper & Brothers, 1951.

Clarke, Arthur C. ed. *The Coming of the Space Age*. Meredith Press, 1967.

Cobun, Peter. "NASA Oldsters Bristle at Star Trek Influence." *Newhouse News Service. Houston Chronicle*, October 6, 1976.

Cocconia, Giuseppe and Philip Morrison. "Searching for Interstellar Communications." *Nature* 184, no. 4690 (September 19, 1959).

Cole, Steve. "Retrospect: Star Fleet Battles." *Space Gamer*, no. 42 (August 1981).

Courtwright, Edgar M. *Space Exploration—Why and How NASA EP-25*. US Government Printing Office, 1965.

Cushman, Marc. *Irwin Allen's Lost in Space: The Authorized Biography of a Classic Sci-Fi Series.* Jacobs/Brown, 2016.

Cushman, Marc, with Susan Osborn. *These Are the Voyages: Gene Roddenberry and Star Trek in the 1970s.* Vol. 1, 1970–1975. Jacobs/Brown Press, 2019.

Cushman, Marc, with Susan Osborn. *These Are the Voyages: Gene Roddenberry and Star Trek in the 1970s.* Vol. 2, 1975–1977. Jacobs/Brown Press, 2020.

Cushman, Marc, with Susan Osborn. *These Are the Voyages: Gene Roddenberry and Star Trek in the 1970s.* Vol. 3, 1978–1980. Jacobs/Brown Press, 2020.

Cushman, Marc. *These Are the Voyages: TOS: Season One.* Edited by Mark Alfred and Susan Osborn. Jacobs/Brown, 2013.

Cushman, Marc. *These Are the Voyages: TOS: Season Two.* Edited by Mark Alfred and Susan Osborn. Jacobs/Brown, 2014.

Cushman, Marc. *These Are the Voyages: TOS: Season Three.* Edited by Mark Alfred and Susan Osborn. Jacobs/Brown, 2015.

Day, Dwayne A. "Blue Skies on the West Coast: A History of the Aerospace Industry in Southern California." *Space Review*, August 20, 2007. https://www.thespacereview.com/article/938/1.

Day, Dwayne A. "Boldly Going: *Star Trek* and Spaceflight." *Space Review*, November 28, 2005. https://www.thespacereview.com/article/506/1.

De Forest, Kellam. "Recollections of My Parents." *Eden: Journal of the California Garden & Landscape History Society*, 17, no. 3, (Summer 2014).

Dick, Karen. "Franz Joseph Timeline." Annotated by Greg Tyler.

Dillard, J. M. *Star Trek "Where No One Has Gone Before:" A History in Pictures.* Pocket Books, 1994.

Dole, Stephen H. *Habitable Planets for Man.* Blaisdell, 1964.

Dole, Stephen H., and Isaac Asimov. *Planets for Man.* Random House, 1964.

Dornberger, Walter J. "Space Shuttle for the Future: The Aerospaceplane." *Rendezvous* 4, no. 1, 1965.

Dornberger, Walter J. "The Recoverable, Reusable Space Shuttle," *Astronautics and Aeronautics*, November 1965.

Drake, Frank. "The Drake Equation Revisited: Part I," *Astrobiology Magazine*, September 29, 2003. https://www.spacedaily.com/reports/The_Drake_Equation_Revisited_Part_I.html.

Drushel, Bruce E. ed., *Fan Phenomena: Star Trek.* Intellect Books, 2013.

Earth Photographs from Gemini III, IV, and V. NASA SP-129. Scientific and Technical Information Division, Office of Technology Utilization, NASA, 1967.

Earth Photographs from Gemini VI through XII. NASA SP-171. Scientific and Technical Information Division, Office of Technology Utilization, NASA, 1967.

Edmonson, Harold A., ed. *Famous Spaceships of Fact and Fantasy.* Kalmbach, 1979.

Engel, Joel. *Gene Roddenberry: The Myth and the Man Behind Star Trek.* Hyperion, 1994.

Fahey, John E. " 'A Tall Ship and A Star to Steer Her By:' Star Trek and Naval History." Official blog of the Naval History and Heritage Command, April 5, 2023. https://usnhistory.navylive.dodlive.mil/Recent/Article-View/Article/3348268/a-tall-ship-and-a-star-to-steer-her-by-star-trek-and-naval-history/#_ftn18.

Faragher, John Mack, ed. *Rereading Jackson Turner: "The Significance of the Frontier in American History" and Other Essays.* Yale University Press, 1998.

Farber, Stephen. "Before the Cry of 'Action!' Comes the Painstaking Effort of Research." *New York Times*, March 11, 1984.

BIBLIOGRAPHY

Fern, Yvonne. *Gene Roddenberry: The Last Conversation*. University of California Press, 1994.

Fontana, Dorothy. "Behind the Camera: Walter M. Jefferies." Edited by Ruth Berman. *Inside Star Trek* 4, (October 1968).

Fontana, Dorothy. "Just Ask." Edited by Ruth Berman. *Inside Star Trek* 9 (March 1969).

Fontana, Dorothy. "The Klingons are Coming!" Edited by Ruth Berman. *Inside Star Trek* 2, (August 1968).

Freeman, Donald. "History of TV Series—Book on 'Star Trek' Makes Good Reading." *Shreveport Journal*, October 31, 1968.

Fulton, Valerie. "Another Frontier: Voyaging West with Mark Twain and *Star Trek's* Imperialist Subject." *Postmodern Culture* 4, no. 3 (1994).

Garr, Teri, with Henriette Mantel. *Speedbumps: Flooring it Through Hollywood*. Hudson Street, 2005.

Gene Roddenberry Papers. UCLA Library Special Collections, Charles E. Young Research Library, University of California, Los Angeles.

Geraghty, Lincoln. "A Network of Support: Coping with Trauma through *Star Trek* Fan Letters," *Journal of Popular Culture 39*, no. 6, 2006.

Gerrold, David. *The World of Star Trek*. Ballantine Books, 1973.

Gess, Alan. *Googie: Ultramodern Roadside Architecture*. Chronicle Books, 2004.

Gibberman, Susan R. *Star Trek: An Annotated Guide to Resources on the Development, the Phenomenon, the People, the Television Series, the Films, the Novels and the Recordings*. McFarland, 1991.

Glaser, Peter E. "A Structural Approach to the Thermal Control of Space Vehicles." In *Proceedings of the Manned Space Stations Symposium: Los Angeles, California, April 20–22, 1960*. Institute of the Aeronautical Sciences, 1960.

Goodman, Eckert. "The Little Model Builders They Grow into Space-Age Giants." *New York Daily News*, February 23, 1959.

Graham, Thomas. *Aurora Model Kits With Polar Lights, Moebius, Atlantis*. 3rd ed. Schiffer Publishing. Ltd., 2008.

Graham, Thomas. *Monogram Models. 2nd ed*. Schiffer, 2013.

Graham, Thomas. *Remembering Revell Model Kits. 3rd ed*. Schiffer, 2008.

Greenberger, Robert. *Star Trek: The Complete Unauthorized History*. Voyageur, 2012.

Grening, David. "Harvey Pack, Legendary On-Air Racing Personality, Dead at 94." *Daily Racing Form*, July 6, 2021. https://www.drf.com/news/harvey-pack-legendary-air-racing-personality-dead-94.

Gross, Edward. "Art Wallace: An 'Obsession' On 'Assignment: Earth.' " *Starlog* 112 (November 1986).

Gross, Edward and Mark A. Altman. *Captain's Logs: The Unauthorized Complete Trek Voyages*. Little, Brown, 1995.

Gross, Edward and Mark A. Altman. *The Fifty-Year Mission: The Complete, Uncensored, Unauthorized Oral History of Star Trek: The First 25 Years*. St. Martin's, 2016.

Gross, Edward, and Mark A. Altman. *The Fifty-Year Mission: The Complete, Uncensored, Unauthorized Oral History of Star Trek: The Next 25 Years*. St. Martin's, 2016.

Gross, Edward and Mark A. Altman. *Great Birds of the Galaxy: Gene Roddenberry & the Creators of Trek*. Boxtree, 1994.

Gurian, Gerald. *To Boldly Go: Rare Photos From the TOS Soundstage: Season One*. Minkatek, 2016.

Gurian, Gerald. *To Boldly Go: Rare Photos From the TOS Soundstage: Season Two*. Minkatek, 2016.

Gurian, Gerald. *To Boldly Go: Rare Photos From the TOS Soundstage: Season Three*. Minkatek, 2017.

Hagerty, Jack, and Jon Rogers. *The Saucer Fleet*. Apogee Books, 2008.

Hagerty, Jack, and Jon C. Rogers. *Spaceship Handbook: Rocket and Spacecraft Designs of the 20th Century, Fictional, Factual and Fantasy*. ARA Press, 2001.

Handsaker, Gene. "Kellam de Forest Runs TV's Reference Service." *Janesville Daily Gazette,* June 7, 1967.

Hardy, Kristen. "Look Up and Learn." *UCLA Magazine*, October 2017. https://newsroom.ucla.edu/magazine/ucla-building-mosaics.

Harvey, Aaron, and Rich Schepis. *Star Trek: The Official Guide to the Animated Series*. Weldon Owen, 2019.

Hernández, José Moreno. *Reaching for the Stars*. Center Street, 2012.

Hilmes, Michele, ed. *NBC America's Network*. University of California Press, 2007.

Hine, Thomas. *Populuxe*. Alfred A. Knopf, 1987.

Hoagland, Richard C. *The Monuments of Mars: A City on the Edge of Forever*. North Atlantic Books, 1987.

Irvine, Matt. *Creating Space: A History of the Space Age Told Through Models*. Apogee Books, 2002.

Jefferies, Richard L. *Beyond the Clouds: The Lifetime Trek of Walter "Matt" Jefferies, Artist and Visionary*. Brown Books, 2008.

Jenkins, Dennis. *Space Shuttle: Developing an Icon, 1972–2013*. 3 vols. Specialty Press, 2016.

Jenkins, III, Henry. "*Star Trek* Rerun, Reread, Rewritten: Fan Writing as Textual Poaching," *Critical Studies in Mass Communication* 5, no. 2 (June 1988).

Jones, Preston Neal. *Return to Tomorrow: The Filming of* Star Trek: The Motion Picture. Creature Features, 2014.

Joseph, Franz. *Star Fleet Technical Manual*. Ballantine Books, 1975.

Joseph, Franz, *Star Trek Blueprints*. Ballantine Books, 1975.

Kamm, Robert W. "Greetings from the NASA Western Operations Office," In *Proceedings of the Science Education in the Space Age*. NASA, NASA, November 1964.

Kapell, Matthew Wilhelm. *Exploring the Next Frontier: Vietnam, NASA, Star Trek and Utopia in 1960s and 70s American Myth and History*. Routledge, 2016.

Kaufman, David. " 'Trek' Into Future; Reagan Win Would Cue 'Death' Exit." *Daily Variety* 132, no. 2 (June 7, 1966): 9.

Kelley, Stephen. *Star Trek: The Collectibles*. Krause, 2008.

Kennedy, John F. Address before a Joint Session of Congress, May 25, 1961.

Kennedy, John F. "If the Soviets Control Space… They Can Control Earth." *Missiles and Rockets* 7, no. 15 (October 10, 1960).

Kennedy, John F. "The New Frontier." Acceptance speech of Senator John F. Kennedy, Democratic National Convention, July 15, 1960. Papers of John F. Kennedy, John F. Kennedy Presidential Library and Museum.

Kerr, Gary. "Restoring a Legend—Part 2, A History of the Starship Enterprise, From Its Construction in 1964 to the Restoration of 2016. *Sci-fi & Fantasy Modeller* 45 (2017): 73–92.

Killian, James. *Introduction to Outer Space*. The White House, March 26, 1958.

Kindryd, Andreea. *From Slavery to Star Trek (Memoir Excerpts from* Code-Switching: A Family's Voyage from Slavery to the Stars*)*. 2022.

Kindryd, Andreea. *From Slavery to the Stars: A Personal Journey*. 2022.

Kmet, Michael. "Script Clearance and Research: Unacknowledged Creative Labor in the Film and Television Industry." *Mediascape: UCLA's Journal of Cinema and Media Studies*, August 28, 2012.

BIBLIOGRAPHY

Kmet, Michael. "The Truth about Star Trek and the Ratings." *Star Trek Fact Check*, July 5, 2014. https://startrekfactcheck.blogspot.com/2014/07/the-truth-about-star-trek-and-ratings.html.

Konecci, Eugene B. and Neal E. Wood, "Design of an Operational Ecological System." *Proceedings of the Manned Space Stations Symposium, Los Angeles, California, April 20–22, 1960*. Institute of the Aeronautical Sciences, 1960.

Kraft, Bill. *Maybe We Need a Letter from God: The Star Trek Stamp*. Gorham, 2013.

Lee, R.A. "Value of Studio Libraries: Closings Called False Economy." *Weekly Variety* 260, no. 8, (January 3, 1973).

Liebermann, Randy. "Frederick C. Durant (1916–2015)." *Space Review*, November 2, 2015. https://www.thespacereview.com/article/2856/1.

Leslie, Stewart W. "Spaces of the Space Age." *Smithsonian Air & Space Magazine*, August 2013. https://www.smithsonianmag.com/air-space-magazine/spaces-of-the-space-age-332231/.

Lichtenberg, Jacqueline, Sondra Marshak, and Joan Winston. *Star Trek Lives!* Corgi, 1975.

Lohr, Steve. "Promoters Hitchhike on *Star Trek*." *Press and Sun-Bulletin*, January 27, 1976, 46.

McCullars, William S. "Starship '72: The Forgotten Detour of the NCC-1702." *Star Trek Communicator: The Magazine of the Official Star Trek Club*, no. 120 (December 1998/January 1999).

McDonald, N. Datin, and Richard C. Datin, Jr. *The Enterprise NCC 1701 and the Model Maker*. CreateSpace, 2016.

Meares, Hadley. "How the Aviation Industry Shaped Los Angeles," *Curbed Los Angeles*, July 8, 2019. https://la.curbed.com/2019/7/8/20684245/aerospace-southern-california-history-documentary-blue-sky.

Newell, Catherine L. *Destined for the Stars: Faith, The Future, and America's Final Frontier*. University of Pittsburgh Press, 2019.

Newitt, Paul M. *StarFleet Assembly Manuals*. CultTVman, 2004.

Okuda, Michael, and Denise Okuda. *The Star Trek Encyclopedia: A Reference Guide to the Future. 2 vols. Rev. and exp. ed.* Harper, 2016.

Orwig, Gail and Raymond Orwig. *Fantastic Serial Sites of California: Science Fiction, Horror and Fantasy Locations, 1919–1955*. McFarland, 2022.

Orwig, Gail and Raymond Orwig. *Where Monsters Walked: California Locations of Science Fiction, Fantasy and Horror Films, 1925–1965*. McFarland, 2018.

Pack, Harvey. "Producer of *Star Trek* Struggling to Keep Space Ship From Crashing." *Wichita Eagle*, November 26, 1967.

Parkin, Lance. *The Impossible Has Happened: The Life and Work of Gene Roddenberry*. London: Aurum, 2016.

Parrish. Wayne R. "A New Age Unfolds." *Missiles and Rockets* 1, no. 1 (October 1956).

Pearson, Roberta and Máire Messenger Davies. *Star Trek and American Television*. University of California Press, 2014.

Penley, Constance. *NASA/Trek: Popular Science and Sex in America*. New Left Books, 1997.

Peterson, Bettelou. "Bettelou Peterson Answers your TV Questions." *Detroit Free Press*, January 18, 1970.

Peterson, Karen and Fan Service. "Artist Creates 'Life' Forms." *Lima* News, January 30, 1976, 7.

Poe, Stephen Edward. *A Vision of the Future: Star Trek Voyager*. Pocket Books, 1998.

Porter, Jennifer E. and Darcee L. McLaren, ed. *Star Trek and Sacred Ground: Explorations of Star Trek, Religion, and American Culture*. State University of New York Press, 1999.

Prelinger, Megan. *Another Science Fiction: Advertising The Space Race 1957–1962*. Blast Books, 2010.

Rabitsch, Stefan. *Star Trek and the British Age of Sale: The Maritime Influence throughout the Series and Films*. McFarland, 2018.

Reagin, Nancy R., ed. *Star Trek and History*. John Wiley & Sons, Inc., 2013.

Rioux, Terry Lee. *From Sawdust to Stardust: The Biography of DeForest Kelley, Star Trek's Dr. McCoy*. Pocket Books, 2005.

Roberts, Robin. *Sexual Generations: "Star Trek: The Next Generation" and Gender*. University of Illinois Press, 1999.

Roddenberry, Gene. *Star Trek is…* First draft, March 11, 1964.

Roddenberry, Gene. *The Star Trek Writers Guide*. 3rd rev., April 17, 1967.

Rohan, Barry. "It's a Model Company, The No. 1 Vehicle Producer." *Detroit Free Press*, March 19, 1978. Labor and Urban Affairs Archives, 512, 48–27 #824, AMT Corp., 1978, Walter P. Reuther Library, Wayne State University.

Root, M. W., "Structural Concept Satellite Space Station." Douglas Aircraft Company, Inc., SM-35661, March 20, 1959; also "Structure Design Criteria for Manned Satellites." Douglas Engineer Paper No. 908.

Root, M. W. "Structure Design Criteria for Manned Satellites." Douglas Engineer Paper 908.

Roughton, Randy, "Squadron 19 Returns to the 'Starship,'" United States Air Force Academy Strategic Communications. https://www.usafa.edu/squadron-19-returns-to-the-starship/?fbclid=IwAR3TmTtDlBV47F-YRZVaFQXtyatpBCdwK3Mz3JRQjJ5fqZj0VAzoRcHR-Z4.

Sackett, Susan. *Letters to Star Trek*. Ballantine Books, 1977.

Scheimer, Lou. *Creating the Filmation Generation*. TwoMorrows, 2012.

Schmitt, Joan. "Male Call." *Los Angeles Citizen News*, January 30, 1965.

Seaforces.org. "CVN-65 USS Enterprise History 1961–2012." https://www.seaforces.org/usnships/cvn/CVN-65-USS-Enterprise-history.htm.

Shales, Tom. "What's the Smithsonian Trying to Tell Us?" *Honolulu Advertiser*, November 1, 1974, 63.

Shuit, Doug. "*Star Trek*: Still Luring a Galaxy of Aficionados." *Los Angeles Times*, June 27, 1972.

Sisk, John. "The Six-Gun Galahad." *Time*, March 30, 1959.

Skorpus, Susan. " 'Vision' Quest: Reno Writer Publishes Second Behind the Scenes 'Star Trek Book.'" *Reno Gazette*, March 8, 1998.

Skotak, Robert. *Ib Melchior: Man of Imagination*. Midnight Marquee, 2000.

Sloman, Ernest. "Looking for a Pearl Peeler?" *TV Guide* 13, no. 32 (August 7, 1965): Issue 646.

Smithsonian Institution Archives. National Air and Space Museum, Office of the Director, NASM Files, 1971–1978 and undated, Record Unit 000306.

Smithsonian Institution Archives. National Air and Space Museum, Dept. of Space Science and Exploration, Subject Files, c. 1960–1986, Record Unit 000348.

Smithsonian Institution Archives. National Air and Space Museum, Department of Astronautics Correspondence, 1965–1984, with materials dating from c. 1953. Record Unit 000398.

Solow, Herbert F. and Robert H. Justman. *Inside Star Trek: The Real Story*. Pocket Books, 1996.

Solow, Herbert F. and Yvonne Fern Solow. *Star Trek Sketchbook: The Original Series*. Pocket Books, 1997.

Sullivan, Joseph T. "Science Drama to Debut Sept. 15." *Boston Herald*.

Swanson, Glen E. "To Boldly Go Where No *Model* Has Gone Before… The USS *Enterprise*, *Star Trek* and the AMT Corporation." *Michigan History* 105, no. 1 (January/February 2021).

Swanson, Glen E. "How *Star Trek* Helped NASA Dream Big and How NASA Helped *Star Trek* Stick Around." *Air & Space/Smithsonian* 35, no. 7 (February/March 2021). https://www.smithsonianmag.com/air-space-magazine/how-star-trek-helped-nasa-dream-big-180976753/

BIBLIOGRAPHY

Swanson, Glen E. "The Making of an Enterprise: How NASA, the Smithsonian and the Aerospace Industry Helped Create *Star Trek*." *Space Review*, September 7, 2021. https://www.thespacereview.com/article/4240/1.

Swanson, Glen E. "In Memoriam: Kellam de Forest, Who Gave Us Stardates and the Gorn." *The Space Review*, January 25, 2021. https://www.thespacereview.com/article/4110/1.

Swanson, Glen E. "The New Frontier: Religion in America's National Space Rhetoric of the Cold War Era." In *The Mutual Influence of Religion and Science in the Human Understanding and Exploration of Outer Space*. Edited by Deana L. Weibel and Glen E. Swanson, 19–36. MDPI, 2021.

Swanson, Glen E. " 'Space, the Final Frontier:' *Star Trek* and the National Space Rhetoric of Eisenhower, Kennedy, and NASA. *Space Review*, April 20, 2020. https://www.thespacereview.com/article/3923/1

Swanson, Glen E., ed. *"Before This Decade Is Out…" Personal Reflections on the Apollo Program*. NASA SP-4223. History Office, Office of Policy and Plans, NASA, 1999.

Thomas, Bob, " 'Eggheads' Like TV's *Star Trek*," *Lancaster New Era*, May 6, 1967, 9.

Tilotta, David and Curt McAloney. "Promoting TOS," *Lost Star Trek History*, May 10, 2016. https://www.startrek.com/article/promoting-tos.

Tilotta, David and Curt McAloney. *Star Trek: The Lost Scenes*. Titan Books, 2018.

Tipton, Scott. *Star Trek Vault: 40 Years from the Archives*. Abrams, 2011.

Torgeson, Ellen "*Star Trek* Is Science Fact, Says Its Creator." *TV Times*, November 16, 1967.

Trimble, Bjo. *On The Good Ship Enterprise: My 15 Years with Star Trek*. Donning Books, 1982.

Van Hise, James. *The Man Who Created Star Trek: Gene Roddenberry*. Pioneer Books, 1992.

Van Treuren, Richard G. "*Star Trek* Miniatures Part Two: The Other Space Ships," *Trek: The Magazine For Star Trek Fans*, 6 (November 1976).

Webb Jr., Alvin B. "Some Spacemen May Be Dentists of Sorts." *Dental Times*, March 15, 1966.

Weitekamp, Margaret A. *Space Craze: America's Enduring Fascination with Real and Imagined Spacecraft*. Smithsonian Books, 2022.

Weitekamp, Margaret A. "Two Enterprise: Star Trek's Iconic Starship as Studio Model and Celebrity." *Journal of Popular Film and Television*, 44 (1), 2–13. https://doi.org/10.1080/01956051.2015.1075955.

Wells, Helen T., Susan Whiteley, and Carrie E. Karegeannes. *Origins of NASA Names*. NASA History Series NASA SP-4402. NASA Scientific and Technical Information Office, 1976.

Westwick, Peter J., ed. *Blue Sky Metropolis: The Aerospace Century in Southern California*. University of California Press, 2012.

Whitfield, Stephen E., and Gene Roddenberry. *The Making of Star Trek*. Ballantine Books, 1968.

Witbeck, Charles, "TV Topics: New Series Pits 400-Man Craft Against Space," *Buffalo Evening News*, September 3, 1966.

Wright, James W. "TV's *Star Trek*: How They Mix Science Fact with Fiction," *Popular Science* 191, no. 6 (December 1967).

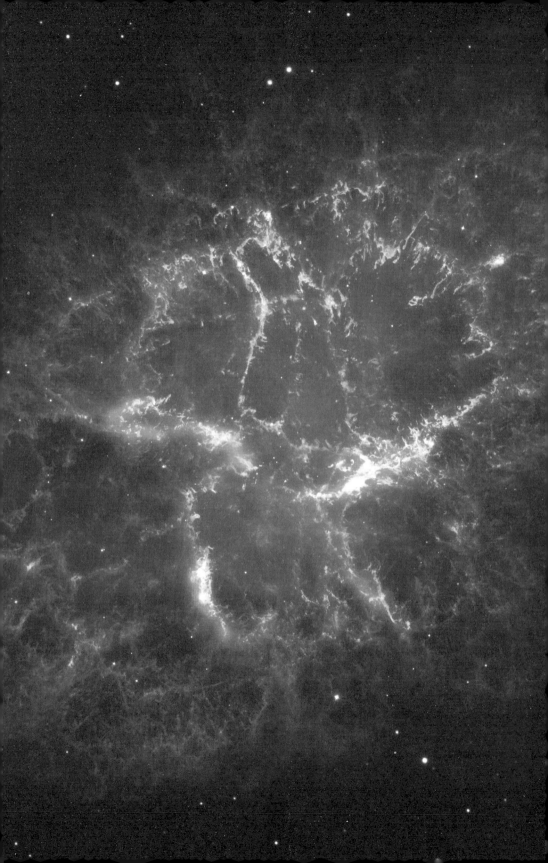

INDEX

ABC, 113
Abrams, J. J., 116
Academy Cadet Squadron 19, US Air Force, 85–86
Academy of Television Arts & Sciences, 202n55
aerospace community, 167
Aerospace Corporation, 95
Aerospace Division, of General Electric, 102
aerospace industry, 98
    Draper relation to, 177n59
    Joseph, F., relation to, 140
    Roddenberry relation to, 96–97, 101–102
    in Southern California, 95–96, 104–105, 111
Aerospace Pilot Research School, US Air Force, 85
The African Queen (movie), 44
Agee, James, 44
Air & Space (magazine), 14
Akimoto, George, 107–109, 194n36
Albert C. Martin & Associates, 105
Aldrin, Edwin "Buzz," 113, 160
Alexander, David, 114
Alexiou, Art, 198n24
Allen, Irwin, 196n4
Allen, Paul, 76, 186n79b
"The Alternative Factor," 98
Altman, Kenneth, 146
Aluminum Model Toys Corporation (AMT), 58
    Desilu Studios relation to, 59, 61–62
    Roddenberry relation to, 68
    Space Cadet relation to, 74–75
    Whitfield, Stephen relation to, 56, 57, 60–63, 69–70, 72
"America in Space," 148
American Cinematographer (magazine), 74, 186n72
American Historical Association, at World's Columbian Exposition, 88
American Institute of Aeronautics and Astronautics, 107, 131
American Institute of Architects, 105
American Institute of Chemical Engineers, 31
American Rocket Society, 131
American Society of Chemical Engineers, 31
AMT. *See* Aluminum Model Toys Corporation
AMT model kits, 13, 14
    of *Constellation*, USS, 65–66
    of *Enterprise*, USS, 55, 58, 60–61, 66, 68–69, 70–71, 76–77, 132, 183n24, 183n33, 185n71

of Klingon ship, 61–65, 66–67, 70, 184n36, 184n38
"Leif Ericson," 75
Analog (magazine), 73–74, 185n71
analog computers, 99
anamorphic format, 119
Anderson, Darrell, 117
Anderson, Howard, 66–67
Andromeda galaxy, 97
The Andy Griffith Show (television show), 121
"Another Frontier" (Fulton), 92
Apollo 1, 123–124, 212n62
Apollo 4, 199n28
Apollo 7, 114
Apollo 8, 212n62
Apollo 11, 133, 153, 160
Apollo 16, 148
Apollo Guidance Computer, 177n59
Apollo program, 96, 170
Apollo-Soyuz Test Project (ASTP), 212n62
Apollo Spacecraft Program Office (ASPO), 212n62
architecture, 104–107
"Arena," 48
Armstrong, Neil, 113, 116, 160
Army Air Corps, 31, 82
Arnaz, Desi, 160
Arts and Industries Building, of Smithsonian Institute, 10, 134, 137–138
Asimov, Isaac, 202n55
aspect ratio, 119
ASPO. *See* Apollo Spacecraft Program Office
"Assignment: Earth," 118, 122, 199n25, 199nn29–30
    Garr in, 201n39c
    Jefferies relation to, 119–120
    Roddenberry relation to, 120, 121
ASTP. *See* Apollo-Soyuz Test Project
Astrobotic, 221n28
Astro-Electronics Division, RCA, 97–98
Astrogator, 100, 192n16
astrostomatology, 84
Atlas-Agena, 120, 200n32
atomic bomb, 26
Aviation Week (trade publication), 87

B-1 flight computer, 99
Bacall, Lauren, 44
Baker, Norman L., 175n38b
"Balance of Terror," 28, 33, 65, 81
Ball, Lucille, 160
Ballantine Books, 55, 142, 143–144, 146
ballistic missiles, 89
Barrett, Majel, 72, 125, 165
Battlestar Galactica (television show), 146

B. Dalton's, 143
Beamish, Christopher, 188n11
Bell Aerosystems Company, 176n38c, 196n49c
Bellows Field, Hawaii, 29
Bernardi, Daniel, 11
Berry, Charles, 114
Beverly Crusher (fictional character), 34
Bewitched (television show), 140
Bezos, Jeff, 170
Black, John D., 158
Blish, James, 55, 81–82
Block, Richard, 160–161
"Blooper Reel," 83
blueprints, for *Enterprise*, USS filming model, 55, 139, 141
Blue Sky Metropolis (documentary), 95
Bluford, Guion, 164
Bogart, Humphrey, 44
Bolden, Charles, 165
Bonanza (television show), 199n25
Bonestell, Chesley, 131
Boneufant, Helene Marie, 189n24
Booklet of General Plans (Joseph, F.), 141
Boston Herald (newspaper), 50
Botany Bay, 110–111
Botany Bay model, 15
Braun, Wernher von, 131, 176n38c, 180n11a
"Bread and Circuses," 174n13
Broadcasting (magazine), 162
Brown, Frederic, 48
Buchanan, Robert S., 115
"buddy-care" kit, 84–85, 190n31
Buffalo Evening News, 158
Bureau of Naval Personnel, 82–83
Burgess, Eric, 210n35
Burton, Lillian K., 194n31
Burton, Mortimer C., 194n31

Caan, James, 194n37
"The Cage," 28, 34, 97–98, 104, 204n1
    NASA relation to, 117
    polar grids in, 100
California Institute of Technology (Caltech), 36, 95–96, 102
Campbell, John W., 185n71
Camp Pendleton, 200n39a
Cannon, James L., 164
Captain Horatio Hornblower (movie), 80
Cernan, Gene, 170
Carpenter, Scott, 116
Carson, Johnny, 190n31
"Catspaw," 209n29a
CB. *See* citizen band
CBS, 113, 196n2
Celestis, 165, 221n28

Cert-1, 221n28
CFI. *See* Consolidated Film Industries
Chaffee, Roger, 123
Chaikin, Andrew, 170
Challenger accident, 164
Chemical Warfare Service Company "B," 31
Child, Julia, 137
children, 33–34
Christies Film and Television Auction, 186n79b
Christopher "Kit" Draper (fictional character), 177n59
Christopher Pike (fictional character), 28, 174n14
citizen band (CB), 163, 220n20
"City on the Edge of Forever," 200n36
Clark, John F., 122
Clarke, Arthur C., 131, 132, 148, 149, 170
Class M planets, 27–28, 38
Close Encounters of the Third Kind (movie), 14, 65, 168
"The Cloud Minders," 118
Cobun, Peter, 220n25
Cocconi, Giuseppe, 24
Cohn, Philip, 202n55
"Cold Hands, Warm Heart," 194n37
Cold War, 87, 131
Cole, Stephen V., 145–146, 214n78
Collins, Michael, 133–134, 147, 160, 205n2, 210n35
Columbia space shuttle, 153, 165
Commander Riker (fictional character), 34
commercials, 161
Connor, James, 164
Consolidated Film Industries (CFI), 82–83
Consolidated Vultee Aircraft, 139
Consolmagno, Guy, 101
Constellation, USS, AMT model kits of, 65–66
"Constitution Class," 187n9
Convair Aerospace, 180n11a
"Convoy," 220n20
Cook, James, 92
Coon, Gene, 48
Copernicus symposium, 209n29a
"The Corbomite Maneuver," 68, 98, 99
Countdown (magazine), 13
"Court Martial," 79, 117, 198n22
Craft, Model, Hobby Industry Magazine, 180n11b
Creifelds, Judith, 72
Crockett, Davy, 87
Crouch, Tom, 138–139, 152
CSG. *See* Slide Graphic Flight Computer
CSG-1A, 99
Cunningham, Walter, 114
curators, 9, 11
Cushman, Marc, 196n1
Cygnus constellation, 98

Daily Variety (magazine), 37, 124–125, 200n39a, 202n55
Dalzell, Bonnie, 138
Daniels, Marc, 115, 119–120
The Danny Thomas Show (television show), 121
Datin, Richard, 109–110, 169
Daugherty, Herschel, 106
Davidson, Karla, 205n2

Day, Dwayne, 91–92, 110
Deatrick, Eugene P., 85
Defense and Space Group Campus, TRW, 106
Defense Space Business Daily (newsletter), 175n38b
Defoe, Daniel, 103
Del Rey, Judy-Lynn, 142, 145
Del Rey Books, 145
Democratic National Convention, 89
Dental Times (journal), 84
Department of State, 133
Desilu Studios, 30, 35, 37, 46, 51, 129
AMT relation to, 59, 61–62
Emmy awards and, 124, 202n55
MSEI relation to, 100
Paramount Pictures relation to, 160
Detroit Free Press (newspaper), 74
Diaz, Franklin Chang, 76
die-cast metal miniatures, 217n105
Disney, 45
Distinguished Public Service Medal, 165
Dole, Stephen H., 26, 27, 37, 174n15
Doohan, James, 46, 115, 126, 165
"The Doomsday Machine," 65–66
Dornberger, Walter, 176n38c
Doster, Alexis "Dusty," III, 138
Douglas, Michael, 126
Douglas Aircraft Company, 26, 96, 194n36
K-7 Space Station relation to, 107, 108–109
Missiles and Space Systems Engineering Department, 195n42
Drake, Frank, 24–25
Drake equation, 25, 26, 38
Draper, Charles Stark, 177n59
Draper Labs, 177n59
Drexler, Doug, 67–68, 146
Dr. Leonard "Bones" McCoy (fictional character), 28, 84–85, 189n29
Dryden Flight Research Center, NASA, 115–116
Duke, Charlie, 148
Dulles International Airport, 152–153
Dumbell Nebula, 98
Durant, Frederick C., III, 67, 130–131, 136, 146, 152, 169
on *Enterprise*, USS filming model, 150
Joseph, F., relation to, 141, 142–143
Lawrence, S., relation to, 207n27
Roddenberry relation to, 132–133, 134–135, 147, 209n29a, 221n27
Durant, Steve, 132, 154

Eames, Charles, 137
Eames, Ray, 137
Earth II (television pilot), 148, 216n96
Earth Photographs from Gemini III, IV, and V, 117–118
eBay, 169
Ehricke, Krafft, 148, 180n11a
Eisele, Donn, 114
Eisenhower, Dwight D., 91–92
"Elaan of Troyius," 65
Emery Freight, 147

Emmy awards, 124, 162, 186n72, 202n55
The Enemy Below (movie), 81
"The Enemy Within," 59
Enterprise, USS (aircraft carrier), 188n17
Enterprise, USS (starship), 73, 82, 138–139, 141–143, 185n55. *See also* filming model
AMT model kits of, 55, 58, 60–61, 66, 68–69, 70–71, 76–77, 132, 183n24, 183n33, 185n71
die-cast metal miniatures of, 217n105
de Forest, K., relation to, 46–47
Jefferies relation to, 110–111
Lynn, H., Jr., relation to, 33–35, 37
in *The Making of Star Trek*, 80–81
at San Francisco Navy Yards, 187n9
"The Enterprise Incident," 65, 66
The Enterprise NCC 1701 and the Model Maker (McDonald), 109
"Enterprise Station," 221n28
Equicon, 141, 142, 207n27
"Errand of Mercy," 104
Ertl, 151
Excelsior, USS, 217n105
Explorer I, 26, 131

Fahey, John E., 80
Falcon 9, 200n32
fandom, 11
Farnon, Robert, 80
"The Federation Trading Post," 67, 146
Field Command Defense Atomic Support Agency, US Air Force, 29
Filmation, 134
Filmation Studios, 162
filming model, *Enterprise*, USS, 9–10, 146–149, 151–154, 164, 211n46, 218n107
blueprints of, 139, 141
on eBay, 169
Roddenberry and, 150, 212n54
Steidinger and, 216n96
Thompson and, 215n94
Weitekamp and, 214n86
film trims, 161, 188n11
First Men on the Moon (Wells), 209n29a
Fitzsimmons, C. John, 65
500F, 199n29
Fletcher, James, 163–164
floaty pen, 150–151, 217n104
Fontana, Dorothy Catherine, 49, 62, 64, 65, 67, 111
Forbidden Planet (movie), 102–103
Ford, Betty, 163, 202n55, 220n20
Ford, Gerald R., 163–164, 220n25
de Forest, Ann, 43–44, 46
de Forest, Kellam, 14, 43–45, 169
Paramount Pictures relation to, 51–52
Roddenberry relation to, 46–47, 48–49, 50–51
de Forest, Lee, 43
De Forest Research Service, 43, 45–46, 51–52, 120, 121

INDEX

Forrester, C. S., 80
"For the World Is Hollow and I Have Touched the Sky," 49
Founders Flight, 165
The French Chef (television show), 137
frontier, 87–89, 92, 93
"frontier thesis," 88
Fulton, Valerie, 92
"The Future of Earth and Mankind," 33
"Future's End," 38

Gagarin, Yuri, 90
Galileo Seven, 59, 61, 75, 134
Garr, Teri, 201n39c
Gary Seven (fictional character), 118, 119, 200n34, 200n36
Geiger counters, 98
Gemini 4, 118
Gemini 11, 113, 196n2
"Gemini 12," 196n4
General Electric, Aerospace Division of, 102
generation ship, 33
Genie, 181n11c
Gerrold, David, 50, 159, 185n55, 209n29a
Get Smart (television show), 83
Gill, Jocelyn R., 198n24
Gilruth, Robert R., 198n24
Goddard, Esther C. Kisk, 123
Goddard, Robert H., 33, 122, 153
Goddard Memorial Dinner, 122–124, 125–127, 129–130, 132, 202n55
Goddard Space Flight Center, 122–123, 125
Golden West College, 147–148, 214n86, 215n94, 216n96
Goldin, Dan, 165
Gomer Pyle, USMC (television show), 121
Googie, 104–105, 194n31
Gorog, William, 163
Graham, Thomas, 181n11c
Gregorian calendar, 47
Gregory, Frederick D., 165
Greg Jein Collection, Heritage Auction, 14–15
Griffith Park Observatory and Planetarium, 97
Grissom, Gus, 123
Gulf and Western, 51
Gulf+Western, 160
Guzman, Pato, 103

Habitable Planets for Man (Dole), 27, 174n15
Haines, Richard, 194n36
handheld calculating tools, 99–100
Hartford Courant (newspaper), 159
Hartley, Jack L., 83–85, 189n24, 190n31, 202n55
Hartley, Le Roy Poston, 189n24
Hartley, Patricia, 84–85
Harvey (play), 56
Harwit, Martin, 217n105
Haskell, Douglas, 194n31
Haskin, Byron, 104
Hawaii, Bellows Field, 29
Hayden Planetarium, 151
Heinlein, Robert A., 50
Heisman, Roger, 207n27
Henderson, Mary S., 217n105
Heritage Auction, Gren Jein Collection, 14–15
"Heritage Edition" kit, 181n11e

HIAA. *See* Hobby Industry Association of America
History of Rocketry and Space Travel Gallery, 150–151
Hoagland, Richard, 163, 202n55
Hobby Industry Association of America (HIAA), 182n12b
Hood, Bob, 122
Hood, James, 69
Hood, Robert H., Jr., 202n42
Horatio Hornblower (fictional character), 80
Hubble Space Telescope, 138
Human Engineering Group, 26–27
Humphrey, Hubert, 123–124, 125, 129
Hunter, Jeffrey, 28–29, 174n14
Huston, John, 44

IBM, 209n32
individualism, 88, 93
Inside Star Trek (newsletter), 64–65, 111
Inside Star Trek (Solow and Justman), 71, 159
International Astronautical Federation, 131
International Space Station, 170
Introduction to Outer Space, 91–92
Irvine, Mat, 180n11a

James T. Kirk (fictional character), 80, 81, 87, 118, 190n38
  in "Court Martial," 198n22
  "new frontier" relation to, 90
  in "Tomorrow Is Yesterday," 82
  in "Where No Man Has Gone Before," 129
Jaren, Ann, 57
Jason, Ben, 124–125
Jean-Luc Picard (fictional character), 92
Jefferies, Walter Matthews "Matt," Jr., 35, 47, 59, 68, 110–111, 209n29a
  "Assignment: Earth" relation to, 119–120
  at Dryden Flight Research Center, 115
  Klingon ship relation to, 62–64, 66–67
  Roddenberry relation to, 85, 101–102
  Space Cadet relation to, 74–75
  in World War II, 82
Jein, Greg, 14–15, 65–66
Jenkins, Dennis, 220n25
Jet Propulsion Laboratory
  Caltech, 95–96, 102
  NASA, 113
Johnson, Peggy L., 74
Johnson Space Center, 13
Johnston, S. Paul, 130, 132–133, 134, 206n20a
Jones, David R., 30
Jones, Peter, 95
Joseph, Franz, 139, 142, 151, 209n29a
  Roddenberry relation to, 140–141
  Star Fleet Technical Manual, 55, 143–146
Joseph, Karen, 139–140, 144–145
Journal of Popular Film and Television, 214n86
Julian calendar, 47–48
"Jupiter 2," 196n4
Jupiter C, 181n11e

Justman, Robert "Bob" Harris, 61, 68, 159
  at Dryden Flight Research Center, 115
  Inside Star Trek of, 71, 159
  US Navy relation to, 82, 188n15
Jutland (game), 214n78

K-7 Space Station
  Datin relation to, 109–110
  Douglas Aircraft Company relation to, 107, 108–109
  Prelinger and, 195n39
Kaiser, Henry K., 160
Kaiser Broadcasting Corporation, 160–161
Katz, Marvin, 63, 204n1
Katz, Oscar, 103
KBHK station, 161–162
Kellam de Forest Collection, 52
Kelley, DeForest, 85, 114
  Celestis relation to, 165
  at Dryden Flight Research Center, 115–116
Kellogg, Thomas W., 60
Kelly Field, 29
Kennedy, Greg, 209n32
Kennedy, John F., 89–90, 92
Killian, James R., 91–92
Kimble, David, 146
Kirby, Russell A., 136
Kirkland Air Force Base, 29
Kitty Hawk, USS, 83
Klein, Paul L., 125
Kline, Bob, 186n77
Klingon Bird of Prey, 151, 217n105
Klingon ship, 61–65, 66–68, 70, 73, 184n36, 184n38
Kmet, Michael, 45, 204n1
"Know the Missiles" poster, 181n11e
Konecci, Eugene B., 195n42
Kubrick, Stanley, 170

Lansing, Robert, 118
LASER. *See* Light Amplification Stimulation Emission Radiation
Laslie, Brian, 86
Laugh-In (television show), 83
Launch Complex 13, 200n32
Launch Control Center, 119
Lautner, John, 194n31
Lawrence, Dick, 147
Lawrence, Stephanie F., 207n27
LAX. *See* Los Angeles International Airport
"Leif Ericson," 75
Lesney Corporation, 75
Letters to Star Trek (Sackett), 209n29a
letter-writing campaign
  to NASA, 163, 221n27
  to NBC, 159–160, 219n10
Ley, Willy, 180n11a, 181n11d
Library of Congress, 205n2
licensed products, 13, 55–56
Licensing Corporation of America, 58
licensing fees, 157
Lieberman, Randy, 132
The Lieutenant (television show), 200n39a
Life (magazine), 180n11a
Life in the Universe? exhibit, 10, 137, 172, 211n46
Mariner 10 in, 197n6
  at NASM, 138–143, 151
  Paramount Pictures relation to, 149–150

Roddenberry relation to, 142
Light Amplification Stimulation Emission Radiation (LASER), 35
Ligon, Elvin S. Jr., 181n12a
Lil Dan'l (Akimoto), 107–108
Lincoln Enterprises, 14, 66
   film trims of, 161, 188n11
   Joseph, F., relation to, 140–141
Lindberg, Charles, 132
Lockheed, 95–96
Lockwood, Gary, 200n39a
long-duration interstellar voyages, 33
Los Angeles
   Griffith Park Observatory and Planetarium in, 97
   space race relation to, 96
   University of California, 36, 47, 52, 107, 194n36
Los Angeles Citizen News, 28
Los Angeles Conservancy, 105
Los Angeles International Airport (LAX), 147
Low, George, 142, 164, 212n62
Low, Mary, 142
Lowman, Paul, 198n24
Lucas, George, 9
Lucasfilm, 217n105
Lunar Orbiter missions, 200n32
Lunar Prospector mission, 221n28
Luna Tranquility, 221n28
Lundin, Victor, 104
Lynn, Dennis, 31–32, 36, 39
Lynn, Harvey Prendergast., III, 31, 39
Lynn, Harvey Prendergast., Jr., 14, 30–31, 43, 51, 52, 168
   "The Cage" relation to, 34
   Roddenberry relation to, 32–33, 35–36, 37, 40
   "space shuttle" and, 175n36b
   Star Trek: The Animated Series relation to, 39

M2-F2, 115, 197n13
MacCormick, Margaret, 44
Maddalena, Joseph, 186n79b
Mailcall (pamphlet), 159
The Making of Star Trek (Whitfield, Stephen), 13, 14, 23, 55–56, 73–74, 76
   "The Cage" in, 204n1
   Drake equation in, 25
   Enterprise, USS in, 80–81
   Joseph, F., relation to, 140
   medikit in, 189n29
   Roddenberry in, 38
   Smithsonian Institute relation to, 135–136
   Starfleet in, 187n1
   Star Trek Writers/Directors Guide relation to, 71–72
Malkin, Myron, 164
Mandel, Geoffrey, 146
Manhattan Project, 26
manifest destiny, 87
Manned Spacecraft Center, NASA, 114, 198n24
Manned Space Stations Symposium, 109, 110
A Man on the Moon (Chaikin), 170
Mantee, Paul, 104, 177n59
"The Man Trap," 196n1
Marine Corps, 200n39a
Mariner 2, 197n6
Mariner 5, 113–114, 197n6
Mariner 10, 197n6

Marrow, Byron, 49
Marsh, Jack, 164
Marshall, Don, 200n39a
MASER. *See* Microwave Amplification Stimulation Emission Radiation
Massachusetts Institute of Technology (MIT), 177n59
Mattingly, Ken, 148
May, Leonie, 30–31
McAloney, Curt, 188n11, 192n17
McCall, C. W., 220n20
McCall, Robert, 151, 154, 205n2
McClay, Howard, 185n65
McCullars, William S., 214n86
McDivitt, Jim, 118
McDonald, Noel Datin, 109
McDonnell Douglas Missile Systems, 122, 202n42
McFadden, Strauss, Eddy, and Irwin (MSEI), 100–101, 192n17
McKinley Rocket Base, 119–120
McMaster, Michael, 146
McNair, Ronald, 164
medikit, 189n29
Melchior, Ib, 103–104
"The Menagerie," 104, 145, 187n1
meteorites, 100–101
MGM, 45, 205n2
Miarecki, Ed, 151
Michigan History (magazine), 14
Microwave Amplification Stimulation Emission Radiation (MASER), 35
military industrial complex, 167
Milkis, Edward, 120
Miller, Stanley L., 137
Mindling, Lou, 142
"Miri," 49–50, 174n13
"Mirror, Mirror," 174n13
Missiles and Rockets (trade publication), 87
Missiles and Space Systems Engineering Department, Douglas Aircraft Company, 195n42
Mission: Impossible (television show), 124, 202n55
MIT. *See* Massachusetts Institute of Technology
Mitchell, Roger, 70–71
MIT Instrumentation Laboratory, 177n59
Model 2586 "Cutie Pie" Radiation Survey Instrument, 98
Monogram, 180nn11a–11b
   Ley relation to, 181n11d
   "US Space Missiles" of, 181n11e
MoPOP. *See* Museum of Pop Culture
Moran, Alberta, 122–123, 125, 126, 202n55
Moran, Pamela, 122–123
Moran, Penny, 124, 125–126
Moreno Hernández, José, 76–77
Moroz, Anne, 123
Morrison, Philip, 24
Movin' On (television show), 220n20
Mr. Spock (fictional character), 114, 118, 123–124, 188n12a
MSEI. *See* McFadden, Strauss, Eddy, and Irwin
"Mudd's Women," 79, 187n7
Mueller, John, 183n24
The Munsters (television show), 58
Muscular Dystrophy Association, 57
Museum of Pop Culture (MoPOP), 207n27

NAACP. *See* National Association for the Advancement of Colored People
Nagler, Ken, 198n24
"The Naked Time," 82, 99
NARA. *See* National Archives and Records Administration
NASA. *See* National Aeronautics and Space Administration
NASM. *See* National Air and Space Museum
National Academy of Sciences, 25
National Aeronautics and Space Administration (NASA), 91, 92, 96, 108, 168
   "Assignment: Earth" relation to, 118–121
   Distinguished Public Service Medal of, 165
   Dryden Flight Research Center, 115–116
   Johnson Space Center of, 13
   letter-writing campaign to, 163, 221n27
   Manned Spacecraft Center, 114, 198n24
   "new frontier" relation to, 90
   Roddenberry relation to, 101–102, 113, 116–118, 122, 126–127
National Air and Space Museum (NASM), 67–68, 69, 129, 138–143. *See also Life in the Universe?* exhibit
   Collins at, 133–134
   Durant, F., at, 131–133, 135, 136
   Enterprise, USS filming model at, 9–10, 150, 151, 154, 164, 169, 214n86, 217n104
   Johnston and, 206n20a
   Preston, R., Sr., at, 130
   Roddenberry at, 137
   "Star Trek: A Retrospective Exhibition", 217n105
   "Where No Man Has Gone Before" and, 47
National Archives and Records Administration (NARA), 188n17
National Association for the Advancement of Colored People (NAACP), 200n39a
nationalism, 87
National Museum of American History, 69
National Research Council, 101–102
National Rocket Club, 122
National Safety Council, 44
National Security Agency (NSA), 118, 200n36
National Space Club, 122, 125, 129
Nation of Speed (gallery), 69
"natural evolution," 174n13
Naval Air Development Center, 85
NBC, 13, 113, 125
   "Assignment: Earth" relation to, 121
   letter-writing campaign to, 159–160, 219n10
   The Lieutenant relation to, 200n39a
   RCA relation to, 126
   Roddenberry relation to, 35
   "Where No Man Has Gone

INDEX

Before" relation to, 129
Necha Butane Products, 31
network technology, 157
"new frontier," 89–90
Newgard, Bob, 160
Newitt, Paul, 140
New York Daily News, 181n12a
New York Times (newspaper), 45, 144
New York Times Book Review, 143
Nichols, Nichelle, 153, 164
   Celestis relation to, 165
   in *The Lieutenant*, 200n39a
Nimoy, Leonard, 85, 123–124, 125, 127, 129, 132. *See also* Mr. Spock
Nimoy, Sandy, 123
"1964 Plan," 134
Nixon, Richard, 153
"No. 921 Space Ship," 60
North American Advanced Space Research Center, 102
North American Nebula, 98
Northrop, 96–97, 107, 115
NSA. *See* National Security Agency
Nuclear-Chicago, Model 2586 "Cutie Pie" Radiation Survey Instrument of, 98

O'Conner, Jerry, 120
Office of Scientific Research and Development, War Department relation to, 26
Ohio State University, 174n17
Okuda, Michael, 38
"The Omega Glory," 133, 160, 219n10
On the Good Ship Enterprise (Trimble), 159, 207n27
"Operation: Annihilate!," 106–107, 194n37
Operation Paperclip, 180n11a
Orbital Sciences Pegasus XL booster, 165
Orion constellation, 97
Orzechowski, Edward A., 117
The Outer Limits (television show), 194n37
OV-101 *Enterprise*, 164, 220n25, 221nn26–27
OV-102 *Columbia*, 221n26

Pacific Design Center, 186n79b
Pack, Harvey, 116, 198n16
Pal, George, 104, 148, 216n96
Palmer method, 139
"Parallel Worlds" concept, 27–28
Paramount Pictures, 9–10
   Desilu Studios relation to, 160
   Durant, F., relation to, 147
   De Forest Research Service relation to, 45–46, 51–52
   Joseph, F., relation to, 141
   Life in the Universe? exhibit relation to, 149–150
Parker, Ted, 113–114
Partial Test Ban Treaty (1963), 194n35
"Patterns of Force," 174n13
Pearce, Joan, 120, 121
Peenemunde, 176n38c, 180n11a
Pentagon, 82, 174n17, 200n39a
Peregrine lunar lander, 221n28
Perlstein, Ed, 36, 117
Peterson, Bettelou, 74
Peterson, Bruce, 115
Petrotti, Yvonne, 31

Pevney, Joseph "Joe," 115
phaser, 35
Pierce, Fred, 182n12b
pinwheel galaxy M-101, 97
Pioneer 10, 210n35
Pioneer 11, 210n35
The Pirates of Orion (animated series), 146
plastic modeling industry, 180n11a
Pleiades, 98
Poe, Bobbi, 56, 72–73
Poe, Edgar Allan, 56
Poe, Reuben, 56
Poe, Stephen Edward. *See* Whitfield, Stephen
polar grids, 100
Ponnamperuma, Cyril, 137
Popular Science (magazine), 73, 113
"Powers of Ten," 137, 210n36
Precision Radiation Model 111 Scintillator, 98
Preliminary Design of an Experimental World-Circling Spaceship, 26
Prelinger, Megan, 195n39
Presidential Science Advisory Committee, 91
Preston, Bannon, 130, 163
Preston, Larkin, 130
Preston, Richard Knowlton, II, 130, 169
Preston, Richard Knowlton, Sr., 129–130, 132
Prickett, Donald Irwin, 29, 174n17
"Prime Directive," 82, 188n12a
Primordial Soup (short film), 137
Probert, Andy, 146
Project Orion, 174n17
Project RAND, 26–27. *See also* RAND Corporation
Prospectus in the Search for Extraterrestrial Civilizations (Drake), 25
Ptak, Roland Daniel, 57
Ptak & Richter Advertising Agency, 57
publicity, 100–101
Publisher's Weekly (magazine), 73
von Puttkamer, Jesco, 220n25

Quest (journal), 13

"rabbit ears," 157
Rabitsch, Stefan, 187n2
radio frequency (RF), 106
radio waves, 24
Ramo, Simon, 105
RAND Corporation, 28, 96
   Lynn, H., Jr., relation to, 32–33, 36–37, 40
   Roddenberry relation to, 38–39, 174n14
   Whitener, J., and, 29–30
RCA, 97–98, 126
Reaching for the Stars (Hernández), 76
Recommendations to the NASA Regarding a National Civil Space Program, 90–91
"The Recoverable, Reusable Space Shuttle," 176n38c
Remora capsule, 196n49c
The Reporter (magazine), 103
Reposh, Nedra, 143
reruns, 157. *See also* syndication
Research And Development. *See* RAND Corporation
Resnik, Judith, 164
Resolution, 92

Retter, Lynn, 123
"Return of the Archons," 188n12a
Rexroat, Eileen-Anita, 206n20a
Reynolds, John, 62
RF. *See* radio frequency
Rhodes, Terry, 83
Richards, Michael (pseudonym). *See* Fontana, Dorothy Catherine
Ride, Sally, 164
RKO Studio library, 45–46
Robbins, Saul, 182n12b
Robinson Crusoe on Mars (movie), 103–104, 177n59
Rocketry and Spaceflight exhibit, 10
Rockwell, Norman, 131, 132
Roddenberry, Gene, 9, 144, 158, 165, 168
   Academy Cadet Squadron 19 relation to, 85–86
   aerospace industry relation to, 96–97, 101–102
   AMT relation to, 68
   "Assignment: Earth" relation to, 120, 121
   "The Cage" relation to, 34, 204n1
   CFI relation to, 83
   on Class M planets, 38
   Collins relation to, 210n35
   Douglas Aircraft Company relation to, 109
   Drake equation and, 25
   Durant, F., relation to, 132–133, 134–135, 147, 209n29a, 221n27
   Enterprise, USS filming model relation to, 150, 212n54
   on *Enterprise*, USS (aircraft carrier), 188n17
   film trims and, 161
   Forbidden Planet relation to, 102–103
   de Forest, K., relation to, 46–47, 48–49, 50–51
   Hartley, J., relation to, 84
   Horatio Hornblower relation to, 80
   Introduction to Outer Space and, 91–92
   Jason relation to, 124–125
   Jefferies relation to, 85, 101–102
   Jones, D., relation to, 30
   Joseph, F., relation to, 140–141
   Klingon ship relation to, 65
   on letter-writing campaign, 159–160
   licensed products relation to, 55
   The Lieutenant and, 200n39a
   Life in the Universe? exhibit relation to, 142
   Lynn, H., Jr., relation to, 32–33, 35–36, 37, 40
   on Mr. Spock, 114
   NASA relation to, 101–102, 113, 116–118, 122, 126–127
   at NASM, 137
   "new frontier" relation to, 90
   OV-101 *Enterprise* and, 164
   on "Parallel Worlds," 27–28
   Prickett relation to, 29
   "Prime Directive" and, 82
   RAND Corporation relation to, 38–39, 174n14
   Robinson Crusoe on Mars relation to, 103–104

237

science fiction relation to, 167
"Series Format" and, 23–24
at Smithsonian Institute, 152
Starfleet relation to, 187n1
Star Trek: The Next Generation and, 33–34
western television shows relation to, 92–93
at Western Test Range, 125–126
Whitfield, Stephen relation to, 71–73, 76
at WorldCon, 61
Roddenberry Collection, Special Collections Library, UCLA, 36, 47
Rogay, Inc., 149
Rohwer War Relocation Center, 107–108
The Rolling Stones (Heinlein), 50
Romulan "Bird of Prey" ship, 65
Romulans, 81
A Rough Sketch for a Proposed Film Dealing with the Powers of Ten and the Relative Size of Things in the Universe (documentary), 210n36
Round 2 LLC, 69, 75
Royal Navy, 79–80
"rules of engagement," 82
Run Silent, Run Deep (movie), 81
Russell, Bryce W., 61

Sackett, Susan, 209n29a
Sagan, Carl, 132
San Francisco Navy Yards, *Enterprise, USS* at, 187n9
Saturn V, 108, 118–121, 199n28, 199n29, 200n34
Scaffidi, Sandra, 198n24
Schirra, Wally, 114
Schlosser, Herbert, 115
Schnaubelt, Franz Joseph. *See* Joseph, Franz
Schneider, Paul, 188n11
Schneiderman, Dan, 113–114
Schoenberg Hall, 194n36
School of Aerospace Medicine, US Air Force, 84
Schwartzchild, Dick, 182n12b
science fiction, 24, 28, 37, 152
Preston, R., Sr., relation to, 130
Roddenberry relation to, 167
Science Fiction Modelmaking Associates (SFMA), 151
"script clearance and research," 45
Search for Extra Terrestrial Intelligence (SETI), 24–25
"Searching for Interstellar Communications," 24–25
"the second variation," 38
Senate, Special Committee on Space Technology of, 90–91
"Series Format," 23–24
SETI. *See* Search for Extra Terrestrial Intelligence
SFMA. *See* Science Fiction Modelmaking Associates
Shatner, William, 85. *See also* James T. Kirk
in *The Outer Limits*, 194n37
in "Where No Man Has Gone Before," 129, 174n14
Shepard, Alan, 89, 116
Shoemaker, Eugene, 221n28
"Shore Leave," 49–50
Shulman, Julius, 194n31
shuttlecraft, 59

SIA. *See* Smithsonian Institution Archives
Silverman, Sarah, 38
Silverstein, Bill, 182n12b
Sinai Peninsula, 199n29
The Six Million Dollar Man (television show), 197n13
Skylab, 212n62
"The Slaver Weapon," 186n77
Slide Graphic Flight Computer (CSG), 99
Sloman, Ernest, 47–48
Sloman, Peter, 47–48
Smithsonian Institute, 152. *See also* National Air and Space Museum
Arts and Industries Building, 10, 134, 137–138
The Making of Star Trek relation to, 135–136
Studies Office of, 217n105
Smithsonian Institution Archives (SIA), 205n2
Solow, Herb, 48–49, 50, 51
Forbidden Planet relation to, 102–103
Inside Star Trek of, 71, 159
on syndication, 159, 162–163
on US Navy, 80
Southern California, 95–96, 104–105, 111, 200n39a. *See also* Los Angeles
Soviet Union, 89–90, 91, 114
Partial Test Ban Treaty with, 194n35
Venera 4 of, 197n6
SPACE (publication), 90, 117
Space Age, 104–105
Space Art International, 131
Space Cadet (television show), 74–75
"Space Fantasy and Social Reality," 217n105
The Space Mural, a Cosmic View, 154
space race, 89–90, 96
Space Research Direction for the Future 1965 (publication), 101–102
Space Review (magazine), 91–92, 110
"Space Seed," 15, 110
"space shuttle," 175n38b
Space Shuttle Developing an Icon, 1972-2013 (Jenkins), 220n25
"Space Shuttle for the Future," 176n38c
space tourism, 170
Space World (magazine), 87
SpaceX, 200n32
Special Collections Library, UCLA, Roddenberry Collection, 36, 47
Special Committee on Space Technology, Senate, 90–91
Speed and Custom Division Shop, AMT, 59–60, 63, 72
Spican flame gems, 50
"Spirit of St. Louis," 153
Sputnik 1, 90, 91, 122
Stamm, Amy, 69
star classification, 27
stardates, 47–48
Starfleet, 79–80, 144–145, 187n1
Star Fleet Battles (game), 145–146, 214n78
Star Fleet Medical Reference Manual, 146

Star Fleet Technical Manual (Joseph, F.), 55, 143–146
"*Star Trek*: A Retrospective Exhibition", 217n105
Star Trek Blueprints (Joseph, F.), 20, 55, 143
Star Trek Communicator (magazine), 148
Star Trek: Enterprise (television show), 116, 119
Star Trek: Exploring New Worlds (exhibit), 66, 207n27
Star Trek Fact Check (blog), 204n1
Star Trek III (movie), 151, 217n105
Star Trek Lives! (Lichtenberg), 135, 150
Star Trek: Lost Scenes (McAloney and Tilotta), 188n11
Star Trek: The Animated Series, 9, 186n77, 192n16
Emmy awards for, 162
Lynn, H., Jr., relation to, 39
Star Trek: The Magazine, 62, 66–67
Star Trek: The Motion Picture (STTMP), 75, 146, 151, 153–154, 192n16
Star Trek: The Next Generation, 33–34, 92
Star Trek: Voyager, 38, 76
Star Trek Writers/Directors Guide, 23
Class M planets in, 27
The Making of Star Trek relation to, 71–72
Star Wars (movie series), 9, 146, 168
"*Star Wars*: The Magic of Myth" (exhibit), 217n105
Statler Hilton Hotel, 162
Steidinger, Scott, 148, 149, 216n96
Steven F. Udvar-Hazy Center, 152–153
Stewart, Patrick, 92
Stine, G. Harry, 73–74, 202n55
stripping, 157–158
Strombecker, 180n11a
STS-128 mission, 76
STTMP. *See Star Trek: The Motion Picture*
Studies Office, of Smithsonian Institute, 217n105
Summary Report of Future Programs Task Group 1965 (publication), 101–102
Superman: The Movie, 146
syndication, 13, 158–159, 161–163, 168

Tartar, 181n11e
Task Force Games #4, 214n78
techno-jargon babble, 48
tektites, 100–101
Texas A&M, 31
Thiel, Walter, 180n11a
Thompson, Craig O., 147–148, 214n86, 215n94, 216n96
Thompson, Ramo and Woolridge (TRW), 105, 106, 107
Thoroughbred Action (recap show), 198n16
Tiedemann, Herb, 198n24
Tilotta, David, 188n11, 192n17
Tinker, Grant, 204n1
"Tomorrow Is Yesterday," 82, 145, 200n36
The Tonight Show Starring Johnny Carson (television show), 85, 190n31
"To Set It Right," 200n39a
To Tell the Truth (television show), 85

INDEX

Toys and Novelties (trade publication), 60, 182n12b
tractor beam, 34, 175n38b
transporters, 58–59
Treaty for the Peaceful Exploration of Outer Space, 120
Trek (magazine), 67, 68
Trifid Nebula, 98
Trimble, Bjo, 159, 160, 163, 202n55, 207n27
"The Trouble with Tribbles," 50, 66, 107
TRW. *See* Thompson, Ramo and Woolridge
TRW Space Park, 105
Tsiolkovsky, Konstantin, 10, 33
Tulare Advance-Register (newspaper), 56
Tupti, Frank, 162
"Turnabout Intruder," 160
Turner, Frederick Jackson, 88
TV antenna, 157
TV Guide (magazine), 125, 202n55
20th Century Fox, 51–52
The Twilight Zone (television show), 161
2001: A Space Odyssey (movie), 170, 205n2
Tyler, Greg, 39

UCLA. *See* University of California, Los Angeles
"U.F.O. Mystery Ship," 75
UFOs. *See* unidentified flying objects
UHF. *See* "ultra high frequency"
"The Ultimate Computer," 66, 80
"The Ultimate Migration," 33
"ultra high frequency" (UHF), 157, 160–161
Underhill, Nick, 182n12b
Underwood, Dick, 198n24
unidentified flying objects (UFOs), 138
United Federation of Planets, 81, 92. *See also* Starfleet
United Launch Alliance, 221n28
Universal, 45
University of California, Los Angeles (UCLA), 107
 Kellam de Forest Collection, 52
 Schoenberg Hall, 194n36
 Special Collections Library, 36, 47
Urey, Harold, 137
US Air Force, 26–27
 Academy Cadet Squadron 19 of, 85–86
 Aerospace Pilot Research School of, 85
 Field Command Defense Atomic Support Agency, 29
 Genie and, 181n11c
 Lynn, H., Jr., relation to, 32
 School of Aerospace Medicine of, 84

in "Tomorrow Is Yesterday," 200n36
US Government Printing Office, 91, 101
"U.S. Missile Arsenal," 180n11b, 181nn11c–11d
US Navy, 26, 79–80, 81, 158
 Durant, F., in, 131
 Justman relation to, 82, 188n15
 in "Tomorrow Is Yesterday," 200n36
US Postal Service, 141–142
"US Space Missiles," 181n11e

Valmassel, Thomas, 68–69
Vanguard, 181n11e
Van Kampen, Hazel, 139–140
Van Treuren, Richard G., 69, 83
Variety (magazine), 103, 162
 Desilu Studios in, 202n55
 Garr in, 201n39c
Vatican Observatory, 101
Vehicle Assembly Building, 119
Vela, 106, 194n35
Velcro, 98–99
Venera 4, 197n6
Verne, Jules, 10, 135, 169
"very high frequency" (VHF), 157
 syndication and, 162
 UHF compared to, 160–161
VideoPlayer (magazine), 75
Vietnam War, 93
A Vision of the Future—Star Trek: Voyager (Whitfield, Stephen), 76
La Voyage a la Lune (Verne), 10
Vulcan, 221n28

Wadi Hadranawt, 118
Wallace, Art, 118, 121–122
War Department, Office of Scientific Research and Development relation to, 26
Warner Brothers, 45
The War of the Worlds (movie), 104, 148
Warwick Hotel, 160
Washington Post (newspaper), 164
Watergate scandal, 153
Weapons Effects and Tests Group Headquarters, Field Command Defense Atomic Support Agency, 29
Webb, James, 122, 123–124
Weitekamp, Margaret, 214n86, 217n105
Wells, H. G., 209n29a
Wesley Crusher (fictional character), 34
West Coast College, 149
Western democracies, 92
Western Operations Office, NASA, 117

western television shows, 88–89, 92–93, 199n25
Western Test Range, 125–126
Westinghouse, 97
WGA. *See* Writers Guild of America
Wheaton, Wil, 34
"Where No Fan Has Gone Before," 190n38
"Where No Man Has Gone Before," 79, 129, 187n7
 NASM and, 47
 Shatner in, 174n14
Whitaker, Walter E., 120–121
White, Ed, 118, 123
Whitener, Jack, 29–30
Whitener, Theresa, 30
Whitfield, Stephen, 13, 14, 23, 80–81, 168, 180n2
 on AMT model kits, 183n33
 AMT relation to, 56, 57, 60–63, 69–70, 72
 on "The Cage," 204n1
 Christies Film and Television Auction and, 186n79b
 on Class M planets, 27
 on Drake equation, 25
 on RAND Corporation, 29
 Roddenberry relation to, 71–73, 76
 Smithsonian Institute relation to, 135–136
 VideoPlayer and, 75
Whitfield, Susan, 76
Whitlock, Albert J., 104
Whitney, Grace Lee, 38–39
"Who Mourns for Adonais?," 99
Winfield, Gene, 60
Wings After War (Johnston), 206n20a
Witbeck, Charles, 158
Wood, Neal E., 195n42
Woolridge, Dean, 105
The World of Star Trek (Gerrold), 159
World Science Fiction Convention (WorldCon), 61, 124–125
World's Columbian Exposition, American Historical Association at, 88
World War II, 26, 31, 82, 139
WPIX-TV, 162
Wright, Frank, 185n65
Wright, James W., 113
"Wright Flyer," 153
Writers Guild of America (WGA), 121–122

X-15, 116

Yablans, Frank, 147
Young, John, 148

Zisfein, Melvin, 137, 138

THE AUTHOR

I was the kid who was always thinking about rocketing to other planets, whether I was watching television or playing with my Major Matt Mason action figures. My wagon was a moon rover and my backyard Hadley Rille. I loved the space program and had a childhood full of moon landings, *Star Trek* and *Star Wars*. After graduating from high school, I followed in my father's footsteps and became a teacher after attended Western Michigan University where I earned my undergraduate teaching degree in history, mathematics and physics. In 1991, I purchased my own offset printing press, lugged it into my parent's garage, and started producing *Quest: The History of Spaceflight Quarterly*. Little did I know then that the publication would still remain in print today. The success of *Quest* helped me acquire *Countdown*, a publication that covered the space shuttle program. These experiences helped me land a job as a technical writer in the aerospace industry. After earning my master's degree in Space Studies from the University of North Dakota, I moved to Houston where I became the chief historian at NASA's Johnson Space Center. After leaving NASA, I began my own space book and memorabilia business called Liftoff Books. Since that time, I moved back to Grand Rapids, Michigan, and built a house next to the home where I grew up. I also got married, had a son and lived a life of research, writing and travel, almost all of it relating to the exploration of outer space with brief detours into the history of renewable energy and Scandinavian polar expeditions. Today, I continue to explore the final frontier of space with my wife (a cultural anthropologist) and son (a budding author) while caring for my mom in the house that my dad built in 1957 and where I was first introduced to Mr. Spock on television that started my life's trek. Glen E. Swanson can be reached at https://www.glenswanson.space/

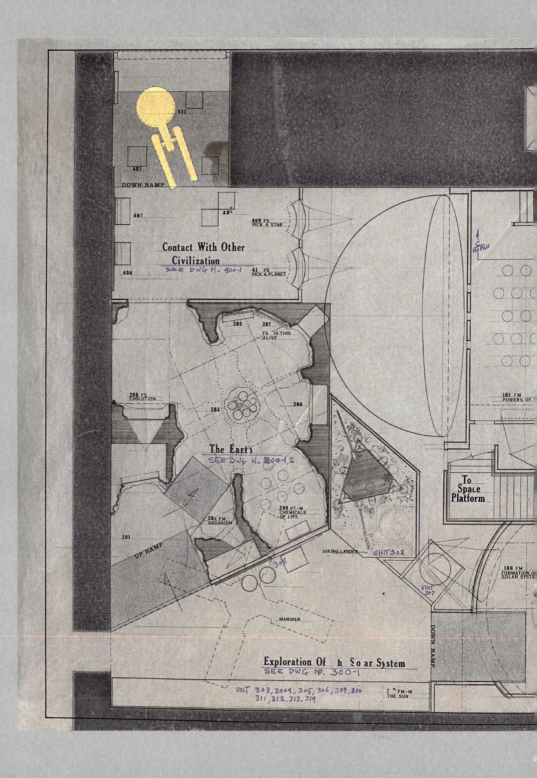